The Endless Voyage
Study Guide

FOURTH EDITION

Paul A. Billeter
The College of Southern Maryland

Robert R. Given, Ph.D.
Marymount College

Prepared by

IN-TELE-COM
INTELLIGENT TELECOMMUNICATIONS

BROOKS/COLE
CENGAGE Learning

Australia • Brazil • Japan • Korea • Mexico • Singapore • Spain • United Kingdom • United States

BROOKS/COLE
CENGAGE Learning

IN·TELE·COM
INTELLIGENT TELECOMMUNICATIONS

The video, print, and web-based course *The Endless Voyage* was designed and produced by INTELECOM Intelligent Telecommunications

ISBN-13: 978-1-4390-4552-7
ISBN-10: 1-4390-4552-6

Brooks/Cole
20 Davis Drive
Belmont, CA 94002-3098
USA

Cengage Learning is a leading provider of customized learning solutions with office locations around the globe, including Singapore, the United Kingdom, Australia, Mexico, Brazil, and Japan. Locate your local office at: **www.cengage.com/global**

Cengage Learning products are represented in Canada by Nelson Education, Ltd.

To learn more about Brooks/Cole, visit **www.cengage.com/brookscole**

Purchase any of our products at your local college store or at our preferred online store **www.ichapters.com**

Printed in the United States of America
1 2 3 4 5 6 7 13 12 11 10 09

Contents

Introduction

From the observations of Lt. Pelham aboard the H.M.S. *Challenger* to the mapping of oceans by satellites and undersea vehicles, the study of the ocean is as rich and dynamic as the ocean itself. *The Endless Voyage* will bring you face to face with exotic denizens of the deep and the peculiarities of their watery world. It will take you on virtual field trips to leading research institutions from Woods Hole, Massachusetts to La Jolla, California. It will introduce you to scientists and academicians involved in the science and study of oceanography in a wide range of disciplines who share their ideas and insights with you.

This guide is one part of the total package available to you when you embark upon the 26 half-hour videos that comprise *The Endless Voyage*. Each video episode will introduce you to academicians, researchers, physical oceanographers, chemists, marine biologists, tsunami specialists, geologists, and a host of others—people whose contributions are helping to interpret current understanding of the Earth's ocean and shape the our environmental future.

These video episodes, in conjunction with the guidance of your instructor, are closely integrated with Tom Garrison's text, *Oceanography: An Invitation to Marine Science.* The seventh edition of this popular text provides a richly illustrated foundation of information upon which to base your study. An interactive web component enhances independent learning.

Each of the half-hour video episodes is accompanied by a complementary lesson containing the following features to help you master your study of oceanography:

— An Overview summarizes each lesson's main topics.

— Learning Objectives identify the major concepts, ideas, and factual data that you should recall and understand after viewing the video and reading the required selections from the text. Many test questions are derived from these objectives.

— Assignments that link the video lesson with related sections of the text.

— Video Focus Points help you follow and analyze information in the video and integrate the information with your readings.

— Text Focus Points are intended to guide your reading of the selections for each assignment.

— Key Terms and Concepts help you to focus on the words and ideas important to understanding the language of oceanography as you work through each lesson.

— A Self Test enables you to check your understanding of the material in the video and text assignments.

— Supplemental Activities provide opportunities for further examination of the issues raised by the video and readings in the text.

Lesson 1

The Water Planet

Overview

Oceanography is the scientific study of the ocean. But what is science? Science is a method of problem solving. It is a body of accumulated ideas, facts, *hypotheses*, *theories*, and *laws*, and the processes used to discover them. Science is based on facts collected by experience or observation in the natural world. It cannot be based on faith or subjective opinion.

Because all living things are composed mostly of salty water, our own origins likely have much in common with the ocean's. So we'll start there by addressing one of humankind's most enduring questions, "Where do we come from?"

The universe seems to have begun its present existence 13 to 14 billion years ago in a cataclysmic event called the big bang. Some cosmologists argue that this happened at a dimensionless point, or singularity. But dimensionless or not, there was a single starting point from which the universe is still expanding. Scientists predicted that if the big bang theory is correct some remnant of this ancient explosion should exist. In 1965 engineers at Bell Laboratories who were studying microwave telecommunications stumbled upon the afterglow of thc big bang in the form of microwave radiation. As predicted, this leftover radiation from the big bang is found virtually everywhere in the universe.

Perhaps a million years after the big bang, the universe cooled enough for atoms of hydrogen—the smallest, simplest, and most abundant type of atom in the universe—to form. About one billion years after that, collapsing masses of hydrogen started to form the first stars. Our own sun formed about 5.5 billion years ago as clouds of hydrogen condensed upon each other, pulled together by their gravitational attraction. As this mass of hydrogen shrunk it heated to enormous temperatures, triggering nuclear *fusion*. During this thermonuclear reaction the sun's hydrogen atoms combined, forming larger helium atoms and releasing a steady stream of energy. Our sun is presently in this stable condition, emitting energy at a relatively steady pace, and should continue to do so for another five billion years.

The formation of the planets began about five billion years ago. The heavier elements that make up the planets, moons, and comets are produced in the nuclear fusion and explosive deaths of older stars. When our own *nebula* was struck by such a mass of energy and matter, it caused our nebula to spin. In addition, it deposited the heavier elements, which coalesced to form the orbiting planets by *accretion*. During this process the dust and atoms of the swirling nebula clumped together and grew, attracting matter by gravity as they orbited. The planets closer to the sun consist of heavy elements, which are solid at higher temperatures. Mercury, the planet closest to the sun, is made mostly of iron. Farther from the sun, water, magnesium, aluminum, oxygen, and silicon collected to form Earth and its neighbors. Still farther out, the gas giants like Jupiter and Neptune coalesced from methane and ammonia gases, which can only be solid at the low temperatures of the outer solar system.

The early Earth was mostly liquid, heated by its own *gravitational energy* and the heat of radioactive decay. In this heated liquid state *density stratification* occurred. Iron migrated inward to form Earth's core while the silicates and other lighter matter moved outward to form the mantle and crust. About 4.6 billion years ago as Earth was cooling and the surface was hardening, a planet about the size of Mars collided with it, exploding some of its material into space. These remnants of the collision went into orbit around Earth and underwent accretion, forming the moon.

The steam *outgassing* from volcanoes, along with extraterrestrial icy comets bombarding Earth from above, supplied the water that was to become the ocean. The steam that condensed and fell on the hot young Earth as rain immediately vaporized and rose back into the atmosphere to repeat the cycle. The weather report for Earth approximately 4.5 billion years ago: 25 million years of continuous heavy rain. Thus the oceans formed.

The oceans are Earth's dominant feature, covering 71 percent of the planet's surface. Although water is a common substance, it is not an ordinary one. Among its uncommon characteristics is a high latent heat, meaning that it takes a lot of heat (calories) to change water's temperature (degrees). So it is not only the distance from the sun and the composition of the atmosphere that create the moderate conditions necessary for Earth to have so much water in liquid form. The very presence of so much water itself contributes to the maintenance of Earth's moderate conditions. Water's many extraordinary characteristics will be further discussed in future lessons.

Earth's atmosphere formed as gases belched forth from volcanoes and otherwise escaped from where they were trapped within the planet. Along with water vapor (H_2O), the early atmosphere contained carbon dioxide (CO_2), nitrogen (N_2), and traces of methane (CH_4) and ammonia (NH_3). But it contained no gaseous oxygen (O_2). This hostile oxygen-free environment is where life on Earth began.

Early in Earth's history, the major elements needed to construct life—carbon (C), hydrogen (H), oxygen (O), and nitrogen (N)—were present. Massive amounts of salty water were present, and energy in the form of lightning, volcanic heat, and ultraviolet radiation from the sun was also present. Somehow from this mix life arose. The possibility that the evolution of life arose from the simple chemicals present on primitive Earth through a process called *biosynthesis* was first suggested in the early twentieth century. Laboratory experiments have since produced most of the important amino acids—the building blocks of proteins—and the simple sugars and other organic compounds necessary for the construction of the complex molecules of life.

A variety of hypotheses about the origin of life have been proposed. Some argue that life began in shallow tide pools; some say it began in the deep ocean near hydrothermal vents; others argue that it formed on beds of clay or under ice. All the hypotheses have some merit. No one has yet created life in a laboratory, but it is certain that the types of chemicals needed can be produced under conditions thought to be similar to Earth four billion years ago. It is also true that organic materials can land on Earth with meteorites, posing the possibility that the chemicals that spawned life on Earth arrived from elsewhere. If that is true it simply displaces the question of how life started from Earth to some other world. Divine creation by a supernatural entity is outside the realm of the natural sciences to investigate.

Earth is the planet of water and life. Its dominant feature is the ocean. The study of the ocean and its numerous subdisciplines is conducted by oceanographers and marine scientists. Physical oceanographers study the waves, tides, heat, and other physical features of the ocean. Chemical oceanographers study the chemistry of seawater, biological oceanographers study the ocean's life, and geological oceanographers study its geology. Others design better ships, rigs for drilling at sea, and satellites that can study the entire ocean in a few passes around Earth. Some work to help take advantage of the bounty of resources that can be harvested from the ocean including food, minerals, medicines, and even fresh water. Others vigilantly study over-fishing, pollution, and other assaults on the enormous, yet finite, ocean. Their work overlaps in the multi-disciplinary world of oceanography.

Every voyage begins with a first step. This lesson introduces oceanography, encourages your curiosity, and invites you to take that first step and join this endless voyage.

Focus Your Learning

Learning Objectives

On successful completion of this unit of study, you should be able to:

1. Define oceanography and list and briefly describe at least five branches of this science.
2. Discuss science as a way of understanding the universe and accumulating knowledge. List the steps in the scientific method and compare and contrast the terms *hypothesis*, *theory*, and *law* as they are used in science.
3. Discuss the big bang theory of the origin of the universe including at least two types of evidence that support this theory.
4. Explain how and when the sun, solar system, Earth, and oceans formed.
5. Discuss current ideas on how life arose.

Assignments

This lesson is based on information in the following text and video assignments. The key terms, focus points, and practice test are intended to help ensure mastery of the material presented in this unit of study.

Text: Chapter 1, "Origins," pages 1–27

Video: Episode 1, "The Water Planet"

Video Focus Points

Review these points before watching the video assignment for this lesson to help you focus on key issues.

— Earth is the water planet. Characteristics of water such as its latent heat, the fact that ice floats, its transparency and its density contribute significantly to both the nature of Earth and nature of life on Earth.

— The universe originated about 14 billion years ago in a cataclysmic event termed the big bang. Two significant observations in support of the big bang theory are the fact that the universe is still expanding and the detection of "leftover" microwave radiation emanating from all parts of the universe.

— Stars and galaxies began to form about a billion years after the big bang, as hydrogen atoms, attracted to each other by gravity, formed cloudy nebula that later condensed into stars.

— The sun formed from a cloud of gas and dust that collapsed into itself by gravitational attraction, forming a nebula. As a protosun, it began to radiate heat and light. Finally, it was transformed into a star as the temperature rose to a level at which the nuclear reaction of hydrogen fusion began.

— Other matter in the spinning disc around the protosun also began to coalesce by gravitational attraction, producing the planets.

— Earth's early atmosphere contained mostly carbon dioxide and nitrogen, and oxygen was totally absent. Oxygen began to accumulate in the atmosphere about 3.5 billion years ago with the evolution of photosynthesis, which produces oxygen as a waste product.

— Water accumulated on Earth to form the oceans. The water probably came from within Earth, outgassing from the hot interior, and from icy comets raining in from space.

— Carbon dioxide, nitrogen, and water contain the four major atoms—C, H, O, and N—needed to make biological molecules.

— There are several plausible hypotheses on the origin of life and the conditions under which life might have originated. None is yet proven.

Text Focus Points

Review these points before reading the text assignment for this lesson to help you focus on key issues.

— Viewed from space, water dominates planet Earth. Nearly three-quarters of the planet is covered by water and more than 97 percent of this water is in the oceans and seas, which will be called the world ocean or simply the ocean in this course.

— Oceanography is a multidisciplinary science drawing on physics, chemistry, geology, biology, and other branches of natural science in an attempt to understand the ocean.

- Science is a systematic method of problem solving. The broad aspects of science are discussed. These distinguish science from other approaches to explaining our universe such as supernatural explanations and art, literature, and the other humanities.
- The scientific theory of the origin of the universe is the big bang theory. Matter expanding from the big bang eventually produced the stars and their planets and the galaxies.
- The condensation theory explains how stars and planets are believed to form. Observations of other stars at various stages of formation and destruction present a general picture of the lifecycle of stars.
- The nebula that produced the sun was struck by the remnants of an older star causing it to spin. The planets formed in the spinning exterior of the early sun. Earth is among the four rocky inner planets. The gas giants, like Jupiter, are farther out in the solar system.
- Earth's layers—core, mantle, crust, ocean, and atmosphere—sorted by density stratification with the densest core on the interior and the least dense atmosphere outermost.
- The water that formed the ocean came from outgassing from Earth's interior and by the gravitational attraction of icy comets from space.
- The early atmosphere of Earth was composed of carbon dioxide and nitrogen with significant amounts of water vapor. It had no oxygen.
- Biosynthesis—the early stages in the production of the organic molecules required for life—is being actively studied, but scientists debate which hypothesis best explains life's origin.

Key Terms and Concepts

A thorough understanding of these terms and concepts will help you to master this lesson.

Accretion. The formation of a larger mass by the accumulation of many smaller masses.

Biosynthesis. The creation of life when simpler molecules combine to produce more complex biomolecules like proteins, carbohydrates, and DNA. Many scientists believe life began on Earth by some form of biosynthesis.

Cosmology. In science, the astrophysical study of the origin of the universe.

Density stratification. The sorting of the materials in a body by their density. The denser materials accumulate in the deeper or lower regions and the less dense rise to the outer or upper regions.

Fusion. A nuclear reaction in which atomic nuclei fuse to produce larger atoms. The atomic fusion of smaller hydrogen atoms to produce larger helium atoms is the thermonuclear reaction that powers the sun.

Gravitational energy. Refers to energy, primarily heat, generated as particles are attracted together by gravity. The compression of matter generates heat.

Hypothesis. A suggested solution to a problem based on observations, measurements, and logical insight. A hypothesis may be tested or currently untested. After testing it may be accepted or rejected.

Nebula. A diffuse cloud of gas and dust. The earliest nebulae following the big bang were composed of hydrogen. As stars formed, passed through their lifecycles, and exploded the other heavier elements produced in the stars were scattered into the universe. Along with hydrogen, these became parts of later nebulae.

Outgassing. The outward migration of gases trapped in the Earth as their lower density causes them to rise from the interior to the exterior. It is one source of the water vapor that produced the ocean and the gases that produced the atmosphere.

Theory. A scientific explanation of some characteristic or process of the natural world. Unlike hypotheses, theories must be strongly and consistently supported by empirical evidence.

Test Your Learning

After working through these questions, check your answers against the key at the end of this book. If you answered any of the questions incorrectly, review the relevant sections of the text and video episode.

Multiple Choice

1. The big bang occurred about ________ years ago.
   a. 14 thousand
   b. 14 million
   c. 14 billion
   d. 14 trillion

2. The gaseous cloud from which the sun formed is called a:
   a. protostar
   b. nebula
   c. nova
   d. supernova

3. The presence of large amounts of water on Earth's surface ________ global temperatures.
   a. increases
   b. decreases
   c. moderates
   d. does not affect

4. The protosun was not considered a star because:
   a. it was too cold for nuclear fusion
   b. it still contained excessive helium
   c. nuclear fission rates were greater than nuclear fusion rates
   d. it had not passed through the supernova stage

5. Which of these was not present in Earth's primitive atmosphere when life arose?
   a. water
   b. oxygen
   c. nitrogen
   d. carbon dioxide

6. The dense rocky planets of the solar system, like Earth, are found:
   a. closer to the sun
   b. farther from the sun
   c. spread throughout the solar system
   d. Earth is the only rocky planet, the others are composed of frozen gases

7. Which of the following is/are suggestive that simple, self-replicating molecules arose in the early ocean?
   a. All life depends on saline water in its cells to dissolve and transport chemicals.
   b. The compounds that are the building blocks of life have been produced in laboratory experiments.
   c. The main chemical requirement for producing these molecules seems to be the absence of free oxygen.
   d. Both a and c

8. As Earth formed in the early solar system the present layers of core, crust, and mantle started to separate by a process called:
   a. density stratification
   b. gravitational sorting
   c. sedimentation
   d. accretion

9. Background microwave radiation is important evidence in support of the ________ theory.
   a. condensation
   b. big bang
   c. solar evolution
   d. galactic fusion

10. The major sources of the water that formed the ocean are thought to be outgassing and:
    a. accretion of molecular water from space
    b. solar winds
    c. capture of icy comets
    d. photosynthesis

11. Wächtenhäuser proposed the idea that life may have begun:
    a. in warm tropical tide pools
    b. on the undersurface of sea ice
    c. between grains of pyrite silt and clay on beaches
    d. in deep-sea hydrothermal vents

12. Most of the matter of the known universe consists of:
    a. neutrons
    b. dust and debris
    c. hydrogen
    d. carbon

Short Answer

1. Consider the following terms:
   a. curiosity
   b. experiments
   c. hypothesis
   d. law
   e. observations and measurements
   f. theory

   Rearrange these terms in order from the one with the least amount of supporting evidence to the one with the most, and explain each term. Briefly describe one specific example of each of these terms that was discussed in the reading or video for Lesson 1.

2. Although it is a teleological, non-scientific statement, one might say "Stars have died so we (humans etc.) may live." Outline the major phases in the life of a typical star and explain how the death of ancient stars contributed to your life.

3. Explain why a biosynthesis similar to that which produced simple organic building blocks that enabled the early steps in the evolution of living organisms cannot occur today.

Supplemental Activities

1. You are a graduate student from a distant galaxy who has studied Earth and written your Ph.D. thesis about it. You are now presenting a seminar on this curious little planet and all of the figures below are in your slide show. Which of the following figures in the text would you use as the most typical image of Earth? Defend your choice by comparing the pros and cons of each figure as a truly typical view of Earth. Figure 1.1a (p. 4); Chapter 6 introduction photo (pp. 154–155); Figures 8.2b (p. 203); 10.10 (p. 274); 14.13 (p. 393) and 18.13 (p. 501).

2. In 1953 Stanley Miller attempted to test the idea of biosynthesis in the laboratory. By conducting a search on "Stanley Miller" you will find detailed information on this experiment. After reading about the experiment, go to the Internet (www.carlsagan.com) and find Carl Sagan's Baloney Detector by clicking on "Carl Sagan" at the top of the page. Suppose a professor proclaimed "Miller's experiment proves how life began on Earth." You disagree. Attempt to evaluate the professor's proclamation using Sagan's first nine detection tools. Then choose any three of the "common fallacies of logic and rhetoric" that you think you can apply to the professor's statement and explain why each of the three applies.

3. We often describe science as a systematic process of problem solving, yet any scientist will freely admit that serendipity, the phenomenon of stumbling onto valuable discoveries by accident, is a significant aspect of science. Both X-rays and penicillin were discovered serendipitously. A major piece of evidence that supports the big bang theory came from a discovery by Edwin Hubble, which supported an earlier hypothetical idea of Georges LeMaitre. Similarly, a discovery made by Arno Penzias and Robert Wilson supported an earlier hypothetical idea of Robert Dicke. Go to the Physics and Astronomy section of the PBS Website at (http://www.pbs.org/wgbh/aso/databank/) and explore the systematic vs. serendipitous aspects of these two discoveries supporting the big bang theory.

Lesson 2

First Steps

Overview

How did we get where we are, and who led us? The origins of human endeavor are often in the most primal animal behaviors—curiosity and the need to survive. Early humans probably lived on the ocean shores or at least frequented them to reap the bounty of the ocean's resources. As population increased, curiosity and the need for additional resources gave man the need to investigate what lay beyond. His mental resources and manual capabilities gave him the means.

The first accounts of ocean *voyaging*, or traveling the ocean for a purpose, are from the Mediterranean. Travel here is recorded for purposes of exploration, trade, warfare, and food. Very early records have the Phoenicians (200 B.C.) leaving the Mediterranean through the Straits of Gibraltar, to the shores of Britain. The Greeks ventured through the Straits (900–700 B.C.) and explored the Atlantic. As they traveled this huge body they recorded mass movement of water flowing like a giant river. What we now know as currents they called the "river" *okeanus* (thus ocean). Other cultures were also voyaging. The Polynesians had been traveling for several thousand years around the islands off Southeast Asia and Indonesia, and the Chinese were establishing trade routes along their inland waterways, some of which led to the Pacific Ocean.

It became necessary to keep records of the routes traveled in the form of drawings, or *charts*, as well as accounts of the hazards encountered and commercial possibilities. The better these records were, the faster and safer the next voyage would be. In pursuing their systematic and organized study of the ocean, these early voyagers were the first oceanographers. By the third century B.C. many of these records and charts were being archived at the Library of Alexandria in Egypt. Often called the first university, this library was used for reference by travelers and scholars from around the world.

For the next thousand years, beginning with the fall of the Roman Empire and the onset of the *Dark Ages*, intellectual development was practically at a standstill. The Polynesians continued their extraordinary travels and island colonization in the Central and Eastern Pacific, and are a high point in the chronology of marine voyaging. Four separate Asian and Indonesian cultures, collectively called Oceanians, explored some ten thousand islands across ten million square miles of ocean. They were skilled at using the stars as well as sunrise and sunset, wind and current analysis, water taste and smell, and other ocean phenomena to navigate effectively. During the so-called Dark Ages in Europe—enabled by their fast, stable ships—Scandinavian Vikings swept the European coasts raiding and pillaging. Although they sought trade rather than conquest, the Chinese were also voyaging during this time and perfecting their navigation techniques and ship building skills. They also contributed a number of important marine innovations, including the magnetic *compass*, the central rudder, compartmentalized hulls for safety, and more efficient sail designs. The early marine voyagers truly saw beyond the commonplace and led other generations forward.

During the Age of Discovery and the *Renaissance*, marine science continued to progress and, fortunately, so did the record keeping. Prince Henry the Navigator is notable here, not only for sponsoring much of the exploration done in the mid-1400s but because he established a center for the study of marine science and navigation at Sagres, Portugal. Master mariner Christopher Columbus ostensibly discovered the New World while searching for trade routes to India and Japan. His discovery got more credit than was due, because Columbus was also a great storyteller and returned with many tales and trinkets, supposedly from these "new" lands. In fact, Columbus never actually sighted North America! With the continued expansion of British sea power, exploration and colonization was done to establish land stations for vessel repair and re-supply.

Although information had been gathered for centuries, the idea of scientific oceanography—voyaging for scientific research—is generally acknowledged to have begun with the 1768 voyage of the British ship *Endeavor*, captained by James Cook. Cook was sent by the British Royal Society to study the transit of Venus across the sun, thus attempting to verify the gravitational theories of Isaac Newton and Edmund Halley as they applied to cometary movement. On completion of that task, he was authorized to do some exploring before returning home. Cook explored and charted around the Great Barrier Reef in Australia and New Zealand, and was probably one of the first to accurately determine and record the locations of his findings. It was easy to find one's *latitude* at sea, but to determine *longitude* (east-west position) required a very accurate timepiece that could withstand conditions at sea. In 1760 John Harrison perfected such a timepiece—the *chronometer*—redefining the concept of accuracy in navigation. Cook made several voyages accompanied by various scientists, collecting specimens and using some of the first crude devices for obtaining samples from the sea floor. Cook was truly a giant in the history of marine science, but it would be nearly 100 years before the first expedition totally devoted to science would be launched.

During the time of Cook's explorations the United States was slowly becoming involved in marine science. In the late 1820s the United States Exploring Expedition was conceived, with the unusual mission of determining if there were holes in the planet at the north and south poles that would allow ships to actually explore Earth's interior! The idea of exploration with a polar component was attractive, but it promised to be expensive and at that time there was no tradition of federal government funding for science. It took some time to secure funding, but ten years later, in 1838, the Expedition began under the leadership of Lt. Charles Wilkes. While its primary mission was military, there were a number of scientists aboard, and it was their work that established natural science as an acceptable endeavor, no longer just an auxiliary to voyaging. Wilkes compiled a great many charts and collected a number of scientific specimens and artifacts. His final report remains a landmark of scientific achievement. The hugely successful Expedition marks the beginning of marine science in the United States. Shortly thereafter, Matthew Fontaine Maury, a naval officer interested in ocean currents as they affected ship travel, became perhaps the first person to be considered a full-time oceanographer. He charted winds and currents for the United States Navy, and has been called the father of physical oceanography. In the early 1830s Charles Darwin, although best known for his theory of biological evolution, made his historic four-and-one-half-year journey on the *H.M.S. Beagle*, gathering natural history and oceanographic samples and data.

By far the most famous and important expedition undertaken to this day is the British *Challenger* expedition. The four-year journey, begun in 1872, was the first sailing expedition totally dedicated to marine science. One of its goals was to test a theory that life could not exist in the deep sea. Many new species were collected, and life in the deep sea was extensively documented, supplemented by water samples and depth and temperature readings. The *Challenger* Report, a 50-volume set of text and magnificent hand-colored illustrations, still stands as a monument in the field of oceanography.

In the twentieth century oceanographic voyages became more technically ambitious and expensive in the quest to explore previously difficult or impossible to reach places. The polar expeditions of Nansen, Peary, Scott, and Amundsen are notable here. Technical breakthroughs in oceanography included the invention of the *echo sounder* for imaging the depth and contour of the sea floor, and the expeditions of the *Glomar Challenger,* which used innovative drilling apparatus to study Earth's crust. However, as Prince Henry realized in the thirteenth century, there was still a need for shoreside institutions to complement the shipboard work. The creation of laboratories at Naples (1800s) and Monaco (1906) were a start, but other countries were also in need of expansion beyond the ships. In the late 1800s a young American student named James Ritter who was working at the Naples laboratory got the idea of developing such a station in the United States.

Through a series of fortunate events, Ritter interested the wealthy Scripps publishing family in the project. In 1912, with a generous endowment from the family, the Scripps Institution of Oceanography was dedicated at La Jolla, California. A similar effort was later undertaken in Massachusetts, encouraged by the concern of the National Academy of Sciences about the reputation of the United States in marine science. Woods Hole Oceanographic Institution was established in 1930, and while remaining independent of affiliation with a college or university, it works closely with the neighboring Marine Biological Laboratory and the Massachusetts Institute of Technology. In 1949 a third major United States oceanographic research institute was founded—the Lamont-Doherty Earth Observatory of Columbia University. Through all of this, the United States government was becoming increasingly involved in oceanographic exploration in both the military and civilian sectors. Through NASA, the use of satellites to conduct oceanographic research has become a major factor in increasing knowledge of the oceans from a previously impossible perspective.

Oceanography is now considered "big science," from manned deep submersibles to satellite imaging and the Ocean Drilling Project. The secret of this success is its interdisciplinary nature and the collaboration of military, civilian, and academic entities in the United States with their international counterparts. While some oceanographers still consider it a "small science" with comparatively few scientists, oceanography remains one of the most exciting and productive endeavors ever undertaken by mankind.

Focus Your Learning

Learning Objectives

On successful completion of this unit of study, you should be able to:

1. Compare and contrast the early voyages of the Polynesians, Vikings, Greeks, and Chinese, especially their motivations, vessels, and seafaring skills.
2. Recognize the importance of record-keeping and cartography to voyaging, and the historical role of the Library of Alexandria.
3. List some of the major contributions to early voyaging of Prince Henry, Magellan, and Columbus.
4. Recognize the role of Captain James Cook in the history of marine science, and be aware of his three major voyages.
5. Describe the United States Exploring Expedition, its objectives and accomplishments, and its importance in bringing the United States into the field of oceanography
6. Describe the *Challenger* expedition and its major contributions in marine science.
7. Chronicle the rise of land-based oceanographic institutions, from the early beginnings in Naples and Monaco to the major facilities now operating in the United States.
8. List the disciplines in which marine research is being done today, as it relates to recent advances in data gathering and technology.

Assignments

This lesson is based on information in the following text and video assignments. The key terms, focus points, and practice test are intended to help ensure mastery of the material presented in this unit of study.

Text: Pages 28–61 in Chapter 2, "History"

Video: Episode 2, "First Steps"

Video Focus Points

Review these points before watching the video assignment for this lesson to help you focus on key issues.

— The early voyagers laid a foundation for further exploration and advances in navigation and vessel design.

- The voyages of Captain James Cook were vital to early marine science, cartography, and discovery.
- The United States Exploring Expedition led the United States into marine science.
- The British *Challenger* expedition was the most important marine expedition ever undertaken in terms of the scope of scientific disciplines involved and the way that future marine science would be done.
- Despite the success of seagoing expeditions, shipboard work has its limitations. The need to expand into land-based facilities and laboratories was recognized in the development of such important facilities as Scripps Institution of Oceanography in California and Woods Hole Oceanographic Institution in Massachusetts.
- After World War II there was a renewed interest in the marine sciences and technology, resulting in advanced deep submergence vehicles, satellite imaging, and highly-sophisticated data analysis methods.

Text Focus Points

Review these points before reading the text assignment for this lesson to help you focus on key issues.

- Although not recognized as such until much later, crude marine science was being done by the early seafarers as they ventured out on voyages of discovery, colonization, trade, warfare, and the search for living space and resources.
- Records of these voyages were made and, where possible, stored in the great Library of Alexandria in Egypt, to be used by researchers and travelers world-wide.
- Navigation using heavenly bodies was used very early, being updated and refined over time. Modifications of many of those techniques are still used today.
- Interest in any kind of science stopped in the Dark Ages, and was revitalized during the Renaissance. Meanwhile, the Polynesians and Chinese continued their voyages and exploration, colonization and trade.
- Prince Henry, Magellan, and Columbus were all notable for their voyages and discoveries, but it was Captain James Cook, on the *Endeavor* in the late 1700s, whose primary mission was science and who began oceanography as a scientific discipline.
- Technological advances continued to be made as necessary for safer and more efficient sea travel. The need to determine latitude and longitude for accurate navigation was addressed, as were the difficulties of obtaining good oceanographic data and natural history samples.
- The United States entered the oceanographic arena with the United States Exploring Expedition in 1838.
- The first truly scientific expedition was that of the H.M.S. *Challenger*. Its trawl samples, records, and other scientific accomplishments still stand as the greatest of all time.
- At the beginning of the twentieth century, polar exploration got much attention through Nansen, Peary, Scott, and Amundsen. Advances in vessel design and technology led explorers to previously inaccessible places.
- Later in the twentieth century, military submarines began gathering oceanographic information, and the capabilities for conducting marine research expanded in both surface and subsurface vessels.
- Land-based oceanographic institutions arose as a complement to vessel operations.
- Satellite imaging for oceanographic purposes began in 1978 with the NASA *Seasat*. That technology continues to expand.
- Highly sophisticated data-gathering techniques continue to evolve, resulting in a network of oceanographic programs with involvement by military, civilian, and academic entities.

Key Terms and Concepts

A thorough understanding of these terms and concepts will help you to master this lesson.

Cartography. The art and science of making charts.

Celestial navigation. The technique of finding one's location on Earth by reference to the apparent positions of stars, planets, the moon, and sun.

Chart. A map that depicts mostly water and the adjoining land areas.

Chronometer. A very consistent clock; does not need to tell accurate time, but its rate of gain or loss must be exactly constant and known so that accurate time may be calculated.

Compass. An instrument for showing direction by means of a magnetic needle swinging freely on a pivot and pointing to magnetic north.

Dark Ages. Term used to describe an early medieval period in Western European history after the fall of the Roman Empire (476–800 A.D.). A period when interest in knowledge and learning practically disappeared, and many artistic and technical skills were lost.

Echo sounder. A device that reflects sound off the ocean bottom to sense water depth. Its accuracy is affected by the variability of the speed of sound through water.

Glomar Challenger. A highly sophisticated "floating laboratory," used in the late 1960s to mid-1980s, for drilling deep into the oceanic crust.

Global Positioning System (GPS). Satellite-based navigation system that provides a geographical position—longitude and latitude—accurate to less than one meter.

Latitude. Regularly spaced imaginary lines on Earth's surface, running parallel to the equator.

Longitude. Regularly spaced imaginary lines on Earth's surface, running north and south and converging at the poles.

Oceanographic institution. A land-based facility dedicated to research in physical and biological oceanography; usually associated with a fleet of research vessels. May operate independently or in affiliation with an academic institution.

Renaissance. The period in Western European history after the Dark Ages (early 1400s to mid-1500s), which saw a rebirth of interest in knowledge and learning and in the development of technical skills.

Satellite oceanography. Ocean research and data-gathering using imagery and other data from specially equipped satellites.

Sounding. Measurement of the depth of a body of water.

Voyaging. Traveling (usually by sea) with a specific purpose.

Test Your Learning

After working through these questions, check your answers against the key at the end of this book. If you answered any of the questions incorrectly, review the relevant sections of the text and video episode.

Multiple Choice

1. What expedition is said to be responsible for bringing the United States into the marine science field?
   a. United States Exploring Expedition
   b. *Glomar Challenger* expedition (1968)
   c. United States Navy *SCICEX*
   d. *Atlantis* expedition (1931)

2. Who is usually acknowledged to have been the first scientific oceanographer?
   a. Matthew Maury
   b. Charles Darwin
   c. Captain James Cook
   d. Alexander the Great

3. Who is considered by many to be the first full-time oceanographer and the father of physical oceanography?
   a. Matthew Maury
   b. Charles Darwin
   c. Capt. James Cook
   d. Alexander the Great

4. This man established what might be considered the first land-based center for the study of marine science and navigation:
   a. Ferdinand Magellan
   b. Prince Henry the Navigator
   c. Prince Albert of Monaco
   d. Fridtjof Nansen

5. Even with the great advances in technology, obtaining one of the following still remains a difficult problem in deep-sea research.
   a. accurate bottom profiles
   b. accurate water column temperatures
   c. accurate locations of sea mounts
   d. adequate biological samples for laboratory analysis

6. Fridtjof Nansen, on the *Fram*, proved that:
   a. polar currents were generally swifter than those at the equator
   b. polar currents were generally slower than those at the equator
   c. accurate water temperatures can be taken under the polar ice
   d. there is no actual Arctic continent

7. Which of the following first calculated the circumference of Earth?
   a. Admiral Zheng
   b. Ptolemy
   c. Alexander the Great
   d. Eratosthenes

8. Which of the following was the first oceanographic institution in the United States?
   a. Woods Hole Oceanographic Institution
   b. Lamont-Doherty Earth Observatory
   c. University of Alaska
   d. Scripps Institution of Oceanography

9. Which of the following was not done on the first *Challenger* expedition?
   a. trawling and netting
   b. measurement of surface chlorophyll
   c. bottom sediment sampling
   d. recovery of manganese nodules

10. Which of the following was a leader in the rise of American sea-power (military expansion onto the oceans), particularly the interdependence of military and commercial control to insure safe lines of supply in wartime?
   a. Robert Scott
   b. Don Walsh
   c. Matthew Maury
   d. Alfred Mahan

11. Returning in 1842 from a four year circumnavigation, the captain and scientists of this expedition prepared a landmark report consisting of 19 volumes of maps, text, and illustrations. Identify the expedition:
   a. U.S. Exploring Expedition
   b. *Challenger* expedition
   c. *HMS Beagle* expedition
   d. *HMS Endeavor* expedition

12. The term oceanography was officially coined by:
   a. Charles Wilkes (*Vincennes*)
   b. Charles Darwin (*Beagle*)
   c. Wyville Thomson and John Murray (*Challenger*)
   d. Fridtjof Nansen (*Fram*)

Short Answer

1. Discuss the role of Claudius Ptolemy (90–168 A.D.) in the discipline of cartography, including the famous Ptolemy's Error and its later consequences.

2. Captain James Cook made three major voyages of discovery and science, all of which were highly successful and productive. Although he had well-trained crews and excellent scientific personnel, much of the credit for the success of these cruises is attributed to the man himself, with his many skills and talents. List and briefly discuss those skills and talents as they applied to these voyages.

3. Although the voyages and accomplishments of Cook, Wilkes, Maury, and Darwin are highly significant in the history of oceanography and marine science, the four-year *Challenger* expedition still stands alone as the most important. List and briefly discuss some of the reasons for that recognition.

Supplemental Activities

1. The United States Exploring Expedition had, as one of its missions, investigating the theory that there are no poles on Earth, but that holes exist there through which one may travel and encounter a whole new world. Admiral Richard Byrd, on his historic North Pole flight, was supposed to have flown into this hole. Preposterous? Look into this concept of a "hollow earth" and be amazed. On the Web you will find a remarkable number of references, including the "Hollow Earth Society"! Look at some of their publications and see how serious these folks are. Write a short (1–2 page) paper about this school of thought, including some of the famous people supposedly involved, what the evidence shows (or does not show). Then write a short paragraph about your personal conclusions: Is it worth more thought? Is it ridiculous and if so, why? You may not become a believer but at least you will know what is going on.
2. One of the missions of the early marine scientists was to collect samples of the plant and animal life in the ocean waters. They used nets of various sizes and configurations, either lowered into the water with a weight or just towed behind the ship. Join those scientists in their quest for new species by making your own net and collecting some of the tiny organisms that live along the shores of your local pond or lake. To make the net, bend the hook of a wire coat hanger out straight. Then bend and arrange the loop part of the hanger into as close to a circle as possible. Cut off the lower portion of a panty hose leg (ankle down to the toe) and attach it to the loop using tape, or sew it on using dental floss or heavy thread. Be sure to leave the toe portion hanging down to collect the specimens. Attach the straightened hook onto a broom or mop handle with tape, and take the whole thing to a nearby pond or lake. Take some plastic bags to put your specimens in. Drag the net along the shore just below the waterline but above the mud (try not to collect the mud, it just makes a mess). After a few passes, empty the toe into the plastic bag, which you have half-filled with water and zip or tie it closed. Using a small magnifying glass or hand lens (available at most toy stores and even some supermarkets) look closely at the creatures swimming around inside. You will see some small crustaceans, worms, beetles, and maybe even a small fish! If you wish to save the material, just put a little rubbing (isopropyl) alcohol in the bag, and look at it later. If you have access to a biology lab, look at your treasures with a dissecting microscope. You will not be able to identify the species, but you will have an idea, on a somewhat smaller and less varied scale, of what the early scientists would have seen in their nets. They usually had to wait until they got back to their land laboratories to take a closer look—it is difficult to use a high-powered microscope on a moving ship! Write up a short field report—name, date, location, weather, time, and your first impressions of what you saw. This activity will give you a good idea what early shipboard scientists would have done and seen.
3. Using the list in the appendix of your text, visit the Website of an oceanographic institution and find out what kinds of research they are doing, how they are gathering the data, and what scientists are involved. If they have a vessel, learn about its size, cruising range, scientific equipment and capabilities, and where they most frequently travel. Often these vessels are available for tours as they visit certain ports. From your research, choose a favorite topic, or a vessel, or a scientist, and write a brief (1–2 page) report, documenting the research being done, how the data are gathered and analyzed, and the disciplines of the scientists involved.

Lesson 3

Making the Pieces Fit

Overview

The currency of science is ideas. As with money, ideas exist in pocket change, small bills, and large bills. The "large bills" of the natural sciences are called paradigms. These major ideas overlie the science and dictate the guiding principles through which scientists of the time view the natural world. A chemist studying salinity does not check every day to make sure that the salt she is studying today is still made of atoms. That all matter is composed of atoms is a paradigm of chemistry that dictates how she views her chemical world. A marine biologist studying salmon does not check each fish to be sure it is composed of cells and that it is more closely related to trout than to tunas. That all living things are composed of cells and that they have come to be what and how they are through a long period of evolution by natural selection are paradigms of biology. Atomic theory, cell theory, and evolutionary theory along with relativity in physics, the big bang in astronomy, and *plate tectonics* in earth science are paradigms of modern science.

Scientists accept these tentative truths until they are convincingly shown to be false. While scientists function assuming current paradigms are true, they understand that every idea is subject to the continuous scrutiny that characterizes the practice of science. Thus, when the data from observation and experiment are better explained by new ideas, the older ones must give way. The origins and development of paradigms demonstrate the nature of scientific study and illustrate the impact of science on sociology and history, and even economics, religious beliefs, and the military.

This course devotes lessons three and four to the paradigm of plate tectonics. Although lesson three focuses on the story of the discovery and development of the theory, and lesson four focuses on the factual details, the facts of science and the process of science are inseparable. Together these two lessons tell the story of a grand idea. And few ideas are more central to the study of oceanography than that of plate tectonics.

The theory of plate tectonics is based on the premise that the surface of the earth is made up of about a dozen large moving plates. When we look at the Mariana Trench, the Pacific Ring of Fire, the global mid-ocean ridges, the Himalaya Mountains, and other features of the earth, we ask not if plate tectonics played some role in their creation and evolution, but rather how it did. After a long history of false starts, the ideas of *continental drift* and *seafloor spreading* combined to form the paradigm of plate tectonics.

The origins of plate tectonic theory can be found in cartography. As seafarers demanded more accurate maps during the ages of Exploration and Colonial Expansion, the remarkable matching of the west coast of Africa and the east coast of South America became obvious. In 1912, Alfred Wegener proposed the idea of continental drift based on the jigsaw puzzle fit of certain continental edges and the existence of the same variety of fossils and geological features on matching sides of the ocean. He hypothesized that all the continents of the earth were at one time combined in a supercontinent called Pangaea and surrounded by a superocean called Panthalassa. Approximately 150 to 200 million years ago Pangea began to break up into the continents and oceans that exist today.

Wegener's theory met with some acceptance. But many of his ideas were rejected, including the mechanism that he proposed could move continents. Another major obstacle was the disagreement about the nature of *isostasy*, the principle explaining how continents are supported from below. There also existed a fundamental disagreement within the scientific community about how the process of scientific study should be conducted, and whether Wegener had practiced good science.

A hypothesis proposing a mechanism that could move continents was advanced in the 1920s. The continents are massive, seemingly unmovable granite structures thousands of miles in length and width, and 60–125 miles thick. But compared to the mantle below, Earth's crust is merely a thin outer layer. If, however, the mantle were fluid, the internal heat of Earth could generate convection cells, which could produce energy sufficient to tear apart and move the pieces of crust. Early study of earthquake waves suggested to some that the mantle is solid. Convection could not exist in a solid mantle, and without convection there could be no continental drift. Wegener died on a research expedition to Greenland in 1930 before his theory of continental drift could be proved, but his idea did not die with him. A landslide of evidence would soon emerge and a historic revolution in the world of geology would be underway.

In 1935 Kiyoo Wadati suggested that volcanoes and earthquakes around Japan are associated with continental drift. In 1940 Hugo Benioff mapped the deep earthquakes in the Pacific region, clearly displaying the Pacific "Ring of Fire." The surprisingly regular and orderly arrangement of earthquakes and volcanoes surrounding much of the Pacific Rim evidenced the border of a large and active Pacific plate.

World War II and the Cold War played a critical role in the development of the next step in tectonic theory. As submarines became vital instruments of war and defense, the need to hide from and root out enemy submarines required detailed knowledge of the oceans. Naval oceanography thus made two significant contributions in the advancement of the theory of plate tectonics. First, the charting of the ocean floor using echo sounding refined the understanding of seafloor structure begun by the *Challenger* expedition in1872. Second, studies of the sea floor using instruments called magnetometers revealed strong evidence in support of seafloor spreading. Just as maps of the oceans and coastlines initiated the idea of continental drift, so maps of the ocean floor (bathymetric charts) propelled science into the investigation of seafloor spreading.

By 1960 the details of bathymetry—the mid-ocean ridge system, trenches, seamounts, and continental shelves—had been examined and catalogued. Using these and other data, Harry Hess and Charles Dietz proposed the idea of seafloor spreading. They suggested that the worldwide system of undersea mountains called the mid-ocean ridges is where new ocean floor is created as plates move apart. As they diverge, they move large crustal plates along with the continents that sit atop of them.

Charting the magnetic features of the ocean crust revealed a distinct pattern of symmetry spreading out from the mid-ocean ridges, demonstrated most clearly in the Atlantic Ocean. Molten crust that exudes at the mid-ocean ridges contains crystals of magnetite, a naturally magnetic mineral. These crystals align in the earth's magnetic field, leaving a permanent record of their "compass" position when the lava hardens. The history of the earth's magnetic characteristics is thus preserved in the seafloor. The magnetic strength, direction, and polarity on either side of the mid-ocean ridges are symmetrical.

Furthermore, evidence collected by the Deep Sea Drilling Project (1968–present) supports the ideas of Wegener, Hess, and Dietz. The ages of bands of seafloor are youngest close to the mid-ocean ridges and progress symmetrically outward to the oldest. The youngest and thinnest sediment deposits overlie the regions closest to the spreading centers and the oldest overlie the regions farthest away, just as their theories would suggest. With the addition of the Hess-Dietz theory of seafloor spreading, the idea of continental drift made more sense. It looked like Wegener had been right!

In 1965 J. Tuzo Wilson integrated Wegener's theory of continental drift with Hess and Dietz's seafloor spreading into the larger idea of plate tectonics, providing the last piece of the puzzle in plate tectonic theory—subduction. As new seafloor is produced, pushing away from a spreading center, the edges of plates elsewhere must collide. When they do, one will often slip under the other, subducting into the mantle where it liquefies. The action of plate boundaries varies with location. Sometimes they diverge (spread), sometimes they converge (collide), and sometimes they slide along a joint parallel to each other (slip).

Many factors such as wind, water, erosion, and glaciation have combined to create the features of the earth's surface on land and sea, but all pale in significance compared to the overriding paradigm of earth science—plate tectonics.

Focus Your Learning

Learning Objectives

On successful completion of this unit of study, you should be able to:

1. Understand Wegener's evidence for the theory of continental drift and how this theory was received.
2. Compare and contrast the classification of Earth's layers based on chemical composition versus their classification based on physical properties.
3. Describe the layered structure of Earth and understand how the structure was determined through the study of seismic waves.
4. Explain the concepts of buoyancy and isostatic equilibrium and their relationship to continental drift.
5. Understand the source of convection and the role it plays in plate tectonics.
6. Describe the roles of Benioff and Wadati, Hess and Dietz, Wilson, and the R/V *Glomar Challenger* in formulating the paradigm of plate tectonics.
7. Compare and contrast the terms continental drift, seafloor spreading, and plate tectonics.
8. Explain the evidence, both direct and indirect, for plate tectonic theory.

Assignments

This lesson is based on information in the following text and video assignments. The key terms, focus points, and practice test are intended to help ensure mastery of the material presented in this unit of study.

Text: Chapter 3, "Earth Structure and Plate Tectonics," pages 62–101, with particular emphasis on pages 62–76

Video: Episode 3, "Making the Pieces Fit"

Video Focus Points

Review these points before watching the video assignment for this lesson to help you focus on key issues.

— Prior to Wegener there was no satisfactory explanation for the large scale geological structure (geomorphology) of the Earth's surface.

— Wegener proposed the idea of continental drift in 1912 and supported it primarily with data based on the jigsaw puzzle fit of the shorelines of continents, as well as geological and fossil similarities on opposite sides of oceans. His idea was not universally accepted and received particular resistance from American scientists because of their different views of isostasy and the way they believed scientific study should be conducted.

— After Wegener's death several pieces of evidence were collected that supported his idea. The work of Wadati, Benioff, Hess and Dietz, and Wilson were among the most significant.

— Research on seafloor bathymetry and seafloor magnetism conducted in association with submarine warfare and evidence collected from the Deep Sea Drilling Project and via satellite also helped to prove Wegener's theory.

— The ideas of continental drift and seafloor spreading were combined with a deeper understanding of crustal movement to produce the theory of plate tectonics. This theory is the major paradigm of modern earth sciences.

Text Focus Points

Review these points before reading the text assignment for this lesson to help you focus on key issues.

— Alfred Wegener's 1915 theory of continental drift, proposed to explain why Earth's land masses look like they once fit together, was dismissed by most geologists.

— Density is defined as the mass per unit volume. This property of matter must be understood to comprehend isostasy and convection, which are central to understanding plate tectonics.

- No one has ever drilled through the crust to the mantle or core. We recognize the properties and dimensions of these layers through indirect evidence, primarily P and S seismic waves created by earthquakes and measured with seismographs.
- Earth is composed of three layers: an inner core, a large mantle, and a thin outer crust. These are sorted by density and distinguished by their chemical composition. The interior core is densest, the mantle is intermediate in density, and the outer crust is the least dense.
- When looking strictly at physical properties, Earth is composed of five layers. The solid outer lithosphere is made from the crust and solid outermost layer of the mantle. The asthenosphere is a partly melted fluid layer of mantle below the lithosphere. Below the asthenosphere is a solid lower mantle, then comes the outer core which is liquid, and the inner core which is solid.
- The most important idea of earth science is plate tectonics. Starting with Wegener's theory of continental drift, and later including the theories of seafloor spreading and subduction, the theory of plate tectonics was created in 1965. The development of and continuous research on plate tectonics has dominated earth science in the twentieth century.
- The origin and development of most of the major geologic features of the continents and oceans, such as mid-ocean ridges, mountain ranges, oceanic trenches, and island arcs, can be explained by plate tectonics.

Key Terms and Concepts

A thorough understanding of these terms and concepts will help you to master this lesson.

Asthenosphere. The hot, fluid region of the upper mantle that lies below the lithosphere. Convection in the asthenosphere creates the movements that drive plate tectonics in the lithosphere above.

Continental drift. The movement of the continents over Earth's surface. Alfred Wegener proposed this idea in 1912 based on the jig-saw puzzle fit of the shores of continents separated by ocean.

Density. The mass per unit volume of a substance. It is often expressed as grams (mass) per cubic centimeter (volume). The density of air is approximately 0.001 g/cm^3; water is 1.00 g/cm^3; granite is 2.7 g/cm^3; basalt is 2.9 g/cm^3; and iron is 7.86 g/cm^3 at the Earth's surface. The density of the iron in the Earth's core is approximately 13 g/cm^3 because it is under tremendous pressure.

Isostasy. Similar to buoyancy, isostasy refers to the way continental masses are supported by "floating" on the fluid (plastic) asthenosphere below.

Lithosphere. The rigid outer layer of Earth composed of the crust and the solid part of the upper mantle.

Paleomagnetism. The magnetic field of a rock, created when naturally magnetic minerals within the rock align in Earth's magnetic field at the time the rock is hardening from molten lava. Once the lava hardens, the fossil (paleo-) magnetic alignment of the rock becomes a permanent record of Earth's magnetic condition at the time the rock formed.

Plate tectonics. The theory that combines continental drift, sea floor spreading, and subduction. It states that the solid outer lithosphere of the Earth is divided into several large plates, which move in relation to each other. The forces that drive this motion are generated by convection in the fluid asthenosphere in the upper mantle.

Seafloor spreading. The theory that new seafloor is created at spreading centers like the mid-ocean ridges where plates diverge. Molten mantle flows up from below, hardening and creating new seafloor and the plates spread apart.

Test Your Learning

After working through these questions, check your answers against the key at the end of this book. If you answered any of the questions incorrectly, review the relevant sections of the text and video episode.

Multiple Choice

1. New seafloor is produced and spreads out from the:
   a. mid-ocean ridges
   b. deep-sea trenches
   c. boundary between continents and ocean
   d. Pacific and Atlantic Rings of Fire

2. The primary evidence used by Wegener to support the theory of continental drift was:
   a. the obvious collision of India with Asia to produce the Himalaya Mountains
   b. the presence of coal deposits with tropical fossils under the ice of Antarctica
   c. laser measurements from geosynchronous satellites showing continental movements
   d. the apparent fit of eastern South America with western Africa

3. Like evolutionary theory, atomic theory, and cell theory, the theory of plate tectonics is called a theory because:
   a. it is generally unproven and is actually a hypothesis
   b. it is one of several equally plausible explanations of the large scale structure of continents and oceans
   c. it is supported by a vast amount of empirical evidence and is the best scientific explanation of the large-scale structure of continents and oceans
   d. the mechanism that drives continental movement has never been discovered

4. The continental masses are often more than three billion years old while the seafloor is rarely older than 200 million years because:
   a. the break up of Panthalassa has destroyed all ancient seafloor
   b. the continents are remnants of ancient mantle before the crust hardened
   c. all the older seafloor exists as a basal layer under present continents
   d. new seafloor eventually subducts and melts in deep-sea trenches; this process rarely takes longer than 200 million years

5. As high continental mountains erode, the low-density mountain root supporting them will:
   a. experience isostatic uplift because the mass of the mountains is decreasing
   b. sink lower into the asthenosphere
   c. subduct and melt because of tectonism
   d. erode at a similar rate to the mountains above

6. Which of the following is true of primary (P) and secondary (S) seismic waves?
   a. P waves are produced by the earthquake and S waves are produced by the aftershocks.
   b. Both travel at the same speed initially, but S waves speed up as they pass through the Earth's core.
   c. P waves travel through Earth nearly twice as fast as S waves.
   d. P waves cannot penetrate through the molten outer core.

7. Which of these series is arranged from least dense to most dense?
   a. air, water, mantle, crust, core
   b. granite, basalt, mantle, outer core, inner core
   c. lithosphere, asthenosphere, outer core, mantle, granite
   d. asthenosphere, granite, basalt, iron, magnetite

8. The paleomagnetic patterns of the seafloor give strong evidence that:
   a. the magnetic poles are constant in strength and location
   b. the polarity of the earth's magnetic field has undergone reversals
   c. the rate of seafloor spreading is about two meters per year and is constant for all oceans
   d. periods of slowest seafloor spreading correspond with glacial periods

9. Core samples from the Deep-Sea Drilling Project reveal that:
   a. the blanket of sediment on the ocean floor is always thickest near the mid-oceans
   b. the oldest rocks of the seafloor are near mid-ocean ridges
   c. the age of seafloor is symmetrical with oldest floor farthest away from spreading centers
   d. Hawaii, Iceland, and virtually all oceanic islands were formed adjacent to subduction zones

10. Which of these is probably most important in driving the movement of plate tectonics?
    a. convection cells in the asthenosphere
    b. convection cells in the outer core
    c. centrifugal force associated with the spin of the Earth on its axis
    d. all of these combine to drive plate tectonics

11. The idea of the supercontinent of Pangaea was proposed by:
    a. Aristotle
    b. Vema
    c. Hess and Dietz
    d. Wegener

12. Benioff and Wadati are most closely associated with the study of:
    a. deep-focus earthquakes
    b. the structure of mid-ocean ridges
    c. hot spots
    d. paleomagnetism

Short Answer

1. Prepare two labeled drawings of Earth sliced in half. On the first, illustrate the layers of Earth based on their composition and on the second illustrate the layers of Earth based on their physical properties.

2. Wegener's idea of continental drift was not widely accepted when he proposed it. Briefly outline the arguments, presented by Naomi Oreskes in the video, between American and European scientists over how the scientific method should be conducted. What are your views regarding this argument?

3. The debate between the ideas of catastrophism and uniformitarianism in the eighteenth century influenced the twentieth century debates about continental drift. Discuss these two contrasting ideas, explain their effect on the continental drift proposal of Wegener, and discuss your views as to whether continental drift is a uniformitarianist or catastrophist idea (or both or neither).

Supplemental Activities

1. You are a lawyer assigned the task of arguing the case in favor of the theory of plate tectonics. Present your case. Begin with an opening argument outlining the types of evidence you will present to the jury to prove Earth is layered and that new crust is forming, moving, and subducting. Call at least eight expert witnesses (dead or alive) and present the evidentiary statements to which you would expect them to testify in support of your case. At least five of your witnesses should be scientists and at least three of your witnesses should be captains of research vessels. Summarize the arguments supporting the theory and end with "I rest my case."

2. The physical property of density and the processes of density stratification and convection are critical to an understanding of plate tectonics and many other aspects of oceanography. In your own words, define "density" and explain clearly the differences among density, mass, and volume. Take a clear glass coffee cup or Pyrex measuring cup and fill it two thirds full with hot coffee (be careful not to spill it on yourself, and do not use plain glass because heat may cause it to expand and crack.) Slowly pour some cold half and half or cream into the coffee and describe and explain what happens over the next five minutes. Observe the mixture from the side and the top occasionally for an hour (preferably while reading Chapter 3 of your textbook!) and explain what you see.

3. Plate tectonic theory is grounded in the study of maps. All flat maps give a distorted picture of Earth and you can greatly improve your understanding of tectonics, oceanography, and Earth's structure by studying a globe. Get a globe that detaches from its base so you are holding a plain sphere, a simple inflatable globe from a toy store is fine, and do the following:
    a. Construct a "harness" that fits loosely around the globe.
    b. Take a manila folder or similar thick paper and cut it into long half inch wide strips.
    c. Using the strips and scotch tape make one belt that circles the globe at the equator and another that circles the globe perpendicular to the first running along the Prime Meridian (0 degrees longitude) up and down over both poles and along the 180 degree line of longitude (roughly the International Dateline) on the other side.

The two paper belts should cross each other at right angles at the North Pole and the South Pole.

d. Tape the two belts together at the poles dividing the globe into four equal wedges. The harness should move freely over the globe; do not make it too snug.
e. Write an "E" at the North Pole and an "X" at the South Pole.
f. Make a handle to move the harness by wrapping a piece of tape around it to form a tab that you can pinch and hold.
g. Starting from the North Pole move down 103° along each strip of the harness and draw a line across each strip (the line will be at 13° south latitude which is the latitude of Brazil, Peru, north Australia, Mozambique, Zambia and Angola).
h. Do the same for 143° down from the North Pole. (143° south of the North Pole is 53° south latitude and runs entirely over ocean except where it crosses the southern most tip of South America).

4. Using the globe, answer the following questions:
   a. Many kids think if they dig a hole through Earth they will come out in China. Place the "E" over New York, Los Angeles, and your hometown, and find out where the "other side" of Earth is. Does the "X" ever fall within China? Name a country where you could live to be opposite China.
   b. Is there any place in the 48 contiguous United States where you could dig straight through Earth and come out on land on the other side? How about Alaska? Hawaii?
   c. Examine figure 3.8 (page 68) in the text to study seismic wave shadow zones. The size of the core was deduced from S wave shadow zones. S waves will be detected around Earth 210° (105° + 105°) from the epicenter of an earthquake and will go undetected over the remaining 150°(360° – 210°).
   d. Place the "E" over the epicenter of the March 27, 1964 Anchorage, Alaska earthquake. Notice that a circle connecting the 105° marks on your harness run roughly through Australia, Indonesia, Sudan, and Brazil. Seismographs north of this line would detect S waves from Anchorage and seismographs south of the line would be in the shadow zone. If we consider the continents to be North America, South America, Eurasia, Africa, Australia, and Antarctica, name the continents where every seismograph on the continent would detect S waves from an Anchorage earthquake. Name the continents where some seismographs would and some would not detect S waves from Anchorage, and name the continents where no S waves from Anchorage would be detected. Answer the same three questions if the epicenter of the earthquake is in Ankara, Turkey; Kobe, Japan; and Santiago, Chile.
   e. If an earthquake occurred in Mexico City, name five countries in which some of the seismographs would and some would not detect S waves.
   f. P waves do pass through the liquid outer core but are refracted (bent) as they go. The refraction produces a shadow zone in a belt 105° to 142° away from the epicenter. If an earthquake were centered in Los Angeles, in which of the following places would seismographs detect P waves: Pakistan, Poland, Iceland, Antarctica, Borneo, Japan, Hawaii, or Vietnam?
   g. If an earthquake were centered in Jakarta, Indonesia, name three place that would detect both S and P waves, three places that would receive neither S nor P waves, and three places that would receive P waves, but not S waves.

Lesson 4

World in Motion

Overview

In science, new answers always give birth to new questions. This is certainly true of plate tectonic theory. The proposals of Hess and Dietz and Wilson in the 1960s concerning plate tectonics did not end the quest begun by Wegener. Rather, they fostered an era of vigorous research and debate that continues today. What happens on the Earth's surface as plates move? Does the evidence accumulated up to 1965 from several places on Earth hold true for all places on Earth? Will new lines of investigation support or refute tectonic theory? Do churning *convection* cells in the asthenosphere below really drag the lithospheric plates above? If so, how exactly does that work? If not, are there different or additional mechanisms to move plates? These are the questions of pure science, the seeking of knowledge simply for the sake of understanding the natural world. Are there ramifications of plate tectonic theory for applied science (using science to improve life, generate profit, or otherwise solve specific human problems)? Perhaps tectonic theory can help us predict earthquakes, volcanic eruptions, or tsunami and thus save lives. Pure science often leads to applied science and applied science can help answer the questions of pure science.

This lesson continues investigation of the evidence that supports tectonic theory and the many still-unanswered questions about it. It focuses on plate boundaries, the places where plates adjoin, and what happens there. Plate boundaries fall into three categories. Places where plates spread apart are called divergent boundaries. Boundaries that collide are called convergent and those that slide past each other in opposite directions, are called transform boundaries.

Divergent Boundaries. The most striking examples of divergent boundaries occur at the mid-ocean ridge. This system of ridges forms a continuous mountain range running under all Earth's oceans. The Mid-Atlantic Ridge, the East-Pacific Rise, and the Carlsberg Ridge in the Indian Ocean are examples. As plates diverge at these ridges (also called spreading centers) molten rock, or *magma*, rises from the mantle below. In some places the magma oozes out forming pillow lavas as the lava solidifies, and becoming new seafloor. In other places there are explosive sub-sea volcanic eruptions that violently build the ridges as they, too, produce new seafloor. Iceland, with its famous volcanoes, hot springs, and flowing lava, is part of the Mid-Atlantic Ridge.

Divergence and rifting are not limited to the mid-ocean ridge. It was, after all, rifting that tore the giant continent of Pangaea apart. A dramatic example of diverging plates on a continent is the East African rift zone. This rift is splitting the eastern part of Africa from the rest of the continent. Much of Ethiopia, Kenya, Somalia, Tanzania, Rwanda, and Mozambique are separating from the main mass of the African continent. Volcanoes, earthquakes, and rift lakes mark this divergent boundary. Lake Tanganyika, Lake Victoria, and Mount Kilimanjaro as well as Africa's most active volcano, Erta Ale, are found on this rift, which joins another spreading center running through the Red Sea and the Gulf of Aden. Here the Arabian peninsula and the African continent are spreading apart as the Red Sea, an embryonic ocean, widens.

Convergent Boundaries. Since it is clear that Earth is not increasing in size, then it must be that as new crust forms at divergent boundaries old crust is being destroyed elsewhere. This happens at convergent boundaries where plates subduct and melt back into the mantle. Where two plates collide one will usually sink, or subduct, under the other. These regions of collision and *subduction* are Earth's most geologically violent places, marked by powerful deep earthquakes and violent volcanoes. Much of the Pacific Ocean is ringed by subducting plates. Volcanoes and earthquakes ring the Pacific rim from New Zealand to Indonesia, along the Philippines and Japan, around Alaska and the west coasts of North, Central, and South America giving it the name, the Pacific Ring of Fire. As a colder, denser plate subducts under another less dense one, both plates bend downward creating an oceanic trench. The Mariana Trench is such a trench, and one of the ocean's deepest features.

There are three major types of convergent boundaries: between an oceanic and a continental plate, between two oceanic plates, and between two continental plates. An example of convergence at an oceanic and continental boundary occurs along the west coast of South America. Here, the Nazca Plate in the Pacific Ocean collides with the South American Plate and the two plates bend downward forming the Peru-Chile Trench. As the colder Nazca plate subducts under South America, its melting spawns the volcanoes that have produced the Andes Mountains.

When one oceanic plate subducts under another oceanic plate, it produces a line of volcanic islands in the ocean rather than a range of volcanic mountains on land. The characteristic curved line of volcanic islands at convergent boundaries is called an island arc. For example, where the Pacific Plate subducts under Alaska and Asia it formed the Aleutian Islands and the Japanese Islands. Most of Indonesia and the Philippine Islands are also island arcs.

Where two continental plates collide, neither subducts. The overlying granite of continental plates has a density of about 2.7 gm/cm^3, so the two masses of granitic continental crust collide, compress, and push up lofty ranges of mountains. As Pangaea broke apart about 200 million years ago, the mass of granite that became the Indian sub-continent moved from its position adjacent to east Africa to collide with southern Asia about 45 million years ago. This relatively recent continent-continent collision created the Himalayas.

Transform (or Transverse) Boundaries. Transform boundaries exist where plates slide past each other. Because Earth is a sphere, spreading centers cannot open in a smoothly linear fashion. If you hold your two hands together, flat and with palms up, they spread evenly when you spread them apart. But if you hold them together and cup them as though you were going to drink water from a stream, as you spread them apart, keeping your pinky tips and wrists together, you'll see that the opening widens faster at its middle. A long, smoothly opening slit on a sphere is not possible. So when two plates diverge they must fracture and shear perpendicular to their length. These fracture zones and *fault* lines are obvious on any map of Earth's plate boundaries. The San Andreas Fault in California is a transform boundary. Volcanoes are rare on transform boundaries but, as anyone who lives in California knows, earthquakes abound.

The most spectacular expression of plate tectonics occurs at the boundaries, but tectonics is hardly inconsequential over the rest of Earth. Much compelling evidence in support of tectonic theory is found independent of the boundaries. The *paleomagnetism* of the continents is also instructive in studying tectonics. While the paleomagnetic stripes of the seafloor are relatively regular and predictable, the paleomagnetism on the continents is rather complex. This suggests that the poles may have wandered about the globe or that more than one north and one south magnetic pole existed at certain times in history. But wandering poles are unlikely and the existence of more than one north and one south pole is impossible. This apparent polar wandering can be explained, however, if the poles stay relatively fixed and the continents move. The paleomagnetism of the continents supports this explanation. Science has even been able to trace the path and rate of continental movement by studying continental paleomagnetism.

At certain places in the mantle, plumes of magma rise below the lithosphere, weaken it, and release heat. Expressions of this heat, including geysers, hot springs, and volcanoes, can be found over these plumes. These places are aptly called *hot spots.* They are found both on land and under the sea; both on plate boundaries and not. Easter Island, the Galapagos Islands, and Iceland are located on hot spots at plate boundaries while Hawaii, Yellowstone National Park, and the MacDonald Seamount are hot spots quite distant from any *plate boundary*.

The Hawaiian hot spot is the longest continuously active hot spot on Earth. As the Pacific Plate moves over this hot spot, a chain of volcanic islands and seamounts erupts. Imagine a lighted candle sitting on a table. If a pane of glass is held parallel to the table and moved slowly over the flame, it will leave a sooty trail on the glass showing its path over the flame. Similarly, the hot spot is presently under Hawaii's three active volcanoes—Mauna Loa and Kilauea on the "Big Island" of Hawaii and the newly forming undersea volcano, Loihi, which is southeast of Hawaii and will become the next island in the chain. The long linear chain of seamounts and islands extending from Loihi, the newest, to Meiji, the oldest, trace out the 70 million year history of the movement of the Pacific Plate over the Hawaiian hot spot. Older volcanoes in the Hawaiian chain have subducted into the Aleutian Trench, where the Pacific Plate sinks and melts. The Meiji Seamount, cold and inactive for 70 million years, will be the next to disappear into the trench. The chain of Hawaiian-Emperor seamounts is more than additional evidence of plate motion, it is a record of the direction and speed of the Pacific Plate.

Tectonic theory is corroborated by still more evidence.

— If new seafloor forms at the mid-ocean ridges and spreads away, the seafloor should be symmetric around the ridges. It is. The paleomagnetic stripes have symmetry of magnetic polarization and age on opposite sides of the ridges. The age of the basaltic seafloor is symmetrical, too, with the youngest basalts at the ridges and getting progressively older moving from ridges to trenches.

— If new seafloor forms at the mid-ocean ridges, the deposits of sediment that build up with time would be thinnest at the ridges and thickest moving away. They are.

— As plates collide, lighter granitic material and small amounts of oceanic crust and sedimentary deposits should scrape, crunch, and deform in various predictable ways leaving evidence of the collision. They do, producing distorted remnants of plate collisions called *terranes*.

Although many questions remain, evidence for the movement of plates is found virtually everywhere it is predicted. Clearly, the oceans are not merely huge puddles in the low spots between the continents. The features and characteristics of the world ocean are dynamic in both space and time, and tectonics is paramount among the forces of change.

Focus Your Learning

Learning Objectives

On successful completion of this unit of study, you should be able to:

1. Describe the three major types of plate boundaries and discuss specific examples of each.
2. Describe the formation, movement, and fate of the Hawaiian Ridge and Emperor Seamounts and explain how these features support the theory of plate tectonics.
3. Discuss the formation of terranes and how they are related to tectonic movements.
4. Compare and contrast deep and shallow earthquakes and describe the Wadati-Benioff Zone.
5. Explain how the patterns of paleomagnetism, seafloor age, and sediment thickness contribute to an understanding of tectonic theory.
6. Summarize the major types of evidence that support the theory of plate tectonics and list at least five significant unanswered questions about this process.

Assignments

This lesson is based on information in the following text and video assignments. The key terms, focus points, and practice test are intended to help ensure mastery of the material presented in this unit of study.

Text: Review Chapter 3, "Earth Structure and Plate Tectonics," pages 61–101, with particular emphasis on pages 76–101

Video: Episode 4, "World in Motion"

Video Focus Points

Review these points before watching the video assignment for this lesson to help you focus on key issues.

— Scientists concentrate much of their research on tectonics at the plate boundaries.

— There are three major types of plate boundaries—divergent, convergent, and transform.

— The East Pacific Rise, the Mid-Atlantic Ridge, and the East African Rift Valley are examples of divergent boundaries where plates are spreading apart.

— The Cascade and Andes mountains and the Japanese and Philippine island arcs were produced by subduction at convergent boundaries where plates are colliding. Oceanic trenches and the Himalaya mountains are also a result of convergence.

— Transform boundaries exist where plates slide past each other. The San Andreas Fault is such a boundary.

— Mantle plumes at hot spots can produce lines of seamounts and volcanic islands as a plate moves over the hot spot, as in the case of the Hawaiian-Emperor Chain.

— Convection in the mantle drives tectonics. Radioactive decay of uranium, thorium, and potassium supplies heat for these convections but the details of the mechanism(s) of plate movement remain in question.

— Several other aspects of plate tectonics are confusing and remain poorly understood.

Text Focus Points

Review these points before reading the text assignment for this lesson to help you focus on key issues.

— The three major types of plate boundaries are further explored. There are two primary types of divergent boundaries, three primary convergent boundaries, and transform boundaries.

— Granitic continental land masses originated from the action of convergent boundaries, where subducting surface materials are heated, compressed, liquefied, and recycled to the surface. Many geologists believe that most of the continental land mass was produced this way.

— Paleomagnetism contributes greatly to the understanding of tectonics. The study of oceanic and continental paleomagnetism has enabled scientists to estimate the rates of plate movement and apparent polar wandering, and demonstrate the periodic reversals of Earth's magnetic field.

— Many of Earth's volcanic islands, seamounts, and guyots are formed by hot spots. Study of these features has helped confirm tectonic theory.

— Tectonic theory helps explain patterns of age and distribution of oceanic sediments and the structure of terranes. Although the break-up of Pangaea began about 200 million years ago, remnants of ancient plate collisions found in terranes, such as ophiolites, provide evidence of tectonic events more than 1.2 billion years old.

— The theory of plate tectonics is not totally explained. Several significant and vexing questions remain unanswered.

Key Terms and Concepts

A thorough understanding of these terms and concepts will help you to master this lesson.

Convection. The physical movement of a fluid (liquid or gas) driven by differences in density within the fluid. Warmer fluid will rise and cooler fluid will sink creating a vertical loop of the material called a convection cell. Convection cells in the mantle contribute to plate movement.

Fault. A fracture in the earth along which movement has taken place

Hot spot. A place in the crust heated from below by a rising plume of hot mantle. Volcanoes, geysers, and other expressions of heat mark hot spots.

Magma. The term for fluid rock when it is found within Earth; above ground it is called lava.

Mantle plumes. An upwelling of hot mantle rock that rises and spreads outward.

Paleomagnetism. The study of ancient magnetic fields.

Plate boundary. Boundary between crustal plates. They may be divergent boundaries where plates spread apart, convergent boundaries where plates collide, or transform boundaries where plates slide past one another.

Radioactive decay. Certain forms (isotopes) of elements convert to other forms, producing radiation in the process. Radiation produced by decaying uranium and other isotopes generates much of the heat that drives convection in the mantle.

Seismic tomography. A method of studying "slices" of the internal structure of Earth. It is somewhat like a CT scan of a medical patient except the images are generated from the passage of seismic waves through different regions of Earth rather than X-rays through different regions of a body.

Subduction. The downward movement of a lithospheric plate into the mantle. Subduction occurs at convergent boundaries producing oceanic trenches, and may carry material as deep as the core in some places.

Terrane. Fragments of seafloor, sediments, islands, or other crustal features transported by plate movements and deposited on continental masses by plate convergence. They may travel great distances before they are deposited and are often of a composition distinctive from the larger mass.

Transform fault. A fracture zone along which the two sides slide horizontally past each other.

Wadati-Benioff zone. More commonly called a subduction zone, a Wadati-Benioff zone is characterized by the formation of trenches, deep-focus earthquakes, and volcanism.

Test Your Learning

After working through these questions, check your answers against the key at the end of this book. If you answered any of the questions incorrectly, review the relevant sections of the text and video episode.

Multiple Choice

1. Which of the following statements about subduction zones is false?
   a. they have active volcanism and can produce island arcs
   b. they are located adjacent to mid-ocean ridges
   c. they are associated with belts of deep-focus earthquakes
   d. their tectonic activity is assisted by sinking lithospheric plates

2. The islands and seamounts of the Hawaiian chain:
   a. are older than those of the Emperor chain
   b. are younger than those of the Emperor chain
   c. are the same age as those of the Emperor chain
   d. in both chains the islands are older while the seamounts are younger

3. Which of these was formed by a convergence of two continental plates?
   a. the Himalaya Mountains
   b. the Mariana Trench
   c. the Aleutian Islands
   d. Chesapeake Bay

4. Which of these was formed by an oceanic-continental convergence?
   a. Rocky Mountains
   b. Mariana Trench
   c. Andes Mountains
   d. Emperor Seamounts

5. Which of these was formed by a divergence?
   a. Himalaya Mountains
   b. Philippine Islands
   c. Mariana Trench
   d. East African Rift

6. Which of the following was formed by transform faulting?
   a. Himalaya Mountains
   b. Mariana Trench
   c. Aleutian Islands
   d. San Francisco Bay

7. Assemblages of oceanic lithosphere obducted onto the edges of continents:
   a. terranes
   b. ophiolites
   c. guyots
   d. seamounts

8. Which of these is not an island arc?
   a. Mariana Islands
   b. Japanese islands
   c. Aleutian Islands
   d. Hawaiian Islands

9. Because of plate tectonics, you would expect to find the thickest layers of ocean sediments:
   a. along the bases of mid-ocean ridges
   b. in the central rift valley of mid-ocean ridges
   c. in the centers of abyssal plains
   d. near subduction zones

10. Of the many aspects of the theory of plate tectonics, which of the following is least understood?
    a. why continental crust is less dense than oceanic crust
    b. the mechanism of plate movement
    c. the reason for the Ring of Fire
    d. the rate at which plates move

11. Which of the following would have the least amount of volcanism?
    a. convergent boundaries between oceanic and continental plates
    b. convergent boundaries between two oceanic plates
    c. transform boundaries
    d. divergent boundaries

12. Earth's strongest and deepest earthquakes are associated with:
    a. subduction zones
    b. spreading centers
    c. terranes
    d. earthquakes in all these places typically register 7–10 on the Richter scale

Short Answer

1. One key to understanding plate tectonics is symmetry. Certain features of the seafloor are symmetrical on opposite sides of spreading centers at mid-ocean ridges. Describe at least three of these symmetrical features.

2. Discuss the relationship among the Hawaiian Islands-Emperor Seamount Chain, the Aleutian Trench, and plate tectonics.

3. Outline the stages of the Wilson cycle. List the motion, features, and an example of each stage.

Supplemental Activities

1. There is evidence that Mars once had tectonic activity, but because of its greater distance from the Sun and its smaller size it has cooled sufficiently to halt its tectonism. Describe three effects a halting of plate tectonics might have on Earth.

2. Using household items, devise ways to demonstrate to an elementary school child the three major forces—slab-sink, ridge-push, and convective drag—that are believed to be responsible for plate tectonics.

3. Both the text and the video discuss some of plate tectonics most baffling mysteries. Choose two of these unanswered questions and describe what scientists would have to do, find, or figure out in order to answer them.

Lesson 5

Over the Edge

Overview

The discovery and study of seafloor contours, *bathymetry*, goes back to the voyagers and was an important part of early cartography. The first "soundings" of the ocean floor were probably made by dropping weighted lines to the seafloor and measuring the length of the line let out. As late as the 1800s, seafarers were still using this primitive system, improved somewhat on the *Challenger* expedition when a powered winch was used to wind in the line.

In the early 1900s, stimulated by the *Titanic* disaster and improved technology, the weighted line was replaced by a crude echo sounder, from which sound waves were bounced off the sea floor and their transit-return times measured. From 1925 to 1927, the German ship *Meteor* profiled the Atlantic, revealing the Mid-Atlantic Ridge system and eventually leading to the study and understanding of plate tectonics.

While echo sounding was a major advance over the weighted line, the technique still lacked accuracy because of alterations in the sound waves caused by variations in water temperature, pressure, and salinity, and by abrupt seabed contour changes. Current technology is considerably improved. One technique currently in use employs a sound (multibeam) system to detect and measure ocean depths and contours directly; another (satellite altimetry) maps the seafloor using the response of sea surface to gravitational attraction from seafloor features.

Current understanding of plate tectonics tells us that Earth's surface is a constantly shifting mosaic of lithospheric plates with the lighter continental lithosphere floating in isostatic equilibrium above the denser ocean basin lithosphere. The important features defining these two regions are based on where they actually meet; the transition between the thick but light granite of the continents and the thin but denser basalt of the deep-sea floor. This transition zone forms the true edge of the continents and divides the sea floor into two distinct provinces—the continents and their submerged outer edges, collectively called the *continental margin*, and the deep sea floor beyond those margins, called the *ocean basins*.

There are two basic types of continental margins formed by tectonic activity. *Passive margins* develop at the edges of diverging tectonic plates; *active margins* occur near the edges of converging plates. These descriptive terms help characterize them. There is little earthquake or volcanic activity around the passive margins (i.e., Atlantic), whereas earthquakes and volcanoes are typical of active margins (i.e., Pacific).

Continental margins can again be divided into two features; the shallow, somewhat flat near-shore shelf and the steeper slope. Moving out from land, through the intertidal zone and into the subtidal, is the *continental shelf*, an extension of the continent composed of the granitic rock of the continental crust. This region tends to mimic continental features, with hills, valleys, canyons, and even mineral and oil deposits. It is difficult to generalize about their depth or distance from shore, since they differ greatly by geographic region. Passive margins tend to have broad shelves, while those at active margins may be steep. Overlying the granitic extension of the continental land mass is a thick (sometimes miles deep) layer of material that has eroded from the continent. The broad shelves of tectonically quiet areas, like those found off Siberia, may be as wide as 800 miles. Tectonically active shelves, like those in the Pacific, may be abbreviated and steep, defined more by faulting, volcanoes, and tectonics than by sedimentation.

There are exceptions to this general passive/active boundary description. A few broad shelves are seen in the Pacific, caused by the same type of sediment-trapping dams that are common in the Atlantic. Due to their gentle slope and proximity to the land mass, continental shelves are greatly influenced by changes in sea level over time, and their topography is changed by rivers and ocean currents. The continental shelves are a prime focus of exploration for natural resources. With an average depth of 250 feet, they are easily accessible for mineral mining and recovery of oil.

At the outer edge of the continental shelf there is a sudden steepening of the sea floor called the shelf break. This transition from the shelf to the *continental slope* occurs at a depth of approximately 140 meters with few exceptions. The slope itself is built of sediments that have been transported across the shelf and over the edge, and may include material scraped from a descending plate during subduction at more active plate margins.

Continental slopes may be incised with valleys and canyons. Canyons are a feature of continental slopes worldwide at both active and passive margins, and can be formed in solid rock or sediments. *Submarine canyons* are most likely the seaward extension of canyons created on the continental shelf during periods of low sea level that continued to form through erosion, landslides, and downhill currents of mud and sand after sea level rose once more. The bottom of the slope is the true edge of the continent.

Along the passive margins of the oceanic crust is an apron of accumulated sediment called the *continental rise* covering the lower part of the slope. The slope is formed mainly of sediments from the shelf, but the rise is the result of *turbidity currents* transporting material from the ocean floor and depositing it along the irregular margins of the basins.

The recently-discovered *hydrothermal vents* are one of the more interesting features of the deep ocean basins, not only for their geology but for the unique and bizarre life forms found there. These vents are formed when seawater moves down through crustal fissures and cracks, coming into contact with the hot rocks associated with seafloor spreading at ridge formations. The heated water dissolves minerals and gases and brings them back to the seabed through convection, often forming large chimney-shaped mineral deposits. These unique formations were discovered in 1977 in nearly 10,000 feet of water. The water temperature there can reach 350°C compared to ambient temperatures of 3–5°C.

Hydrothermal vents represent an entirely new ecosystem supported not by photosynthesis but by geochemical energy from Earth's interior. Scientists estimate that there may be hundreds of new species unique to the vents. With the development of submersible vehicles, both manned and unmanned, and ever-increasing depth capability and sophistication of instruments, the study of these vents has become big science, particularly for biologists and for mineralogists interested in their formation and the potential economic value of the minerals found there.

Another deep-seabed characteristic is the relatively flat, featureless *abyssal plains*. These areas of sediment-covered sea floor lie between the continental margins and *oceanic ridges*. Their apparent flatness is the result of the layers of sediments covering them, often up to 1,000 meters thick. Most of this sediment, even in the deepest basins, appears to be of terrestrial origin, possibly transported from the continent by surface winds or by undersea turbidity currents. If the plains sediments are not thick enough to totally cover the outer ridge features, *abyssal hills*—small, sediment-covered extinct volcanoes or magmatic extrusions—may be evident. Other seabed features that do not rise above the sea surface are *seamounts*—steep-sided volcanic projections more than one kilometer high, and flat-topped seamounts called *guyots*, which once approached or rose above the sea surface, only to be worn down by wave action. *Trenches* are flat-bottomed "creases" in the sea floor that occur at subduction zones. Massive earthquakes and their resulting tsunami are often associated with these tectonically active areas.

Trenches are the deepest places in the crust; the deepest found so far being the Mariana Trench, which is over 11,000 meters deep, and the site of the famous 1960 dive of the bathyscaphe *Trieste*. *Island arcs* are associated with the concave edges of these trenches. They are typically curving chains of volcanic islands and seamounts that form as magma and lava rise and sometimes surface to form chains of islands behind the trench.

New technology has enabled astonishing advances in the ability to visit and document Earth's sea floor. But each new voyage reveals how much is left to learn just to catch up with terrestrial geologists! One next logical step might be the establishment of highly-instrumented *in situ* monitoring stations on the sea floor, gathering data continuously and sending it back on request.

Focus Your Learning

Learning Objectives

On successful completion of this unit of study, you should be able to:

1. Discuss the history of bathymetry from the early voyagers to the scientists of today.
2. Understand the modern techniques and technologies (both shipboard and satellite) used to describe and study the ocean floor and the variety of specially-designed and instrumented vessels.
3. Describe the features and boundaries characterizing the continental shelf, shelf break, slope, and rise.
4. List the various types of continental margins and their relationships to the plate margins.
5. Compare and contrast manned and unmanned undersea research vessels regarding their basic construction, operation, and capabilities.
6. Explain Earth's oceanic ridge system and how it relates to the lithospheric plates.
7. Understand the hydrothermal vent phenomenon—how it works, its biological and physical aspects, and its potential for research and exploitation.
8. Describe the abyssal plain—its extent and boundaries, relief features, and sources of sediment.

Assignments

This lesson is based on information in the following text and video assignments. The key terms, focus points, and practice test are intended to help ensure mastery of the material presented in this unit of study.

Text: Chapter 4, "Oceans Basins," pages 102–129

Video: Episode 5: "Over the Edge"

Video Focus Points

Review these points before watching the video assignment for this lesson to help you focus on key issues.

— New bathymetric technologies have enabled scientists to profile the continental margins and the deep sea floor in great detail.

— In addition to their scientific interest, some offshore and seabed features are considered resources for the recovery of oil and gas, high-grade minerals, and new fuels.

— The processes that occur at plate margins and ridge systems, and how they relate to tectonics, are becoming much better understood.

— The sediments that characterize the continental slopes and rises and the abyssal plains are generally thought to be of continental shelf and terrestrial origin, transported by turbidity currents.

— Submarine canyons are important features of the continental shelf and slope that probably formed when the areas were exposed during periods of low sea level and continued to restructure after submergence.

— Hydrothermal vents are the focus of much attention, both for their physical structure and for the unique biological systems they support.

— The abyssal plain, once described as flat and featureless, actually has a number of relief structures—seamounts, guyots, and abyssal hills.

Text Focus Points

Review these points before reading the text assignment for this lesson to help you focus on key issues.

— The history of bathymetry begins with the crude "soundings" of the voyagers and leads to the famous 1960 dive of the bathyscaphe *Trieste* into the Mariana Trench.

— Equipment and technology used in doing bathymetric studies now includes both shipboard and satellite imaging, and manned and unmanned submersibles.

— Description of the oceanic component of Earth's crust is generally divided into two major regions—the continental margins and the ocean basins.

— The continental margins are further characterized by their relationship to the lithospheric plate boundaries and to the continental land mass—shelf, shelf break, slope, rise, and submarine canyons—and can be either active or passive.

— The true edge of the continents is marked by the transition from the granitic structure of the continents to the basaltic structure of the deep sea floor.

— The deep ocean basins consist of two main features—the oceanic ridge system and the sediment-covered abyssal plain.

— The rift zones associated with oceanic ridges are where new ocean floor is formed.

— Hydrothermal vents are a recently-discovered feature of the ocean basins, and are often associated with ocean ridges in areas of rapid seafloor spreading.

— Hydrothermal vents are of great interest to scientists, not only for their physical structure and unique mineral composition, but for their bizarre ecosystems, which are totally supported by geochemical energy.

— The area of deep ocean floor not occupied by the ridge system is called the abyssal plain—a flat expanse of sediment-covered sea floor, interrupted occasionally by such relief structures as seamounts and guyots.

— Trenches are the deepest places in Earth's crust, and are associated with subduction activity at converging plate boundaries. At the concave edges of the trenches are often found island arcs—curving chains of volcanic islands and seamounts.

Key Terms and Concepts

A thorough understanding of these terms and concepts will help you to master this lesson.

Abyssal hill. Small sediment-covered inactive volcanos or intrusions of molten rocks less than 200 meters (650 feet) high, thought to be associated with seafloor spreading. Abyssal hills punctuate the otherwise flat abyssal plain.

Abyssal plain. Flat, cold, sediment-covered ocean floor between the continental rise and the oceanic ridge at a depth of 3,700 to 5,500 meters (12,000 to 18,000 feet). Abyssal plains are more extensive in the Atlantic and Indian oceans than in the Pacific.

Active margin. Continental margins near an area of lithospheric plate convergence; also called a Pacific-type margin.

Bathymetry. The discovery and study of ocean floor contours.

Continental margin. The submerged outer edge of a continent, made of granitic crust; includes the continental shelf and the continental slope.

Continental rise. The wedge of sediment forming the gentle transition from the outer (lower) edge of the continental slope to the abyssal plain; usually associated with passive margins.

Continental shelf. A gradually-sloping submerged extension of a continent, composed of granitic rock overlain by sediments; usually having features like the edge of the nearby continent.

Continental slope. A sloping transition between the granite of the continent and basalt of the seabed; considered to be the true edge of a continent.

Fracture zone. An area of irregular, seismically-inactive topography, marking the position of once-active transform faults.

Guyot. A flat-topped, undersea volcanic peak. Guyots are seamounts eroded flat by waves when their tops were at the ocean surface. Their subsurface location can be the result of either subsidence or rising sea level.

Hydrothermal vent. A spring of hot, mineral-and-gas-rich seawater found on some ocean ridges in zones of active seafloor spreading. Water temperatures may reach 350°Celsius.

Ice age. One of several periods of low temperature during the last million years; each period lasting several thousand years. Glaciers and polar ice were derived from ocean water, lowering sea levels at least 100 meters (328 feet).

Island arc. An island chain that forms in a characteristic arc adjacent to a subduction zone as a descending plate melts, producing volcanic islands above. The Aleutian Islands are an example.

Methane compounds (methane ice). Methane gas is formed by microbial communities on the continental shelves. Under the right conditions of pressure and temperature it can form a solid compound that may have importance as a fuel resource.

Ocean basin. Deep ocean floor made of basaltic crust.

Oceanic ridge. Young seabed at an active spreading center, often unmasked by sediment, bulging above the abyssal plain; considered to be the boundary between diverging plates. Often called a mid-ocean ridge, even though less than 60 percent of the length exists at mid-ocean.

Passive margin. Continental margins near an area of lithospheric plate divergence; often referred to as an Atlantic-type margin.

Remotely-Operated Vehicle (ROV). An unmanned submersible, either cable-tethered to a ship or unattached and operated electronically. Used to gather data over long periods of time; cheaper and safer to operate than people-occupied vehicles.

Seamount. A conical mountain rising 1,000 meters or more above the sea floor.

Submarine canyon. A deep, V-shaped valley running roughly perpendicular to the shoreline and cutting across the edge of a continental shelf and slope.

Trench. An arc-shaped depression in the deep-ocean floor with very steep sides and a flat, sediment-filled bottom coinciding with a subduction zone; usually found in the Pacific.

Turbidity current. An underwater "avalanche" of abrasive sediments, thought to be responsible for the deep sculpturing of submarine canyons and a means of transport for sediments accumulating on abyssal plains.

Test Your Learning

After working through these questions, check your answers against the key at the end of this book. If you answered any of the questions incorrectly, review the relevant sections of the text and video episode.

Multiple Choice

1. A multibeam system is basically:
   a. a radar system
   b. an echo sounder
   c. a laser system
   d. a light-beam system

2. Satellites can profile seabed features by measuring variations in:
   a. basalt density
   b. granitic rock density
   c. reflected light
   d. sea surface shape

3. The crust under a continental shelf is composed of:
   a. basalt
   b. continental sedimentary rock
   c. granite
   d. limestone

4. Remotely-Operated Vehicles (ROVs) used for deep-sea work have the advantage of being ________________ than are people-carrying vehicles.
   a. cheaper and safer
   b. able to carry more sophisticated equipment
   c. able to stay underwater for longer periods of time
   d. able to cover more area per dive

5. Continental shelves are primarily shaped when:
   a. sea level is high
   b. sea level is low
   c. there are turbidity currents
   d. the rift zones are active

6. A flat-topped seamount is called a:
   a. flat-topped seamount
   b. protrusion
   c. guyot
   d. abyssal hill
7. A continental rise is made of sediment primarily from:
   a. river mouths
   b. sediments transported by turbidity currents
   c. abyssal plains material
   d. the shelf break
8. If the ocean evaporated, Earth's most obvious feature would be:
   a. the abyssal plains
   b. the continental slope
   c. the deep trenches
   d. the oceanic ridge system
9. The superheated water associated with hydrothermal vents is moved by:
   a. conduction
   b. convection
   c. turbidity currents
   d. near-surface mantle plumes
10. The deepest places on Earth are:
   a. island arcs
   b. guyots
   c. trenches
   d. abyssal plains
11. Submarine canyons are most often found on:
   a. continental shelves and slopes
   b. continental rises
   c. transition zones
   d. the sides of seamounts
12. Transform faults are fractures along which lithospheric plates move:
   a. vertically
   b. rapidly
   c. diagonally
   d. horizontally

Short Answer

1. Since their discovery in 1977, deep-sea hydrothermal vents have attracted the attention of many types of scientists. Now, with the study and description of the large seafloor structures called "black smoker chimneys," there may be another player in this drama—commercial interest in those structures as a resource for highly-purified minerals. Think of all the different factions with potential interest in these amazing features. Comment briefly on each and indicate what their particular focus might be. Consider the federal government (both military and civilian), academic and private research (with their wide range of specialties), and more commercial interests.
2. Turbidity currents are blamed for transporting significant amounts of sedimentary material onto and around undersea features. Write a basic description of a turbidity current, then list some of the seabed features and areas where turbidity currents are most active.
3. A general description of the deep ocean floor is a mixture of ocean ridge systems and adjacent sediment-covered plains. The ridge systems draw much interest because of their active rift zones, spreading centers, faults, and vents, making the abyssal plain seem dull and featureless by comparison. But since the abyssal plains and the abyssal hills cover about one-fourth of Earth's surface, they are worthy of some attention. Briefly describe the abyssal plains and the abyssal hills.

Supplemental Activities

1. Your text refers to the idea of comparing seabed features and topography to terrestrial mountains, plains, canyons, and hills. Get a topographic (topo) map of your region from a local recreational supply, map store, or USGS office and look at the features. (An Internet search for USGS maps will also give you information on how to find topographical maps.) A geology textbook or lab manual will have instructions on how to read maps (contours, units of measurement, etc.). Note how they depict the features—canyons, washes, gullies,

hills, and mountains. Then visit a nearby canyon or hill that is on your map, and get a sense of those features and their dimensions—they are essentially the same on land as they are under water, and are measured and mapped in about the same way, although underwater maps include depth. If there is a marine supply store nearby, or a good Internet site (see your text for Hyperlink addresses), look at a hydrographic survey chart/map, and note the similarities and differences in their presentation (fathoms, feet, meters). Also, look at your text (Figure 4.17 on page 114) depiction of the Hudson Canyon. Write a summary comparing and contrasting the terrestrial techniques and presentations with bathymetric charting.

2. Imagine yourself as a marine geologist making your first trip in the deep submersible *Alvin*. Using the magic of electronics, and your imagination, take a virtual dive in *Alvin* by calling up submersible *Alvin* on the Internet. Check out this site: www.ocean.udel.edu/deepsea/dailynews/neatstuff/day/insidealvin.html

   Keep a running journal of your experience, concentrating on your emotions and feelings as you are introduced to this amazing technology. Do you feel claustrophobic, and would this hamper your ability to do good observations? Are you overwhelmed by the instrumentation, or do you see it as a "magic wand" into the deep sea realm? When you reach the virtual bottom on the Internet journey, choose one or more of the activities offered and continue your documentation and journaling.

3. Look at the hand-drawn map of the Atlantic sea floor (Figure 4.22 on page 118) and find the tiny island of Iceland, lying right on the Mid-Atlantic Ridge—an active spreading center! Starting with the comments in the text, expand on what it might be like to be an Icelander. Check out a book about Iceland from the library, or search on the Internet for basic facts about the history, people, and culture of Iceland. What is Iceland's history? Where did the first settlers come from (or does anyone really know)? How did they get there and when? How do the present inhabitants get their power, food, and other supplies? Are there automobiles, trains, and airports? Maybe, with a little ingenuity, you can locate an Icelander (in person or on the Internet) and "chat" with her/him about what it's really like to live on what amounts to an active volcano! Journal your search, including your chat, and keep in touch with your new friend.

Lesson 6

The Ocean's Memory

Overview

The statement has been made that "when we study sediments we see the true face of the ocean floor." Very little of the seabed is totally free of overlying *sediments*, which can vary in thickness from a few meters to several kilometers. These sediments may be permanent features or just passing through, carried by turbidity currents. Sources of seabed sediments range from the remains of once-living organisms to material washed off the terrestrial land mass or blown in from Saharan storms, and even some (microtektites) resulting from extraterrestrial events impacting Earth. Scientists study marine sediments for their composition, formation, distribution, and use as a habitat for marine organisms. Marine sediments are also of economic interest as a source of oil and gas and the mining of *sand* and gravel, manganese and phosphorite nodules, and occasionally even diamonds.

Sediments are also of historical value. They can reveal the recent history of an ocean basin, the changes in climate over time, and the story of *Earth's* shifting *magnetic field*. Seafloor sediments have been documented as a part of marine science since Cook's voyages. He collected samples wherever he went, and carefully described and recorded what he saw. Cook's sampler was a wax-coated lead weight. The *Challenger* used wax-tipped poles and weights, mechanical grabs, and trawl nets to collect both geological and biological samples. Crude coring devices, grab samplers, and scoops have also been tried over time. Today's researchers employ a variety of coring and drilling devices, as well as deep-water cameras and seismic profilers to collect sediments.

The major difference between the collection of marine and terrestrial sediment samples is the modifications made to the vessels and sampling gear to retrieve material from the ocean. The study of marine and terrestrial sediments employ essentially the same techniques and technology. The basic analytical criteria are the appearance of the sediment, particle (grain) size, origin or source of the material, distribution, and location on the continental margin or in the deep sea.

The appearance of both a sediment sample and the site where it was collected can provide valuable information. A photograph of the collection site can reveal whether it is smooth or rippled; associated with a relief feature such as a seamount, ridge, or guyot; and if there is life visible in the area. Color, whether from a photograph or in a core sample, may be an important indicator of the sediment's origin or *mineral* content.

The particle or grain size is measured to determine its probable origin. Particle size is based on a standard scale used by both terrestrial and marine scientists, ranging from the largest (*boulders*) to the smallest-known sedimentary material (*clay*). Since pure, or unmixed, sediments are rare, analysis involves sorting by methods such as passing it through a series of sifting screens and measuring the settling rate. Determining the sorting profile of a sample can often reveal the environmental history of the material.

Another criterion for sediment study is how the material was formed, or its source. The origin of sediments can be broken into four categories—terrigenous, biogenous, hydrogenous, and cosmogenous. It is rare to find a sample derived from a single source. The pattern and composition of various seabed sedimentary strata can preserve a record of both past and present environmental conditions.

— *Terrigenous sediments* are the most abundant type, originating on nearby islands or continents, as the name suggests. Granite—the most common continental rock—breaks down into quartz and feldspar to form sands and clays. These particles are then carried by wind or water (rivers, streams, ocean currents) across the continental shelf and onto the deep-sea floor.

— *Biogenous sediments*, as the name also implies, are of mainly siliceous or carbonaceous composition from the skeletal remains of organisms. These organisms may have originated in fresh water and been carried to sea. They could also have originated in the ocean as tiny planktonic plants or animals, or larger types like mollusks and corals. These sediments are of particular interest as sources of oil and gas.

— *Hydrogenous* (or *authigenic*) *sediments* are minerals that have been dissolved in either fresh water or sea water and then precipitated out under certain conditions. The best-known hydrogenous sediments are manganese nodules, first recorded during the *Challenger* expedition. These, and the less-common phosphorite nodules, are a potential mineral resource.

— *Cosmogenous sediments* are the least abundant but possibly the most interesting. These sediments may be tiny, *silt*-sized micrometeorites from an asteroid or comet fallen through Earth's atmosphere, or from a large asteroid actually impacting Earth. *Microtektites* are the most spectacular of these particles, having the appearance of tiny, tear-shaped oblongs of cosmic glass.

Sediments are also classified by geographic location as either neritic—terrigenous material associated with the continental shelf—or *pelagic*—on the slope, rise, or sea floor. Over 70 percent of all marine sediments (by volume) are associated with continental rises. On the continental margins most terrigenous sediment is carried to the oceans by streams and rivers, to be further distributed by currents and wave action. Larger particles are sorted out and deposited close to shore, while the silts and clays are usually washed to the shelf break and then out to the deep-sea floor. Factors such as glaciation and turbidity currents can also be involved in moving sediments. Under certain conditions, such as under pressure from overlying strata, sediments can lithify, or turn into sedimentary rock. Deep-ocean basin sediments can vary in thickness depending on location, the number of source streams and rivers, and the ability of the local terrain to trap or dam the loose material.

A number of deep-sea sediments have been thoroughly studied and described. *Turbidites* result from the action of turbidity currents carrying material from submarine canyons to the rise and even the abyssal floor. Clays are the finest of the sediments. They are transported and distributed by currents, often taking hundreds or even thousands of years to settle out. One of the more interesting deep-sea sediments is the *ooze*. These biogenous materials are composed of the shells and skeletons of tiny marine organisms, and named for the type whose skeletal remains are most abundant. Hydrogenous sediments are also found on the deep-sea floor, usually formed from the chemical reaction of dissolved minerals with the dominant sediment type of the area or organic artifacts such as shark teeth or microskeletons.

Marine sediments have much to reveal about the history of our lands and oceans. The science of *paleoceanography* has recently arisen as a result of advances in deep-sea drilling and profiling technology. Projects engaged in such study include the Deep Sea Drilling Project (DSDP) and its current successor, the Ocean Drilling Project (ODP) which is benefiting from state-of-the-art technology aboard the JOIDES ship *Resolution*.

Focus Your Learning

Learning Objectives

On successful completion of this unit of study, you will be able to complete the following tasks.

1. Characterize the distribution of the major sea-floor sediment types, and how they can be associated with the more sediment-free relief features.

2. Discuss the role of sediments in constructing a history of recent seafloor formation.
3. List and describe the major criteria used to classify marine sediments.
4. Compare and contrast the sediments of the continental margins and those of the deep-sea floor.
5. Understand the biological and chemical diversity of the biogenous sediments.
6. Discuss the growing commercial interest in marine sediments, including their potential as a resource and recovery methods.
7. Characterize the technologies involved in seafloor sediment study: locating, profiling, recovering, laboratory analysis.
8. Understand the basic processes involved in the formation of oil and gas in marine sediments.
9. Explain how knowledge of marine sediments can aid in describing Earth's historical climatic changes and magnetic field orientation.

Assignments

This lesson is based on information in the following text and video assignments. The key terms, focus points, and practice test are intended to help ensure mastery of the material presented in this unit of study.

Text: Chapter 5, "Sediments," pages 130–153

Video: Episode 6, "The Ocean's Memory"

Video Focus Points

Review these points before watching the video assignment for this lesson to help you focus on key issues.

— Marine sediments are an important resource in piecing together information about former ocean basins, past climates, changes in Earth's magnetic field, and ice ages.

— Marine sediments have been studied by scientists for many years, and interest has been expressed in them recently as oil, gas, and mineral resources.

— Most marine sediments actually originate on land and are carried to the sea by streams, rivers, and winds.

— Biogenous sediments are of interest to science because of the types of organisms that form them, and because they are involved in oil and gas formation.

— Some of the first collections and descriptions of marine sediments were made by early marine science expeditions, using crude bottom dredges and wax-coated weights.

— Present technology employs specially-designed vessels and sophisticated coring and drilling apparatus, along with magnetometers, deep-sea photography, and satellite sensing.

— Paleoceanography is a relatively new science, based on the state-of-the-art sample location and retrieval technology, and concurrent seafloor imaging and description.

Text Focus Points

Review these points before reading the text assignment for this lesson to help you focus on key issues.

— Most of the sea floor is covered by sediment of various types and thicknesses. Analysis of this material, its origin and distribution, can reveal much about the recent history of Earth, formation of the ocean basins, climatic changes, and the orientation of Earth's magnetic field.

— Sediment is generally described as "unconsolidated organic or inorganic particles." Its location on the sea floor, appearance, and its origin and formation are some of the many criteria used to classify and describe it.

— Seabed sedimentary materials are collected and described from core samples, deep-sea photographs, and seismic profiles, then studied in the laboratory for their finer details.

— One of the first steps in analyzing both marine and terrestrial sediments is determining the ratios of the various sediment particle sizes using standard methods and descriptions.

— Sediments are also classified according to their origin—where they came from and how they were formed. These four basic subcategories are named for their material of origin—terrigenous, biogenous, hydrogenous, or cosmogenous.

— Marine sediments tend to vary in thickness, general character, and composition between the nearshore (continental shelf) and the deep ocean floor.

— Sediments associated with continental margins are mainly of continental origin (terrigenous). Nearly three-fourths of all marine sediment is associated with continental slopes and rises.

— Biogenous sediment becomes more common moving seaward from the continental margin. A common biogenous sediment called *ooze* is usually composed of the remains of single-celled drifting organisms whose shells sink to the sea floor.

— Hydrogenous sediments result from chemical reaction between materials dissolved in the seawater and particles on the ocean floor.

— Sediments have been of interest since the early marine science expeditions, but the drillships of today are equipped to do very precise sea-floor description and sample retrieval. Of particular note is the Ocean Drilling Program, using the JOIDES drillship *Resolution.*

— Much new information is being gathered about the history and composition of the ocean basins, giving rise to some re-evaluation of the old concepts about sediment distribution and plate tectonics.

— Along with the science being done, the ability to accurately sample and describe sediments and their locations has renewed interest in their potential as resources for oil, gas, sand, gravel, and minerals such as manganese and phosphorite.

Key Terms and Concepts

A thorough understanding of these terms and concepts will help you to master this lesson.

Authigenic sediment. Sediment formed directly by precipitation from seawater. Also known as hydrogenous sediment.

Biogenous sediment. Sediment of biological origin. Organisms can deposit calcareous (calcium-containing) or siliceous (silicon-containing) residue.

Boulder. The largest (coarsest) sediment particle, larger than 25.6 cm (10 inches) in diameter.

Calcareous ooze. Ooze composed mostly of the hard remains of calcium-carbonate-containing organisms.

Calcium carbonate compensation depth. The depth at which the rate of accumulation of calcareous sediments equals the rate of dissolution of those sediments. Below this depth, sediment contains little or no calcium carbonate.

Clamshell sampler. Sampling device used to take shallow samples of the ocean bottom.

Clay. The smallest sediment particle; smaller than 0.004 millimeters in diameter.

Coccolithophore. A very small planktonic alga carrying discs of calcium carbonate, which contributes to biogenous sediments.

Cosmogenous sediment. Sediment of extraterrestrial origin.

Diatom. Earth's most abundant, successful, and efficient single-celled phytoplankton. Diatoms possess two interlocking valves made of silica, which contribute to biogenous sediments.

Earth's magnetic field. Consider Earth as a huge magnet, with North and South (+ and –) poles. The magnetic field would be the forces associated with that North/South orientation, as it affects Earth's surface.

Evaporite. Deposit formed by the evaporation of ocean water.

Foraminiferan. One of a group of planktonic amoeba-like organisms with a calcareous shell, which contributes to biogenous sediments.

Hydrogenous sediment. Sediment formed directly by precipitation from seawater. Also known as authigenic sediment.

Lithification. Conversion of sediment into sedimentary rock by pressure and/or by the introduction of a mineral cement.

Microtektite. Small translucent oblongs of glass, thought to be formed by meteors or asteroids.

Mineral. A naturally-occurring inorganic crystalline material with a specific chemical composition and structure.

Natural gas (methane). A gas formed at sites where the organic materials associated with oil formation are subject to very high temperatures or overcooking.

Neritic sediment. Terrigenous material associated with the continental shelf.

Nodule. Solid mass of hydrogenous sediment, most commonly manganese or ferromanganese nodules and phosphorite nodules.

Oolite sands. Hydrogenous sediments formed when calcium carbonate precipitates from warmed seawater as the pH rises, forming rounded grains around a shell fragment or other particle.

Ooze. Sediment of at least 30 percent biological origin.

Paleoceanography. The study of the ocean's past.

Pelagic sediment. Sediments of oceanic origin.

Piston corer. A seabed-sampling device capable of punching through up to 25 meters (80 feet) of sediment and returning an intact plug of material.

Poorly sorted sediment. A sediment in which particles of many sizes are found.

Pteropod. A small planktonic mollusk with a calcareous shell, which contributes to biogenous sediments.

Radiolarian. One of a group of usually planktonic amoeba-like organisms with a siliceous shell, which contributes to biogenous sediments.

Sand. Sediment particles between 0.062 and 2.00 millimeters in diameter.

Sediment. Particles of organic or inorganic material that accumulate in an unconsolidated form.

Siliceous ooze. Ooze composed mainly of the hard remains of silica-containing organisms.

Silt. Sediment particles between 0.004 and 0.062 millimeters in diameter.

Stratigraphy. The branch of geology concerned with the composition, origin, and age relationships of stratified rocks.

Terrigenous sediment. Sediment derived from the land and transported to the ocean by wind and flowing water.

Turbidite. A terrigenous sediment deposited by a turbidity current; typically, coarse-grained layers of nearshore origin interleaved with finer sediments.

Well-sorted sediment. A sediment in which particles are of uniform size.

Test Your Learning

After working through these questions, check your answers against the key at the end of this book. If you answered any of the questions incorrectly, review the relevant sections of the text and video episode.

Multiple Choice

1. Which of the following could not be studied based on sediment analysis?
   a. early fishing methods
   b. early climatic changes
   c. changes in Earth's magnetic field
   d. Cretaceous extinction

2. Manganese nodules are considered to be ______________ sediment.
   a. terrigenous
   b. hydrogenous
   c. cosmogenous
   d. none of the above

3. Well sorted sediments are:
   a. a specific mixture of several particle sizes
   b. all of one size
   c. also called rubble
   d. formed by turbidity current

4. The largest particle still considered sediment is ________; the smallest is ________.
   a. cobble, silt
   b. cobble, sand
   c. boulder, sand
   d. boulder, clay

5. The study and description of layered sedimentary deposits is called:
   a. sedimentology
   b. seismic profiling
   c. stratigraphy
   d. paleontology

6. At great depths seawater tends to become slightly acidic, and ___________ oozes will not form.
   a. siliceous
   b. terrigenous
   c. radiolarian
   d. calcareous

7. Seabed sediments have given us records of the glacial-interglacial climatic cycles of the last:
   a. 65 million years
   b. 32 million years
   c. 14 million years
   d. 2 million years

8. Much of Earth's history has been lost due to:
   a. conduction
   b. subduction
   c. reduction
   d. influction

9. Deep sea sediment containing at least 30 percent biogenous material is called:
   a. a turbidite
   b. authigenic
   c. an ooze
   d. poorly sorted

10. Analysis of sediments and fossils from the Deep Sea Drilling Project have helped verify the theory of:
    a. stratigraphy
    b. plate tectonics
    c. the sequence of ice ages
    d. ooze formation

11. Evaporites are classified as _______ sediments.
    a. turbidites
    b. biogenous
    c. oolitic
    d. hydrogenous

12. Which of the following has not (yet!) been mined from the deep sea floor?
    a. natural gas
    b. diamonds
    c. gravel
    d. manganese nodules

Short Answer

1. Discuss the growing commercial focus on marine sediments, noting particularly the different types of resource materials involved and the methods and economics of their retrieval.

2. Interest in the composition of the sea floor probably began with the early voyagers as a navigational aid, then became more focused with the first marine science expeditions, and is currently big science due to major advances in sampling and study methods. Document some of the milestones in this sequence of events, culminating with the state-of-the-art technology being used today.

3. The first recorded retrieval and identification of what we now call hydrogenous sediments was on the *Challenger* expedition. Although not considered to be rare, they are thought to cover less than one percent of the total ocean floor, and their chemistry and the mechanics of their formation is still not entirely clear. Summarize what we know about the composition and formation of these sediments.

Supplemental Activities

1. Do a coring exercise using a hand-held coring device and two corks. A coring device can have a number of variations, depending on available materials (metal pipe, PVC pipe, clear plastic tubing). It should be between 15 and 18 inches long, and about 1½ inches in diameter. The thinner the walls, the better (clear plastic tubing works best). You will need two stoppers to close the ends (rubber corks from a school laboratory, or "cork" corks wrapped with plastic wrap for water tightness). One of the stoppers should have a hole punched through it large enough to see a little light through (this helps create a suction when removing the tube from the mud). Find a shoreline (lakeshore is best, or a quiet inlet of a stream or river) where there seems to be a good accumulation of silt or mud (sand and gravel do not collect well). Push and rotate the tube into the mud, down to where you can still get a grip on the tube. Push

the stopper with the hole into the top of the tube. Holding your thumb over the hole (creating a suction), gently pull and rotate the tube out of the sediment. Just as the lower end of the tube leaves the mud, push a stopper into that end, while removing your thumb from the top stopper. This releases the pressure in the tube and allows the bottom stopper to be inserted. Wash the mud from the outside of the tube, taking care to always keep the tube upright. If you used clear plastic tubing you may be able to see the different layers (strata) of sediments. If you wish to go further and examine the composition of the different layers, it will be best to freeze the whole apparatus, full of mud, for a couple of days (again keeping it as vertical as possible). After it is frozen, remove the stoppers and allow the outside of the tube to warm up a little. This will free the core sample and allow it to slide from the tube. If it is stubborn you can help it along using the handle of a kitchen utensil to push with. Now the strata will be easier to see and describe. Going further still, you may wish to slice the core longitudinally to see the stratification, or crosswise to separate the different-appearing layers for further study. Be aware, however, that they will quickly begin to melt, so be ready to place them into separate paper cups for further study. Write a report describing the procedure and collecting site, and sketch the appearance and description of the different strata. Try to reconstitute a history of the site, based on what you know about sediment transportation (heavy particles sink first, lighter and small ones have been carried farther and take longer to settle). This might reveal something about the possible source of the material, over time, and the agents that transported it (water, wind, even ice). Use your imagination.

2. On page 134 of your text Table 5.1 depicts the standard sediment particle sizes and their settling rates, which is used to determine the composition of a sample. You can observe this in a simple experiment, using a teaspoon of sediment and a glass of water. Beach sand is not good (too clean), and regular dirt usually has too much organic material. Obtain a handful of sand from a construction site, or a building supply store, or even from a child's playground or sandbox. How to test the settling rate? Use a large (100 ml or more) graduated cylinder from a school laboratory, or just a tall, clear-sided drinking glass with no fluting or pattern or etching on it. This experiment must be set up in a location where the glass will not be moved or jostled for at least a week. If you wish to actually measure the settling rate (in inches or millimeters per day), use the graduations on the cylinder or tape a clear plastic ruler to the outside of the glass. This will enable you to keep track of the top of the settling sediment layers (mainly the silt and clay) as it leaves the top of the water column and settles to the bottom. When all of the clay has settled (it may be a week or so) you should be able to see through the glass of water as well as you could before you added the sand. To begin, fill the glass or cylinder with water to within about an inch of the top. Use bottled water—tap water is often cloudy with its own sediment. Add about a teaspoon of sand to the water and stir it well. The larger particles (probably the sand and maybe some pebbles) will sink to the bottom immediately (check the table in the text, page 134). The silt and clay will remain suspended (and thus cloud the water) and will settle out according to their ratios in the sample. You may be able to document the differential settling as the silt separates from the clay by noting differences in the clarity of the water. When the water clears to its original before-sand clarity, the clay has all settled. Document the procedure, recording the approximate lengths of time it took for each particle size to settle out, and compare it to the standards on the chart.

3. The video mentions that sediment analysis may be useful in chronicling the history of Earth's magnetic field using the presence and abundance of particles of magnetite, which can be found in both terrestrial and marine sediments. Magnetite looks like tiny elongate crystals (they are often called "iron filings") which can be attracted with a magnet. Using a common toy magnet, drag it through sand or dirt (garden soil is not good—it has too much organic matter). Sand found in dry stream beds works best; and there may even be visible black streaks along the bottom of the bed, usually composed of concentrations of pure magnetite. The particles will stick to the poles of the magnet, and can be easily seen with the unaided eye (although using a hand lens makes it more interesting). Sample a number of different types of sand (stream, beach, sandbox) and journal your findings, noting the comparative abundance of magnetite in each type. To fur-

ther quantitate your results, choose a likely sediment and drag your magnet through it for a certain number of seconds, then do an "eye-ball" quantification of your yield. Do this in different kinds of sediments. Would any of them be worth mining for iron? You may even have seen some of these filings in children's magnetic games for sketching and doodling using the particles and a small magnet.

Lesson 7

It's in the Water

Overview

The practice of science is limited to observable phenomena, and everything observable is either matter or energy. Since matter and energy are the subjects of physics and chemistry, it could be argued that all the natural sciences are branches of physics and chemistry. Biology is the physics and chemistry of living things. Astronomy is physics and chemistry applied to planets and stars. And oceanography is physics and chemistry applied to the oceans. Lesson 7 begins at the level of molecules and Lesson 8 proceeds to large scale aspects of the physics and chemistry of the oceans.

Water is the stuff that makes the oceans and it is an extraordinary substance with some unusual chemical properties.

— It exists naturally on Earth in all its three phases—solid, liquid, and gas.

— Its solid form (ice) floats on its liquid form (water).

— It absorbs and loses great quantities of heat with only small changes in temperature.

— It is a remarkably good solvent.

As with any substance, the properties of water are related to the properties of its molecules. The most important features of the water molecule (H_2O) are its polarity and the ability of separate water molecules to form hydrogen bonds.

Polarization occurs if a separation of positive and negative charges exists. Because the H-O-H of a water molecule are not in a straight line and because the negative electrons of the three atoms are not equally shared, a water molecule has polarity. The oxygen atom in the middle has a slight negative charge and the two hydrogen atoms, each arranged at an angle of 105° with the oxygen, have slight positive charges. The chemical bonds that hold the oxygen atom to the hydrogen atoms within a single water molecule are called polar covalent bonds. They are covalent because the three atoms are sharing electrons and they are polar because the electrons are shared unequally creating polarity.

Hydrogen bonds form between water molecules. A *hydrogen bond* forms when a positively-charged hydrogen atom of one water molecule is attracted to a negatively-charged oxygen atom of another. Individual hydrogen bonds are weak and temporary. Their lifetime is measured in trillionths of a second and they are ten times weaker than the polar covalent bonds within a water molecule. But because a vast number of hydrogen bonds exist in any drop, lake, or ocean of water, hydrogen bonding is one of water's most important features. These bonds make water molecules stick to one another, inhibiting their movement. This has a dramatic effect on the thermal properties of water—the properties related to temperature and heat.

Heat and temperature are not the same. Heat, measured in calories, is a form of energy produced by the vibration of atoms and molecules. The amount of heat is related to both how fast the molecules are moving and how many molecules there are. Temperature, measured in degrees, is related only to how fast the molecules of a particular substance are moving. Thus a bathtub of water at 120°F has much more heat than a teacup of water at 120°F because it has many more molecules. But the teacup and the bathtub have the same temperature because the molecules in each are moving at the same speed.

If heat is added to a substance its molecules move faster and its temperature rises; if heat is removed they move slower and the temperature falls. The amount of heat needed to raise the temperature of one gram of a particular substance 1°C is called its *heat capacity* and the heat capacities of different substances vary widely. Water's heat capacity is one calorie per gram per degree Celsius. Thus, one calorie[1] of heat will cause one gram (about 20 drops) of water to rise 1°C. This is about the highest heat capacity of any substance on Earth and it is a consequence of water's hydrogen bonds inhibiting molecular motion. It is why air and sand at the beach heat rapidly and water heats slowly. Water moderates temperature fluctuations because it must absorb a lot of heat to increase in temperature and it must lose a lot of heat to decrease in temperature. Earth, with its huge amount of water, has much more moderate temperatures than it would have if it were waterless.

Hydrogen bonding also affects the evaporation of water. To change one gram of water from liquid to vapor requires the addition of 540 calories. This is water's *latent heat of vaporization*, or the amount of heat needed to change one gram of liquid water to one gram of gaseous water. This very high latent heat of vaporization is also a consequence of hydrogen bonding. To evaporate, the molecules must spread apart from each other and all the hydrogen bonds must be broken. Every gram of water that evaporates absorbs 540 calories and every gram of water that condenses back to liquid releases 540 calories. This also has an important moderating effect on Earth's climate.

Like most substances, water gets denser as it cools. But water reaches its maximum density at 4°C and expands slightly as it cools down to 0°C. And, as anyone who has ever put a can of soda in the freezer and forgotten about it knows, water expands when it freezes. Ice is less dense than liquid water, so it floats. Without this unusual property, icebergs would sink, the oceans would solidify, and Earth would be a much colder and largely uninhabitable planet since all life requires liquid water.

Water's latent heat of fusion is 80 calories per gram, which means that in order to freeze one gram of water to ice at 0°C (32°F), 80 calories of heat must be removed. Similarly, to melt that gram of ice back into water requires the addition of 80 calories. This is not as dramatic as the latent heat of vaporization, but significant amounts of heat are absorbed or lost as ice thaws and water freezes. And this is yet another property of water that moderates Earth's temperature. The moderating effect that water has on Earth's climate is called water's *thermostatic effect*. Water's high heat capacity, its high latent heat of fusion, its high latent heat of vaporization, and the fact that ice floats all contribute to this thermostatic effect and make Earth a very climatically mild planet with an abundance of liquid water, which permits an abundance of life.

Temperature also affects the density of water. Global warming causes sea-level to rise not only because ice melts causing water to run into the ocean, but also because water expands as it warms. The ocean gets larger as warming causes it to expand. Temperature and salinity are the two main factors affecting the density of sea water. Salt water is denser than fresh water and cold water is denser than warm. Denser water sinks into less dense water. The sinking of denser water and rising of less dense water is one of the most important movers of ocean water.

Very cold, dense water forms and sinks in both polar regions. In the north, remnants of the salty Gulf Stream arrive in the north Atlantic near Greenland and cool. The combination of high salinity and low temperature creates a very dense mass of water that sinks into the deep Atlantic and spreads southward. In the south, around the coast of Antarctica, exists the coldest ocean water on Earth. Here, as sea ice forms more water is removed into the ice and more salt stays behind in the water. This "salting-out" increases salinity and this very cold, salty, and dense water is called Antarctic bottom water as it sinks to the seafloor and spreads northward. This *thermohaline circulation* moves great masses of water from the surface to the depths and from the poles toward the equator, and is another important factor controlling global climate. Additionally, the thermohaline circulation carries life-supporting oxygen to the deep ocean, which would be anoxic without it.

Seawater is a *solution*, a homogeneous and permanent mixture of two or more substances. In a solution, the substance present in the larger amount is called the solvent while the substance present in the smaller amount is the solute. In seawater the solvent is, of course, water. Atoms and molecules that have a charge generally dissolve

1. This is not a coincidence. A calorie is defined as the amount of heat needed to make a gram of water rise 1°C.

well in water. Ionic molecules (acids, bases, and salts) and polar molecules (sugars and alcohols) dissolve well in water. Uncharged, or nonpolar, molecules such as fats and oils do not dissolve well in water. The solutes in seawater fall into three broad categories: dissolved inorganic solids (salts), dissolved gases (oxygen) and dissolved organic molecules (proteins). This discussion will focus on the salts and gases.

Salinity is the total concentration of dissolved inorganic solids in seawater. The average salinity of seawater is about 3.5 percent. However, salinity is usually expressed as parts per thousand (ppt or ‰) rather than parts per hundred (pph or %) so the salinity of seawater is expressed as 35‰ rather than 3.5%. Seawater has a lower specific heat, a lower rate of evaporation, and a lower freezing point than pure water. These are water's *colligative properties*, or the properties that are affected by solutes.

There are many ways to accurately measure salinity. Evaporating the water and weighing the leftover salt is probably the least accurate. Water does not evaporate well from seawater, and if the water is heated to speed up evaporation, the salts and water react chemically giving inaccurate measurements. The most efficient and widely-used method of measuring salinity is to use an electronic salinometer to measure the water's electrical conductivity. Water's conductivity, or how well water conducts electricity, increases proportionally as salinity increases.

Another method of measuring salinity takes advantage of the fact that many of the constituents of seawater are present in constant ratios throughout the ocean—a fact known as the *principle of constant proportions*. Therefore, it is possible to measure just one of the solutes and convert this to salinity. Chlorine (Cl–), the most abundant constituent of salinity, is frequently measured to calculate salinity. The chlorine concentration, or chlorinity, is always 54.2 percent of salinity. If the salinity of a sample of seawater is 35‰, the chlorinity will be 18.98‰ ($18.98 \div 35 = 54.2\%$). Whether the salinity has been decreased by rain or increased by evaporation, chlorine will always be 54.2 percent of the total.

The salts of the ocean are leached, or dissolved, from the rocks and soils of the land and carried to the ocean in streams and rivers. But they also originate from other sources. They rain in from the atmosphere, enter from hydrothermal vents and undersea volcanoes, percolate up from groundwater near shore, and dissolve out of underlying sediments.

Salts leave the ocean, too. Because salts remain behind as pure water evaporates from the sea surface, the salinity of the ocean should be continually increasing. But the ocean seems to be in a state of equilibrium. The salts have pathways to exit the ocean that balance the rate at which they enter. One important exit, or sink, for solutes is the conveyor belt of plate tectonics. For example, calcium carbonate ($CaCO_3$) or limestone dissolves from rocks on the land and is carried into the ocean by rivers. Some of the $CaCO_3$ dissolved in the ocean is removed by organisms like oysters, corals, and foraminifera to make shells and skeletons. When these organisms die, their skeletons fall to the seafloor and become part of the sediment lying on a tectonic plate. The sediments are eventually conveyed to a subduction zone where they leave the ocean as they sink into the mantle.

Once a particular substance enters the ocean it remains there for a measurable time called its *residence time*. The residence time for chloride (Cl^-) is 100 million years while iron (Fe^{++}), which reacts quickly with sediments and is rapidly used by living organisms, may reside in the ocean for a mere 200 years. Water molecules have a residence time of about 4,100 years, entering mostly as runoff and precipitation and leaving mostly by evaporation. Solutes with long residence times are called *conservative constituents* and those with short residence times are called nonconservative.

The mixing time, or time it takes the ocean to totally mix, is about 1,600 years. Substances with residences longer than 1,600 years are conservative and fully mixed throughout the ocean. This full mixing is the basis of the principle of constant proportions. Constituents whose residence is less than 1,600 years do not get fully mixed throughout the ocean, so nonconservative constituents are not evenly distributed.

The three most abundant dissolved gases in seawater are nitrogen (N_2), oxygen (O_2), and carbon dioxide (CO_2). Although all living things require nitrogen, it is not usable in its gaseous form by most.

Oxygen is vital for the survival of most organisms. It is about 100 times more abundant in air than in seawater. Oxygen enters the ocean, and its highest levels are found, in the surface layers where it either mixes in from the air or is manufactured by photosynthesis. The deep ocean has higher levels of oxygen than the mid layers because the thermohaline circulation carries oxygen to the depths as this cold, dense, oxygen-rich water sinks from polar surface water to the seafloor. Solids dissolve better in warm water than in cold, but gases—like oxygen—dissolve better in cold water.

Unlike oxygen, there is much more carbon dioxide dissolved in the oceans than in the atmosphere. This important gas is used by photosynthesis to make the sugars that serve as the base for almost all of Earth's food chains.

Because carbon dioxide produces a weak acid when dissolved in water, it plays a critical role in regulating the ocean's acid-base balance. Dissolved in water, carbon dioxide will exist in three forms: carbonic acid (H_2CO_3), bicarbonate (HCO_3^-), and carbonate (CO_3^{-2}). Carbonic acid is acidic while bicarbonate and carbonate are basic. The relative concentrations of these forms of carbon dioxide give seawater an average pH of about 8.0.

Carbon dioxide also affects global warming. Like the windows of a car in the summer, CO_2 allows light energy to pass in freely through the atmosphere but inhibits heat energy from radiating back out. Thus heat builds up and Earth warms when CO_2 increases in the atmosphere. The ocean is a huge storehouse for CO_2, so understanding the uptake and release of CO_2 between ocean and atmosphere is important for understanding and predicting the effects of increasing CO_2 on global climate.

There are two ways to reduce global warming by reducing atmospheric CO_2: stop putting it into the atmosphere or start taking it out. This is why efforts are made to reduce the emission of CO_2 by reducing the burning of fossil fuels in automobiles and factories. In the 1980s scientists at Moss Landing Marine Laboratories proposed an interesting remedy. Since large areas of the ocean around Antarctica and the tropical southern Pacific are deficient in the iron needed for phytoplankton growth, adding iron to these areas could stimulate phytoplankton growth, increasing photosynthesis and the demand for CO_2 in the ocean. The scientists concluded that this would remove CO_2 from the atmosphere and reduce or even reverse global warming. Controlled experiments in the ocean have shown that adding iron can cause phytoplankton to bloom in these regions. This seemingly simple solution is highly controversial and vigorously debated by oceanographers. One thing is certain; an attempt to control global warming by adding iron to vast areas of the ocean would be the largest intentional manipulation of the biosphere ever attempted.

Focus Your Learning

Learning Objectives

After completing this unit you should be able to do the following:

1. Explain the structure and characteristics of water molecules and explain how these contribute to the global thermostatic effects of water on Earth's climate.
2. Understand the factors that regulate the density of seawater and explain the nature and importance of the thermohaline circulation.
3. Detail the sources, composition, and measurement of the ocean's salinity and outline the colligative properties of water.
4. Discuss the principle of constant proportions and compare, contrast, and give examples of conservative and non-conservative constituents of seawater.
5. Compare and contrast the general distribution, concentration, and roles of dissolved carbon dioxide and dissolved oxygen in the ocean.
6. Explain the relationship between carbon dioxide and the greenhouse effect and the interplay of oceanic and atmospheric carbon dioxide in global warming. Explain why some scientists believe that adding iron to the oceans could affect atmospheric carbon dioxide and global warming.

Assignments

This lesson is based on information in the following text and video assignments. The key terms, focus points, and practice test are intended to help ensure mastery of the material presented in this unit of study.

Text: Pages 154–168 in Chapter 6, "Water and Ocean Structure"; pages 182–199 in Chapter 7, "Ocean Chemistry"; and in Chapter 9, "Circulation of the Ocean," the section "Thermohaline Circulaton Affects All the Ocean's Water," pages 250–258

Video: Episode 7, "It's in the Water"

Video Focus Points

Review these points before watching the video assignment for this lesson to help you focus on key issues.

— Water has several unusual properties because its molecules are polar and form hydrogen bonds.

— Water has a dramatic effect on global climate because it absorbs large amounts of energy as it thaws from ice, as it heats, and when it evaporates.

— Because water expands when it freezes, ice floats. Without this unusual characteristic, ice would sink, the oceans would freeze, and Earth would be a much colder place.

— Water is an excellent solvent and seawater is a chemical soup. The chemicals dissolved in the ocean enter from land run-off and glaciers, from the atmosphere, from hydrothermal vents and volcanoes, from ground water, and from sediments.

— Chemicals dissolved in seawater leave the ocean by various pathways called sinks.

— The two main ways to measure salinity are to measure water's conductivity or chlorinity.

— Salinity and temperature are the two major features of seawater that determine its density. The thermohaline circulation caused by the sinking of cold, salty, dense polar water is one of the major mechanisms that moves and mixes ocean water.

— Ocean circulation, rates of photosynthesis, burning fossil fuels, and deforestation are some of the things that affect the dynamic interaction of oceanic and atmospheric carbon dioxide. Carbon dioxide is a greenhouse gas and affects global warming and cooling cycles.

Text Focus Points

Review these points before reading the text assignment for this lesson to help you focus on key issues.

— Water is both a common and unusual chemical. The unusual properties of water result largely from its polarity and hydrogen bonding.

— Because of hydrogen bonding water has a very high heat capacity, a very high latent heat of fusion, and a very high latent heat of vaporization. Additionally, water expands when it freezes so ice floats. The thermal properties of water are responsible for moderating global climate. These effects are called global thermostatic effects.

— Because water is polar it dissolves other molecules with electrical charges including ionic molecules such as salts and polar molecules such as sugars. Non-polar molecules such as oils do not dissolve well in water.

— The total concentration of dissolved inorganic solids in a sample of seawater is called its salinity. These chemicals enter the ocean from weathering (erosion) of land, outgassing from undersea volcanoes, rifts and hydrothermal vents, dissolution from sediments, and washing in from the atmosphere.

— The length of time a particular chemical remains in the ocean is called its residence time. Residence times for the various chemicals that make up seawater vary greatly.

— The movement of water in the ocean causes the ocean to completely mix about every 1,600 years. Chemicals such as sodium and chlorine whose residence times exceed the mixing time are called conservative constituents because they are homogeneously mixed throughout the sea and their proportion in any sample of seawater is always constant. Non-conservative constituents have short residence times and their concentrations vary from place to place.

— Nitrogen, oxygen, and carbon dioxide are the most important gases dissolved in the ocean. Oxygen mixes in from the atmosphere and is manufactured by photosynthesis in the upper sunlit layer of the ocean. Carbon dioxide is one of the raw materials used in photosynthesis and plays a major role in regulating the ocean's acid-base balance.

Key Terms and Concepts

A thorough understanding of these terms and concepts will help you to master this lesson.

Colligative properties. The properties of pure water that are altered by dissolving something in it. For example, the freezing point of pure water is lowered when another chemical is dissolved in it.

Conservative constituent. The solutes in seawater whose relative concentration is always the same. In contrast, non-conservative constituents such as oxygen, iron, nitrate, and other biologically or geologically active chemicals can vary widely in their concentration in seawater.

Greenhouse effect. The trapping of heat as light energy penetrates the atmosphere and is converted to heat. The light moves easily in through the transparent air but the heat does not easily move out. Certain gases such as carbon dioxide and methane increase the greenhouse effect and contribute to global warming.

Heat capacity. The amount of heat energy (calories) needed to raise one gram of a substance one degree Celsius. The heat capacity of water (1.0) is very high. For example, it is 17 times higher than that of silver (0.06).

Hydrogen bonds. The weak bonds that hold separate water molecules to each other as the positive (hydrogen) region of one water molecule is attracted to the negative (oxygen) region of another. Although they are relatively weak and temporary, their large number greatly restricts the motion of water molecules and has profound effects on the nature of water.

Latent heat of fusion/vaporization. The amount of heat required to make one gram of a substance change state (from solid to liquid or liquid to gas). The latent heat of fusion of water is 80 calories per gram, meaning it takes 80 calories of heat to thaw one gram of ice. The latent heat of vaporization of water is 540 calories per gram, meaning 540 calories of heat are needed to change one gram of water to vapor.

Principle of constant proportions. Also known as Forschhammer's Principle, this principle states that the proportions of the conservative constituents of seawater are constant regardless how much the sample has been diluted by rain or concentrated by evaporation. Chlorine, for example, is always 54.2 % of the total salinity of seawater.

Residence time. The average length of time a particular chemical resides in the ocean. The solutes of seawater enter and leave through several pathways. Conservative constituents have long residence and non-conservative constituents have short times.

Salinity. The total dissolved inorganic solids in a sample of seawater. Salinity is generally around 35 ‰ (3.5 %). The two most common ways to measure salinity are to measure the electrical conductivity or the chlorinity (chlorine concentration) of a sample. Salinity is proportional to both.

Solution. A uniform and permanent mixture of two or more chemicals. The chemical present in the largest concentration is the solvent and the chemicals present in smaller proportions are called the solutes. In seawater, water is the solvent and the salts are the solutes.

Thermohaline circulation. A large scale mixing of the oceans driven by the sinking of cold dense water near the poles.

Thermostatic effects. Compared to most chemicals, water absorbs large quantities of heat with minimal change of temperature. The presence of large amounts of water on Earth, therefore, moderates global climate.

Test Your Learning

After working through these questions, check your answers against the key at the end of this book. If you answered any of the questions incorrectly, review the relevant sections of the text and video episode.

Multiple Choice

1. Hydrogen bonds are:
   a. weak and temporary in liquid water
   b. weak in ice, moderate in liquid water, and strong in water vapor
   c. polar and covalent
   d. weak in water with a low pH and strong in water with a high pH

2. Ice is
   a. more dense than water
   b. less dense than water
   c. equal in density to liquid water at 0°C and more dense at lower or higher temperatures
   d. equal in density to liquid water at 0°C and less dense at lower or higher temperatures

3. The salty water near Greenland that becomes cold and dense and sinks is part of the:
   a. Abyssal Gyre
   b. Thermohaline Circulation
   c. Arctic Convection Cycle
   d. Arctic Advection Cycle

4. Which atom(s) in a water molecule usually have a slight positive charge?
   a. hydrogen
   b. oxygen
   c. both
   d. neither

5. In addition to mixing in from the air, oxygen enters seawater:
   a. through the respiration of animals
   b. as a result of decomposition of plant and animal remains
   c. as a by-product of photosynthesis
   d. through oxidation of metal ions in seawater

6. Which of the following are nonconservative ocean chemicals?
   a. sodium and potassium
   b. sulfate, chloride, and sulfide
   c. oxygen, phosphate, and nitrate
   d. calcium and magnesium

7. The properties of water affected by dissolving salts in it are called:
   a. osmotic properties
   b. steady state properties
   c. ionic properties
   d. colligative properties

8. The major pH-buffering agents in the ocean are:
   a. manganese/magnesium
   b. phosphate/nitrate
   c. oxygen/ozone/peroxide
   d. carbon dioxide/bicarbonate/carbonate

9. The average salinity of the world's oceans is about ________ (‰ = parts per thousand)
   a. 0.35‰
   b. 3.5‰
   c. 35‰
   d. 350‰

10. Which of the following is a true statement about gases dissolved in seawater?
    a. nitrogen is present in a higher percentage than oxygen or carbon dioxide
    b. dissolved oxygen levels are lowest in the deepest regions of ocean
    c. gases dissolve better in warmer water than in colder water
    d. production of CO_2 by photosynthesis increases pH in the photic zone

11. Which of these is true about residence time?
    a. it is inversely proportional to mixing time
    b. conservative constituents have long residence times and non-conservatives have short ones
    c. it is not affected by activity at hydrothermal vents
    d. iron, nickel, and other ferromagnetic metals have the longest average residence times

12. In iron deficient regions, adding iron to the oceans might reduce global warming by:
    a. increasing photosynthetic use of carbon dioxide
    b. increasing photosynthetic production of carbon dioxide
    c. increasing photosynthetic use of oxygen
    d. increasing photosynthetic production of oxygen

Short Answer

1. Explain the reasoning that makes some scientists believe that adding iron to certain regions of the ocean can reduce global warming.

2. Figure 7.8 on page 194 of your text illustrates the general distribution of carbon dioxide and oxygen in seawater. Explain why these two curves are roughly opposite. Explain also why the concentration of dissolved oxygen is least

at around 500 meters and increases in deeper water.

3. Discuss how hydrogen bonding contributes to the thermostatic effects of water.

Supplemental Activities

1. Experiment with some of the properties of water in your kitchen. You'll need water, a dropper, food colorings, ice, salt, wax paper, aluminum foil, and a variety of glasses, a glass bowl, and an ice cube tray. Do the following activities and attempt to explain what you observe in each.
   a. Sprinkle a few drops of water on a piece of wax paper and a few on a napkin. Compare, describe, and explain.
   b. Create a thermohaline circulation. Mix some very salty water in a small glass and dye it dark red. Fill a glass bowl with warm tap water and float an ice cube in one corner. Slowly place drops of the salty red water on the ice cube. Describe and explain.
   c. Study surface tension. Cut a piece of foil about 1 square inch, pinch it into a very tight ball, and drop it into a glass of water. Take another piece of foil about one half inch wide and two inches long and, using scissors, cut tiny slivers of foil and let them fall into the water. Compare and explain what happens to the ball and the slivers.
   d. Study sea ice. Dissolve 3 teaspoons of salt in a cup of water and add 5 drops of red dye. Pour this into several compartments of an ice cube tray. Fill several more with tap water. Place them in the freezer and observe them every hour until they freeze. Compare and explain.
   e. Study the expansion of ice. Fill a plastic bottle to the very top with tap water and cap it. Place it the freezer over night. Observe and explain what happens.
2. The iron hypothesis explores how ocean phytoplankton may affect global climate. An equally interesting hypothesis suggests that global warming may actually trigger a mini-ice age in Europe and North America. Use the Internet to explore this gloomy prospect, then explain how the melting of Arctic ice could thwart the thermohaline circulation. Discuss the causes and predict the effects of such an event on global climate, economy, and agriculture.
3. The calories we use to measure the energy in foods are kilocalories or 1,000 calories. Thus a 300 kilocalorie doughnut actually has 300,000 calories of energy in it. One calorie raises the temperature of one gram of water 1°C. This is water's heat capacity, 1calorie/1gram/1°C. Suppose you released all the energy in one doughnut by burning it. Calculate what that energy would do to:
   a. a liter (1,000 grams or about one quart) of water at 0°C
   b. 300,000 grams of water (about 75 gallons or a half filled bath tub) which starts at 35°C
   c. 300,000 grams of gasoline starting at 35°C (the specific heat of gasoline is 0.50)
   d. 300,000 grams (660 pounds) of silver (the heat capacity of silver is 0.06)

Lesson 8

Beneath the Surface

Overview

The ocean is not merely a large puddle of seawater lying in the low places between the continents. Like the land and the atmosphere, the ocean has a complex structure in both space and time. This lesson examines some aspects of the ocean's structure and the characteristics of light and sound in the ocean.

Several of the ocean's properties change with depth. Some, such as pressure and light, change gradually and continuously while others, such as temperature, can change abruptly. These changes create a complex environment with a distinct and dynamic structure and are important to understanding oceanography. Graphs like those in Figures 6.18 and 6.19 on pages 170 and 171 of your text illustrate how a particular feature of the ocean—in this case temperature—changes with depth. The vertical axes of these profiles always represent depth.

Less dense water floats on more dense water and the ocean often exhibits a horizontal layering or *stratification* resulting from differences in density at different depths. Salinity and temperature are the two most important factors affecting seawater density and creating stratification. The zone of transition between density layers is called a *pycnocline*. A pycnocline layer caused by temperature differences is called a thermocline and a pycnocline caused by salinity differences is called a halocline. On a large scale, temperature is the primary stratifying factor in the oceans. Salinity is generally important on a more localized scale. For example, it is particularly important in estuaries where fresh water rivers encounter the salt water of the oceans or where freezing, evaporation, or precipitation is significant in altering salinity.

In most places the ocean is divided into three temperature zones—a surface zone, a middle thermocline zone, and a deep zone. About 80 percent of the ocean is the deep zone. The water there is always cold, ranging from $-1\,°C$ to $3\,°C$. The surface zone or mixed zone is the layer warmed by the sun. It contains about two percent of the ocean's water and can extend to a depth of up to 1,000 meters, although it is usually about 150 meters. It is well mixed by the action of waves and currents. The thermocline zone lies between the deep zone and the surface zone and holds about 18 percent of the ocean's volume. Moving down through this zone, density increases rapidly as temperature decreases. The thermocline is a variable oceanic feature. Such factors as latitude and water clarity affect how much and how deeply sunlight penetrates into the sea and, therefore, how much the water warms. Polar oceans have no warm surface zone or thermocline because they never warm enough. Horizontal currents such as the warm Gulf Stream or the cold California current, vertical currents such as upwellings, and other factors that affect density, like evaporation and precipitation, also affect the nature of the three layers in specific locations.

The temperate oceans have the largest seasonal variation in temperature. With some significant exceptions, the polar oceans are always cold and the tropical oceans are always warm. In the temperate oceans a secondary thermocline develops with the seasons. It forms as the surface warms rapidly in the temperate zone's spring and breaks apart in the autumn as the surface cools and storms remix the water. The production of food by photosynthesis can only occur in the sunlit upper layer of the ocean and the upwelling of water from below brings the nutrients needed for the growth and survival of phytoplankton. The prevention of this upward mixing of nutrients into the sunlit zone by the thermoclines can retard oceanic photosynthesis. Conversely, the upward mixing of nutrient-rich water when a seasonal thermocline breaks down can enhance photosynthesis.

In a sense, the ocean is also layered by light penetration. Solar radiation is the major agent affecting ocean temperatures but the most dramatic effect of light in the sea is the impact it has on biology. Sunlight powers photosynthesis creating food and oxygen. Additionally, sunlight and the light produced by bioluminescent organisms are the basis for vision and the roles that sensing light plays in biology. Sunlight diminishes as it passes through seawater creating three zones defined by their light intensity. The upper sunlit layer of the ocean is called the *photic zone* while the deeper, perpetually dark region that forms the greatest mass of the ocean is called the *aphotic zone*. The photic zone is further divided into the well-lit euphotic zone and the poorly lit disphotic zone. The euphotic zone is the narrow uppermost layer of the ocean where photosynthesis occurs. The disphotic or twilight zone is a shadowy region with diminished light. It eventually phases into the lightless aphotic zone. The light zones are more dynamic and variable than the temperature zones because seawater can store heat but it cannot store light. Seasons, latitude, cloud cover, surface conditions, and water clarity all affect light penetration. And the photic zone disappears each night.

Oceanographers have found it useful to subdivide the ocean into five layers defined by various aspects of light, temperature, and other features. These layers are the epipelagic, mesopelagic, bathypelagic, abyssopelagic, and hadopelagic zones. The *epipelagic zone* corresponds to the euphotic zone. It is the surface water and is generally considered to extend down to about 200 meters. Below this is the mesopelagic zone, which corresponds to the disphotic or twilight zone and extends down to about 1,000 meters. The bottom of the mesopelagic zone also corresponds roughly with the lower reaches of the main thermocline. The bathypelagic, abyssopelagic, and hadopelagic zones are the deep zones. They are dark, cold, and under high pressure. The bathypelagic extends roughly between 1,000 and 4,000 meters. The abyssopelagic zone lies between 4,000 and 6,000 meters and is approximately the depth where the Arctic and Antarctic bottom waters have sunk from the polar oceans. These water masses are identifiable by their specific properties including temperature, salinity, and oxygen content. The very deepest ocean regions, deeper than 6,000 meters, form the hadopelagic zone. These are the regions of the deep-sea trenches at the tectonic subduction zones. The deepest (11,022 meters) is the Mariana Trench in the western Pacific.

As light enters the ocean and forms the photic zone two things happen to it—it is scattered and absorbed. It is scattered in a multitude of directions as it bounces off water molecules, bubbles, suspended particles of silt and clay, and even plankton and fish. It is absorbed by water and other molecules. When light is scattered it hits something and bounces away as light. When it is absorbed by a molecule, it causes the molecule to vibrate and the light energy is converted to heat energy. This is how water is heated by light and how the surface layer is created.

The various wavelengths, or colors, of light are absorbed differentially by the sea. Red is absorbed most rapidly and 71 percent of red light is absorbed in the first meter of ocean. At a depth of only four meters 99 percent of the red light has been absorbed and converted to heat. Yellow and orange are also absorbed rather quickly while blue and green penetrate the deepest, with blue not being absorbed until a depth of 254 meters. Light that is scattered is visible; light that is absorbed is converted to heat or infrared and is, therefore, invisible. The sea looks blue because the largest percentage of red, yellow, and orange light are absorbed rapidly, while the greatest percentage of blue and green are scattered. Where substantial amounts of phytoplankton are present, the water looks green because chlorophyll does not absorb green light, but scatters it.

Light is important to vision. The vision of ocean creatures is adapted to the nature of light in the sea. Some photosynthetic plankton have eyespots that detect light intensity, allowing them to swim toward the light. Many zooplankton can also detect light and swim toward it because that's where phytoplankton, their primary food source, congregates. Swarms of small fish and other animals migrate to the euphotic zone at night to feed on the abundant food found there, then retreat to the cover of darkness before the sun rises and reveals them to visual predators. Many deep-sea animals lack eyes entirely, relying on other senses in the virtually lightless habitat. Some have eyes that are sensitive to the blue light that penetrates to their depth and unresponsive to red light. Some animals in the aphotic zone have very large eyes to capture the biological light, or *bioluminescence*, that is produced by many organisms in this perpetually sunless zone. Bioluminescence can attract mates and it can also attract predators or prey. Scientists who study the deep oceans with manned and unmanned submersibles report a continuous, confusing, and startling abundance of flashes, streaks, and clouds of bioluminescence.

Sound behaves very differently in the water than it does in the air, creating some interesting differences in the acoustic properties of the ocean. Like light, sound scatters and is absorbed as it spreads from its source. But sound is unlike light, heat, and other forms of electromagnetic radiation in that it can travel through matter but cannot pass through a vacuum. High frequency sounds are absorbed more quickly than low frequency sounds. The passage of sound through water is more efficient than the passage of light, so many marine animals rely more heavily on hearing than vision. Sound travels fastest in solids, slower in liquids, and slowest in gases. It travels about five times faster in water than in air. Its speed is affected by both temperature and pressure, traveling faster where water is warm or pressure is high. Because temperatures are generally higher at the surface and pressure is higher at the bottom, the profile of the speed of sound in water is the backward S-shaped curve illustrated in Figure 6.24 on page 176 in the text. In general, the speed of sound increases moving down through the warm surface layer because pressure is increasing rapidly. The speed then decreases as the temperature decreases rapidly through the thermocline layer, and then increases again as pressure increases while temperature remains constant through the deep layer. Because of the opposite relationship between depth and temperature and depth and pressure, the sound velocity profile has a bulge of high velocity at a depth of about 80 meters and a bulge of low velocity at about 600 to 1,200 meters. These bulges create two interesting sound regions in the ocean: the sofar layer and the shadow zone.

As sound passes through regions where temperature and pressure alter its speed, it will *refract*, or bend. Sound always refracts toward a region of lower speed. The refraction of sound produces the *sofar layer* at the depth of minimum velocity. As sound spreads out from this depth it passes into regions of higher velocity both above and below. As it moves into these two regions of higher velocity, the sound is refracted back toward the depth where it was produced. Refraction traps the sound energy in the sofar layer, so sound can travel great distances through the ocean at this depth. It is thought that blue whales can communicate across entire oceans in the sofar layer and low frequency sounds have been produced and detected by scientists through this layer over distances greater than 11,000 miles! Researchers are actively studying ocean temperatures and global warming by monitoring the speed at which sound travels through the sofar layer.

The high velocity bulge at a depth of about 80 meters creates a zone that is the opposite of the sofar channel. This depth is called the *shadow zone* and also results from the refraction of sound waves toward lower velocity layers. The shadow zone is just above a pycnocline layer. Sound velocity above this depth is slower because of lower pressure and sound velocity below this depth is slower because of lower temperature. Clever submarine commanders can avoid detection by sonar by hovering at this depth. As the pings of sonar from a ship at the surface head toward the hiding submarine they are refracted away into the slower velocity water above and below the submarine, thus never reaching and bouncing off the submarine.

Light travels better through air and sound travels better through water. Understanding the relationships between these two phenomena and the physics, chemistry, and biology of the oceans provides a better understanding of the oceans themselves and the organisms that inhabit them. Detailed knowledge of the acoustic nature of the oceans is a critical requirement to naval operations and has important applications to maritime navigation, salvage, commerce, fisheries, and an ever-growing assortment of oceanographic research initiatives.

Focus Your Learning

Learning Objectives

After completing this unit you should be able to do the following:

1. Describe the general density stratification of the oceans and explain why it exists and how it differs in tropical, temperate, and polar oceans.
2. Compare and contrast the euphotic, disphotic, and aphotic zones of the ocean.
3. Compare and contrast the epipelagic, mesopelagic, bathypelagic, abyssopelagic, and hadopelagic zones.
4. Discuss how the quantity and wavelength of light changes as it passes through seawater and list some ways this affects marine life.
5. Draw a profile illustrating how the speed of sound changes with depth in the ocean. Explain how these changes create the sofar layer and the shadow zone.
6. Explain how and why sound is used to study global ocean temperatures. Explain what sonar is and describe how humans use it.

Assignments

This lesson is based on information in the following text and video assignments. The key terms, focus points, and practice test are intended to help ensure mastery of the material presented in this unit of study.

Text: Chapter 6, "Water and Ocean Structure," pages 168–181 and page 364 in Chapter 13, "Life in the Ocean"

Video: Episode 8, "Beneath the Surface"

Video Focus Points

Review these points before watching the video assignment for this lesson to help you focus on key issues.

— The ocean is not a monolithic mass of water, but has a distinct structure.

— Most of the world ocean has a three-layered structure based on density. The density differences result from variation in temperature and produce a thin warm surface layer, a massive deeper cold layer, and a zone of rapid cooling between the two called the thermocline.

— The ocean is also divided into three light zones: the well lit euphotic zone, the shadowy disphotic zone, and a lightless aphotic zone. All oceanic photosynthesis takes place in the euphotic zone.

— Based on temperature, density, and light, oceanographers divide the ocean into five zones: the epipelagic, mesopelagic, bathypelagic, abyssopelagic, and hadopelagic zones.

— Both the quantity and color of sunlight change as it passes through seawater. These changes affect the adaptation of marine organisms for photosynthesis and vision. Additionally, biologically produced light, or bioluminescence, plays many roles in the biology of certain marine organisms.

— Sound is also affected as it passes through seawater. Its velocity is affected by temperature, pressure, and salinity. It is scattered, absorbed, and refracted. The behavior of sound in water creates the sofar layer and the shadow zone and allows oceanographers to use the passage of sound through water to measure global ocean temperature.

— The acoustic properties of seawater have been exploited in naval warfare and sound and sonar are employed for communication and sensing in many marine organisms.

Text Focus Points

Review these points before reading the text assignment for this lesson to help you focus on key issues.

— The ocean is often divided into a surface or mixed zone, a pycnocline zone, and a deep zone. This stratification is created by density differences that, in turn, are caused by temperature and/or salinity differences.

— When density stratification is pronounced, vertical mixing between deep water masses and surface water masses is prevented.

— Temperature differences caused largely by latitude, and salinity differences caused by evaporation, precipitation, runoff, and freezing will affect stratification. These may inhibit or enhance vertical water movements and oceanic mixing.

— As light passes through seawater it is either absorbed or scattered. The absorption of light by water molecules is the major way ocean water is heated. The depth of penetration of light creates a sunlit upper layer, or photic zone, and a dark lower layer, or aphotic zone.

— The photic zone rarely exceeds 600 meters deep and is usually less than 100 meters. This thin photic zone is where all of the ocean's photosynthesis takes place, so it is the most important biological zone of the ocean.

— Sound is also absorbed and scattered in seawater but it travels much farther through the oceans than does light.

— The speed of sound is faster in water with warmer temperatures and higher pressures. Because sound waves refract (bend) toward water of slower sound speed, two interesting acoustic zones are created in the ocean: the sofar layer and the shadow zone.

— Research has been conducted using the relationship between sound velocity and water temperature to study global ocean temperature. The various types of sonar utilize the passage of sound through water for military, industrial, exploration, geological, archaeological, and salvage operations.

Key Terms and Concepts

A thorough understanding of these terms and concepts will help you to master this lesson.

Aphotic zone. The zone of the ocean that is too deep for the penetration of sunlight and is perpetually dark. The deep zone accounts for most of the ocean.

Bioluminescence. Light produced by organisms. Fireflies are an example. This phenomenon is very common in the oceans.

Epipelagic zone. The upper layer of the ocean corresponding roughly with the photic zone. Although it is only about 200 meters deep, all ocean photosynthesis occurs here so it is the major region of food and oxygen production. Below it lie the mesopelagic zone (200–1,000 m), the bathypelagic zone (1,000–4,000 m), the abyssopelagic zone (4,000–6,000 m) and the hadopelagic zone (greater than 6,000 m).

Photic zone. The upper layer of the ocean that is shallow enough for the penetration of sunlight. The upper part of the photic zone is the euphotic zone. This is where there is enough light for photosynthesis to occur. The lower part of the photic zone is the disphotic zone. This "twilight" region has enough light for vision but not enough for photosynthesis.

Pycnocline. A transition zone between ocean layers of different densities. If the pycnocline separates water of different temperatures it is called a thermocline and if it separates masses of different salinities it is called a halocline.

Refraction. The bending of light or sound waves. This occurs when the waves move at an angle (other than 90°) from a medium of one density to a medium of a different density. Light refraction produces the broken straw effect when a straw is placed in a glass of water and sound refraction produces the sofar layer and shadow zone of the ocean.

Shadow zone. An ocean region where sound velocity is high. Because velocity is lower both above and below this zone, sound waves refract away from it. This enables a submarine to "hide" from sonar in this layer. It is generally found at a depth of about 80 meters.

Sofar layer. A region in the ocean where sound waves travel long distances, it corresponds to the layer of minimum sound velocity. Sound generated there refracts back into the layer and so is conserved. Its depth is usually about 1,200 meters.

Stratification. Layering. In this chapter it refers to the density layering of the ocean caused by water masses of different temperature or salinity.

Three-layered ocean. Most of the ocean is divided into three density layers created by temperature differences. The thin, warm, upper layer is the mixed or surface zone, the transition zone where temperature drops rapidly is the thermocline (pycnocline) and the cold, dense zone below is called the deep zone.

Test Your Learning

After working through these questions, check your answers against the key at the end of this book. If you answered any of the questions incorrectly, review the relevant sections of the text and video episode.

Multiple Choice

1. Which of these groups is arranged in correct order?
   a. epipelagic, mesopelagic, bathypelagic, hadopelagic, abyssopelagic
   b. epipelagic, bathypelagic, mesopelagic, abyssopelagic, hadopelagic
   c. hadopelagic, abyssopelagic, bathypelagic, mesopelagic, epipelagic
   d. hadopelagic, abyssopelagic, mesopelagic, epipelagic, bathypelagic

2. The ________ is the region of maximum sound transmission in the ocean.
   a. shadow zone
   b. sonar zone
   c. Doppler layer
   d. sofar layer

3. Which of the following pairs are roughly synonymous?
   a. epipelagic and mesopelagic
   b. epipelagic and euphotic
   c. thermocline and halocline
   d. pycnocline and hadopelagic

4. The bending of sound or light waves is called:
   a. deflection
   b. diffraction
   c. reflection
   d. refraction

5. Sound waves in seawater:
   a. travel five times slower than sound waves in the air
   b. travel faster in colder water than in warmer
   c. travel faster than sound waves in the air
   d. are affected by pressure but not by salinity

6. Which of these regions generally lacks a thermocline?
   a. polar
   b. temperate
   c. tropical
   d. the thermocline is present at all latitudes except at upwelling zones

7. The boundary between the thermocline and the deep zone usually lies at a depth of about:
   a. 100 meters
   b. 1,000 meters
   c. 2,000 meters
   d. 3,000 meters

8. Which of the following colors of visible light is absorbed least by seawater?
   a. red
   b. infrared
   c. blue
   d. violet

9. Researchers are attempting to monitor fluctuations in global ocean temperature by:
   a. measuring the speed of light transmission through the pycnocline
   b. measuring speed of sound transmission through the sofar layer
   c. collecting data from a series of floating and submerged buoys deployed worldwide
   d. using infrared sensors on a series of 24 satellites in geosynchronous orbit

10. The sinking of large amounts of surface water to great depths occurs:
    a. in all polar oceans
    b. primarily in the southern polar oceans
    c. primarily at mid-latitudes via submarine canyons
    d. in temperate oceans in July and August in the northern hemisphere and January and February in the southern hemisphere

11. The shadow zone is:
    a. at the boundary of the pycnocline and the deep zone
    b. at a depth where sound travels rapidly
    c. deeper than the sofar layer
    d. shallower than the sonar layer

12. A halocline would be most likely to occur in:
    a. the region associated with the Antarctic circumpolar current
    b. the tropical epipelagic zone
    c. the sofar layer
    d. regions where rivers enter the sea

Short Answer

1. Describe where the sofar layer and the shadow zones are located and explain how refraction of sound waves creates these two peculiar ocean regions.
2. Describe and define the following ocean zones: epipelagic, mesopelagic, bathypelagic, abyssopelagic, and hadopelagic. Prepare a cross-section illustration of these zones.
3. Compare and contrast the salinity, temperature, and precipitation balance in the surface waters of tropical, temperate, and polar oceans.

Supplemental Activities

1. You are aboard a research vessel in the tropical Pacific over the Mariana Trench collecting oceanographic data. It is noon on a clear summer day and you lower an array of instruments overboard to collect data at 1 meter, 100 meters, 1,000 meters, and 10,000 meters. The instruments measure temperature, salinity, red light intensity, blue light intensity, chlorophyll concentration, and sound velocity. Predict, compare, and explain how the data from the four depths would differ.
2. Study density stratification in water by performing the following simple experiments at home. Explain what happens in each mixture. Prepare a mass of dense water by mixing four tablespoons of sugar or salt into a pint of tap water colored with green food dye and placing it in the refrigerator overnight. The next day, prepare a mass of low-density water by heating a pint of water on the stove to about the temperature of hot soup. Color it yellow. Fill about two-fifths of the glass with the cold water and very carefully pour about the same amount of hot water on top of the cold. Describe what happens and observe your model ocean over a period of several hours. Can you create stratification? How long does the stratification last?
3. You are hired by the defense department to design an outreach program for elementary school children that explains research on monitoring the global ocean temperature using the ATOC project, which ended several years ago but whose findings are available for study. Go to the Acoustic Thermometry of Ocean Climate (ATOC) Website (http://atoc.ucsd.edu/). Choose the parts of the project that would be of most interest to school children and outline a 40 minute presentation/demonstration that would excite and educate them.

Lesson 9

Going to Extremes

Overview

Earth is a planet of moderate temperatures because of the thermostatic effects of water and yet the ocean's polar and tropic zones represent dramatic extremes. Comparing these very different ocean regions helps to clarify how the physical and chemical principles discussed in the previous lessons apply to the world's ecosystems.

Primary productivity is the driving force of all ecosystems. Productivity refers to the rate of growth of the organisms in a food chain; primary productivity is the rate of growth of the organisms at the foundation of any food chain. Photosynthesis is the basis of *primary production* in most ecosystems, and the ocean is no exception. Single-celled microscopic drifting phytoplankton are responsible for most oceanic primary production. Using light energy to manufacture the sugars that are the fuel which sustains life, these tiny organisms support the entire food chain above them.

The raw materials necessary for photosynthesis—carbon dioxide and water—are readily available in ocean water. Two major resources are required to fuel the process—light and nutrients. Sunlight is available only near the surface of the ocean. Nutrients are restricted to the deeper regions in many parts of the ocean. Without both light and nutrients, photosynthetic productivity is low or impossible. These, then, are the factors that create the conditions that define the patterns of primary production in ocean ecosystems.

Nutrients are chemicals dissolved in the water. Photosynthesizing organisms absorb these chemicals from the water and use them to build proteins, DNA, and other molecules within their cells. When these organisms die or are consumed, the nutrient molecules, trapped within cells or incorporated in the fecal pellets of the herbivores that eat the phytoplankton, fall to greater depths. Here the cells and feces are decomposed by bacteria and the nutrient molecules are released back into the water. Thermoclines—often a permanent feature of the well-lit tropical regions—stratify the ocean, creating a barrier to the mixing of nutrients back up into the photic zone. Polar oceans are generally too cold to form thermoclines, so nutrients are always available in the upper layers, but light is often scarce. High polar latitudes are marked by six months of total darkness followed by six months of light. It is often true that tropical oceans have abundant light but few nutrients, while polar oceans have abundant nutrients but little light.

It is generally true that harsh ecosystems have low *biodiversity* and less harsh systems have high diversity, but the coral reef ecosystem has a very high biodiversity. The polar ecosystems are harsh and have fewer species than the coral reef. But the few species that can tolerate the harsh conditions are often very abundant. Coral reefs generally have a large number of different species but relatively small populations of each species.

Coral reefs have long been a paradoxical place to biologists. Clear blue water generally signifies a low level of nutrients, resulting in a low level of phytoplankton. Such a place should not be teeming with life. But coral reefs are highly productive ecosystems in nutrient-poor waters. They teem with colorful life in the clear, warm, shallow waters of Hawaii, the Caribbean, and Australia. Where are the primary producers to support this ecosystem? The answer lies in symbiosis. The coral species that build the massive *calcium carbonate* (limestone) skeletons that form the reefs have algae living within their tissues. These symbiotic algae are a type of dinoflagellate called *zooxanthellae*. As they photosynthesize, they produce food for the *coral polyps*. They also produce oxygen and create conditions that seem to favor the rapid skeletal growth of the coral. In return, the coral animal produces carbon dioxide and nutrients, and provides protection for the zooxanthellae. A tight and efficient cycling of nutrients between the coral polyp and its symbiotic algae is the key to the high level of productivity of the reef.

Tropical reefs are built largely by organisms that secrete calcium carbonate. Along with coral animals, calcareous green algae (*Halimeda*) and coralline red algae (*Lithothamnion*) are simultaneously primary producers and major reef-builders. In many Pacific reefs coralline red algae are actually more important than corals in constructing reefs. Additionally, as they die and erode they produce the calcium carbonate sand that is the major substrate around the reef and may serve to strengthen and cement the reef as well.

Coral reefs are not universal in shallow tropical oceans; they lie beneath the surface of only about two percent of tropical oceans. They are common around open-ocean tropical islands such as Hawaii. Where coral reefs exist adjacent to continents they are generally on the western sides of oceans, like Australia's Great Barrier Reef and reefs in the Caribbean. Because of the general circular flow of major surface currents (see Figure 9.8 on page 239), the eastern sides of oceans are dominated by cold currents that flow away from the continents, while the western sides of oceans are dominated by warm North and South Equatorial Currents that flow westward toward the continents. Major coral reefs exist where these surface currents have flowed across entire oceans. Having traveled above a permanent thermocline, they are nutrient-poor, have low productivity, and are clear blue. On the opposite sides of the ocean the cold currents move away from the continents, drawing water up from the depths. This nutrient-rich deep water feeds a flourishing upwelling community with abundant phytoplankton, zooplankton, massive schools of tiny silvery fish like anchovies and sardines, and the larger fish that feed on them. This is the environment off the coasts of southern California, Peru, and West Africa.

Whereas tropical ecosystems are regulated by nutrient availability, polar ecosystems are regulated more by the availability of light. While the tropical home of coral reefs have only minor differences of light and temperature throughout the year, the seasonality of the polar region is stark. When the sun is shining, polar regions exhibit some of the highest productivity in the oceans. Without a thermocline, nutrients wash up from below more freely. Vast regions of continuous strong upwelling make the polar oceans, particularly the Southern Ocean, highly productive during their short summer growing seasons (June, July, August in the Arctic and December, January, and February in the Antarctic). During those months the already-present nutrients combine with the arrival of strong sunlight to cause a seasonal bloom of phytoplankton (see Figure 14.13 on page 395 of your text) that feeds the higher levels of the food chain including zooplankton, invertebrates, fish, sea birds, seals, and whales.

The three months of summer feast is followed by nine months of lightless famine. With no primary productivity, polar animals must cope with severe scarcity. Some, like the great filter-feeding whales and many bird species, migrate to warmer latitudes. Other animals remain during the harsh winter and drift into periods of low metabolic activity. Antarctic *krill* store lipids during the bloom season and then survive on these stores until the next summer.

Both polar and tropical regions are of considerable interest for economic reasons. The reef ecosystem and surrounding waters are important fishing grounds providing food and other marine products. Like rain forests, the biodiversity of coral reefs make them a potential source of chemicals with biomedical uses. Acyclovir, the first antiviral medication approved for humans, was isolated from a Caribbean sponge, and a variety of promising pharmaceutical chemicals have been isolated from marine species. Meticulously cleaned pieces of coral skeleton have been used as a matrix for bone regeneration.

Exploiting the profitability of coral reefs is not always a benign endeavor. Reefs are destroyed for building materials and dredged for sand and gravel (beach mining), which stirs up silt and sand and kills reefs by reducing light and covering the sensitive coral polyps. The cutting of reefs into bricks of limestone called "living rocks," along with the poisoning of large patches of reef in order to narcotize colorful reef fish for the aquarium trade, produces many more dead animals to be discarded than live ones to be sold. Reefs are dynamited to catch fish and to cut channels for boat passage. Coral reefs suffer world wide from *coral bleaching* disease and white band disease, both seemingly related to pollution and other human-induced ocean disturbances. An alarming 27 percent of the world's reefs are already destroyed and it is estimated that fully 70 percent of the reefs near population centers such as those in the Caribbean, Gulf of Mexico, and Mediterranean will die in the next 50 years. Natural disasters such as hurricanes periodically damage reefs, and the damage is intensified if the reef is already weakened by pollution and disease.

Although farther from centers of human activity, the polar regions are of great interest as well. Overfishing has found its way to virtually all parts of the sea and the polar oceans are no exception. Trawlers dragging their nets destroy swaths of ocean bottom as they haul in substantial quantities of fish for sale. Their nets also ensnare a variety of other animals, including slow-growing polar sponges and unwanted fish. These unwanted fish and invertebrates, euphemistically called bycatch, are the collateral damage of commercial fishing. The reduction of Earth's ozone layer resulting from chlorofluorocarbon (CFC) air pollution is most dramatic over the poles and is responsible for the passage of greater levels of DNA-damaging ultraviolet radiation through the atmosphere to earth's surface. Skin cancers, cataracts, and crop damage are among the effects.

Because they are hard, rather permanent, and very sensitive to environmental changes, coral reefs can be used in the studies of paleoclimatology and *paleoecology*—ancient climates and ancient environments. The chemical composition of their skeletons vary depending on the temperature when they were living, so cores can be taken from the fossil skeletons of ancient corals and analyzed for growth patterns, somewhat like reading the growth rings of trees. Such studies provide clues to ancient climatic conditions, nutrient levels, sedimentation rates, and a variety of other features of past oceans and their adjacent land.

Much current research in the polar regions is associated with ocean circulation and global climate. It is at the poles that the very cold, salty masses of dense seawater sink, driving thermohaline circulation as the water spreads throughout the ocean basins and causing *upwelling* elsewhere. This flow is an important factor regulating global climate. A current hypothesis suggests that global warming could melt polar ice, cause runoff of fresh water into the polar oceans, and shut down the conveyor because the fresh water would not sink. In the north Atlantic, this mass of low salinity water could prevent the remnants of the Gulf Stream from flowing as far north as usual and initiate a mini-ice age in Europe and North America. There is some evidence this has happened in the past. It is a curious hypothesis that global warming could initiate global cooling if the ice-melt shuts down the thermohaline conveyor belt.

An understanding of the principles of oceanography that account for the differences between polar and tropical oceans provides a useful springboard to the further study of ocean science, including topics like meteorology, biology, resources, and others that follow in the remaining lessons.

Focus Your Learning

Learning Objectives

After completing this unit you should be able to do the following:

1. Define primary production and compare and explain the patterns of production in polar and tropical oceans.
2. Explain the structure of coral animals and coral reefs and explain why coral reefs are much more common on the western sides of oceans than on the eastern sides.
3. Define biodiversity; compare and contrast biodiversity in polar oceans and on coral reefs.
4. Compare and contrast the Arctic and Antarctic polar regions; compare and contrast western tropical oceans and eastern tropical oceans.
5. Explain the symbiotic relationship between zooxanthellae and the coral polyps they inhabit.
6. List and give examples of several ways organisms cope with the long dark season of polar oceans.
7. Discuss several uses, abuses, and problems specific to coral reefs and to polar ecosystems.

Assignments

This lesson is based on information in the following text and video assignments. The key terms, focus points, and practice test are intended to help ensure mastery of the material presented in this unit of study.

Text: Read "Surface Water Moderates Global Temperatures" pages 164–168; "Seawater Flows in Six Great Surface Circuits" page 239; "Wind Can Cause Vertical Movement of Ocean Water" pages 244–247; "Thermohaline Circulation Affects All the Ocean's Water" pages 250–257; "Coral Reefs Can be Built by Coral Animals" page 332; "Primary Productivity is the Synthesis of Organic Materials" page 348; "Phytoplankton are Autotrophs" pages 380–381; "Lack of Nutrients and Light can Limit Productivity" pages 390–391; "Phytoplankton Productivity Varies with Local Conditions" page 392; "Stinging Cells Define the Phylum Cnidaria" pages 407–408; "Coral Reefs Are Earth's Most Densely Populated and Diverse Communities" page 447; "Coral Reefs are Stressed by Environmental Change" pages 504–505; "The Protective Ozone Layer Can be Depleted by Chlorine-containing Chemicals" pages 509–510

Video: Episode 9, "Going to Extremes"

Video Focus Points

Review these points before watching the video assignment for this lesson to help you focus on key issues.

— Various oceanographic principles explored in earlier lessons are integrated and applied to a comparison of polar and tropical ocean ecosystems.

— Light, nutrient availability, temperature, and other factors interact in predictable ways to explain the differences of the polar and tropical ocean realms.

— Biodiversity of the coral environment is high while that of the polar environment is low.

— The extended dark season of the polar oceans results in little or no primary productivity. Polar organisms employ a variety of strategies to cope with the season of scarcity.

— The structure and composition of coral skeletons respond differently to different conditions of temperature, light, sedimentation, and other factors. Analysis of core samples of ancient reefs can reveal historical environmental patterns helpful in understanding ancient climates and environments.

— Both polar and tropical oceans provide a wealth of resources that are sometimes utilized rationally and sometimes abused and overexploited.

— Scientific interest in the polar regions often focuses on the role they play in generating the oceanic thermohaline circulation and the effect of this on global climate.

Text Focus Points

The text readings for this unit are several short sections drawn from a wide variety of chapters. Use them along with the overview to embellish and clarify the video material.

— The production of organic food molecules or primary production is usually a result of photosynthesis. Sunlight and nutrients are the two major factors regulating rates of oceanic photosynthesis. In polar oceans light is generally the more important of these two limiting factors while in tropical seas nutrient availability is more critical.

— Coastal upwellings and gyre circulation patterns create different conditions in western and eastern tropical oceans. The tropical upwelling regions such as those in the eastern Pacific near Peru and California are quite different from the tropical reef regions in the western Pacific such as those near the Philippines or Australia's Great Barrier Reef.

— Coral reefs flourish in nutrient-poor seas. The symbiotic relationship between hermatypic coral animals and their zooxanthellae is a key to understanding the reef ecosystem.

— Although both ecosystems can be highly productive, the harsh conditions of the polar ocean limit biodiversity while the constant warmth and availability of sunlight favor the high biodiversity of the tropical reef ecosystem.

— Several factors including coral bleaching disease, white-band disease, overfishing, sewage, sedimentation, and hurricanes are interacting to cause a rapid decline of Earth's coral reefs.

— CFC atmospheric pollution contributes to destruction of the ozone layer, particularly over the polar regions. This causes increased human skin cancers, crop damage, declines in phytoplankton productivity and damage to other species exposed to the increased ultraviolet radiation formerly absorbed by the ozone.

Key Terms and Concepts

A thorough understanding of these terms and concepts will help you to master this lesson.

Biodiversity. The variety of different species in a particular region.

Calcium carbonate. The limestone chemical ($CaCO_3$) that forms the skeleton of corals and many other ocean organisms including clams, snails, calcareous algae, and foraminifera.

Coral bleaching. A condition, often fatal, in which corals expel their symbiotic zooxanthellae.

Coral polyp. An individual coral animal. Polyps are generally cylindrical and have a central mouth surrounded by tentacles.

Gyre. A large circular ocean current that flows around the periphery of an ocean. Gyres in the northern hemisphere flow in a clockwise direction while in the southern hemisphere they flow counterclockwise.

Krill. A large crustacean zooplankter that feeds on phytoplankton and is in turn fed upon by larger carnivorous organisms including many whales. They are an important link in many ocean food chains and are particularly important in polar oceans.

Nutrients. Any substance except CO_2, O_2, and H_2O required for survival and obtained from the environment by an organism. Nitrate, phosphate, silicate, and iron are examples of nutrients required by phytoplankton.

Ozone depletion. A reduction of ozone (O_3) in the upper atmosphere caused mostly by a chemical reaction between the ozone and chlorofluorocarbon (CFC) air pollution.

Paleoecology. The study of ancient environmental conditions and ecosystems.

Plankton. The organisms that drift about in the ocean. They are generally microscopic and divided into zooplankton (animal-like plankton) and phytoplankton (plant-like plankton).

Primary production. The biological synthesis of organic food molecules such as glucose sugar from inorganic molecules. Photosynthesis is the major form of primary production.

Upwelling. An upward moving ocean current that brings cold and often nutrient-rich bottom water to the surface.

Zooxanthellae. Symbiotic, photosynthetic algae that live within the tissues of hermatypic corals and supply the coral polyps with sugars and other food molecules derived from the algae's photosynthesis.

Test Your Learning

After working through these questions, check your answers against the key at the end of this book. If you answered any of the questions incorrectly, review the relevant sections of the text and video episode.

Multiple Choice

1. Primary production is most closely synonymous with:
   a. herbivory
   b. solar energy
   c. reproductive rate
   d. photosynthesis

2. The process responsible for the circulation of the global ocean as a whole is:
   a. upwelling
   b. the geostrophic currents
   c. thermohaline circulation
   d. gulf stream circulation
3. Depletion of atmospheric ozone is mostly related to:
   a. global warming
   b. acid rain
   c. excessive atmospheric carbon dioxide
   d. chlorofluorocarbon pollution
4. The skeletal material of coral reefs is composed of:
   a. silicon dioxide
   b. calcium carbonate
   c. calcium phosphate
   d. disodium phosphate
5. Zooxanthellae are:
   a. crustacean zooplankton that are the major herbivores of polar oceans
   b. yellow sponges that produce prostaglandins and acyclovir
   c. symbiotic algae that live within coral polyps
   d. the major diatoms of coastal upwelling zones
6. The area of greatest annual phytoplankton productivity occurs:
   a. in the open tropical oceans
   b. in tropical oceans near continents
   c. in the Northern polar oceans
   d. in the Southern subpolar open ocean
7. Much research in the Arctic polar ocean presently focuses on:
   a. whaling
   b. thermohaline circulation
   c. white band disease
   d. El Niño
8. Coral bleaching seems to be associated with:
   a. excessively high water temperatures
   b. excessive sedimentation
   c. overfishing
   d. excessive algal blooms
9. The thermocline contributes to low rates of productivity in:
   a. coral reefs
   b. most open tropical oceans
   c. the Arctic Ocean
   d. coastal upwelling zones
10. One would expect to find the greatest biodiversity in/on:
   a. the Peruvian upwelling region
   b. an Indo-Pacific coral reef
   c. a Southern Ocean upwelling region during summer
   d. upper layers of open equatorial oceans
11. In general, the two most important nutrients in limiting primary production are:
   a. silicate and chloride
   b. potassium and zinc
   c. carbonate and bicarbonate
   d. nitrate and phosphate
12. A large crustacean zooplankton important in Antarctic food webs:
   a. krill
   b. dinoflagellates
   c. zooxanthellae
   d. isopods

Short Answer

1. Compare and explain the general patterns and causes for the differences in primary production in polar vs. tropical oceans.
2. Discuss man-induced problems contributing to the global decline in coral reef ecosystems.
3. Discuss the causes, extent, and consequences of ozone depletion.

Supplemental Activities

1. Find the following locations on the map in Appendix V, page 534. For each location, integrate the information in Figures 9.8, 9.15, 9.27 and 14.13 to predict the type of ecosystem at each location and outline the oceanographic characteristics of each system.
   a. Location A: the intersection of the Prime Meridian and the Antarctic Circle;

b. Location B: the intersection of the Prime Meridian and the Equator;
c. Location C: the intersection of the Equator and 120°W Longitude; and
d. Location D: the intersection of the Equator and 120° E Longitude (Sulawesi, Indonesia).

2. Several fishing enterprises favor harvesting Antarctic krill for human use and consumption. Predict some of the possible consequences of a large scale krill fishery on the Antarctic ecosystem.

3. Lima, Peru and Cairns, Australia both lie at about 15° south latitude adjacent to the tropical Pacific Ocean. The coastal ecosystems of these two regions are very different. Compare the ecosystems of these two regions and explain how patterns of water movement, thermoclines, nutrients, and sunlight interact to cause the differences.

4. Compare and contrast the relationship between zooxanthellae and hermatypic corals with the relationship between polar phytoplankton and krill.

Lesson 10

Something in the Air

Overview

The word "oceanography" implies the study of the ocean, but even the early voyagers realized the inseparable interaction between the ocean, the atmosphere, and the land making up what is called the whole Earth system. This lesson concentrates on the composition and activities of Earth's atmosphere as a part of the system.

Earth's lower atmosphere is composed mainly of nitrogen and oxygen along with some minor gases and, depending on local conditions and geographic location, various amounts of water vapor. Air has mass which—along with its temperature, water content, and distance above sea level—determine its *density*, an important factor in atmospheric activity. As these components interact (warm air rising, cold air descending, altitude and water content changing), a restless, active "system" results. Add an energy factor and the system begins to circulate.

The source of the energy that powers atmospheric circulation is the sun. Only a miniscule amount of the sun's total radiant energy reaches Earth's surface, and about 30 percent of that is immediately reflected back into space. About 50 percent is absorbed by Earth's land and water and 20 percent by Earth's atmosphere. The 50 percent striking land and water is converted to heat and transferred into the atmosphere by conduction, radiation, and evaporation. The atmosphere eventually radiates this heat back into space, establishing what is called Earth's *heat budget*, and a state of general *thermal equilibrium*. The concept of equilibrium, however, is misleading; the heat budget differs at different latitudes depending on the angle of sunlight reception and reflection, the amount of atmosphere to be penetrated, and other variables. Tropical latitudes constantly receive more solar energy than do polar regions, while the mid-latitudes experience seasonal heating and cooling. All of these factors and more are important in attempting to understand Earth's climate system, realizing that the system can and will continue to remind us of its potential for surprise and devastation. Hurricane Mitch, in 1998, was such a reminder.

Mitch began as a chain of thunderheads over Central Africa, became a tropical depression, and moved on to hit the Central American coast as a category-5 storm with winds up to 170 miles an hour. One of the symptoms of such weather is a fall in the barometer—a measure of the weight of an air mass as it affects the surface of Earth under it. As the barometer dropped (the weight of the air, mass decreased) during Mitch, the sea surface rose under the decreased pressure into a dome estimated to be 20 feet high and tens of kilometers across. Mitch continued west over Honduras and Nicaragua, leaving near total devastation in its path. Meteorologists tracked and documented this series of events, but still cannot explain how or why all of the conditions necessary to create it came together at that particular time and place. Although many of the physical phenomena involved in the weather can be characterized and quantified, its location and intensity are often still a surprise.

Nearly all weather events have in common an interaction between atmosphere and ocean. As mentioned earlier, the amount of solar energy absorbed and re-radiated differs by latitude—more at the equator and less at the poles—suggesting a major heat imbalance. Water moves large amounts of heat from the tropics to the poles, helping to correct that imbalance, but more heat is actually transferred by water vapor than by liquid water. Masses of air move about two-thirds of the heat toward the poles; ocean currents move the other third. The mid-latitudes experience periodic temperature fluctuations, or seasons, based on the tilt of Earth as it rotates around the sun. This uneven solar heating of different parts of Earth also affects the atmosphere by causing air masses to move, or circulate, as *convection currents*, which are created when hot air rises and cold air sinks. But this is not a linear phenomenon. Earth's rotation also plays a part through the Coriolis effect. A simple explanation of *Coriolis effect* is difficult to offer, but it is most important to have some understanding of how it works and how it affects both atmospheric and oceanic circulations. The eastward rotation of Earth on its axis deflects air, water, or any moving object that has mass, away from a linear north-south or south-north course. Coriolis deflection works clockwise in the Northern Hemisphere and counterclockwise in the Southern Hemisphere because the frame of reference is reversed. It is essentially non-existent at the equator.

While the Coriolis effect influences any moving object, it most dramatically affects the movement of air and water, being most apparent at the mid-latitudes as an interaction between ocean water and winds. Air warms and rises at the equator, and cools and falls at the poles, but not in a linear equator-pole-equator circuit. In the Northern Hemisphere that circuit is moved eastward by Coriolis effect; in the Southern Hemisphere it is moved westward. These large masses of moving air are called *atmospheric circulation cells*, and they are named and defined by their locations and characteristics. *Hadley cells*, or tropical cells, are located on each side of the equator, *Ferrel cells* operate at mid-latitudes, and *polar cells* at the poles. All three of these cell types exist in each hemisphere, powered by uneven solar heating and influenced by the Coriolis effect. Another important set of cells, *Walker cells*, are at work in an east-west direction along the equator, overturning the very warm equatorial waters and the cooler subequatorial waters between the western and eastern Pacific oceans.

All of this implies air movement, or wind. Air moves vertically between circulation cells and is weak and erratic. Within circulation cells, air moves horizontally and is strong and dependable. These wind patterns are described and named according to their characteristics and locations. The descriptive term *doldrums* refers to the calm equatorial areas where the two Hadley cells converge; scientists call this the *Intertropical Convergence Zone*, or *ITCZ*. It is at the rising part of the Hadley circulation that the Northern and Southern Hemispheric trade winds meet. Because the trade winds have been moving over very warm air, they are forced to rise when they meet, forming a strong and turbulent circulation cell with accompanying heavy rain as the humid air rises and loses moisture. The exact position of the ITCZ determines the location and intensity of the seasonal wind patterns known as *monsoons*, which are characterized by high winds and heavy rain. Monsoon climates are usually associated with Africa, Asia, and northern Australia, but there are also so-called American monsoons, which move up from the Gulf of Mexico into the Central United States bringing rain, wind, and thunderstorms.

Other areas where wind patterns can be characterized are between the Hadley and Ferrel cells in the *horse latitudes* where the atmospheric pressure is high and there is very little wind. These were feared by early mariners because they could becalm a ship for weeks. The *trade winds* are the dependable, usable surface winds of the Hadley cells as they move from the horse latitudes to the doldrums. *Westerlies* form as air moves from the Ferrel cells into the polar cells. Early mariners learned to use the characteristics of these wind patterns to great advantage.

An *air mass* is a large body of air with consistent temperature and humidity, thus density, throughout. Air masses are influenced by the land or water over which they form. While air masses can move within or between circulation cells, their different densities usually prevent them from mixing as they meet. The boundary where they collide is called a *front,* and is the point of origin for many storms. *Storms* tend to be regional, characterized by strong winds and often rain.

One of the most common types of storm is the *cyclone*—a huge rotating mass of low-pressure air in which winds converge and ascend. The term cyclone can include a number of different kinds of storms, but all have some similar components. *Tropical cyclones* are masses of warm, humid, rotating air, and are common to most tropical oceans. Large tropical cyclones are called *hurricanes*, while those with winds of less than 150 miles per hour are tropical storms or tropical depressions. From above they appear as spirals of clouds hundreds of kilometers in diameter surrounding a calm center, or eye. Tropical cyclones are not the result of colliding fronts, but form within a single warm, humid air mass.

An *extratropical cyclone* forms at the boundary of polar and Ferrel cells as a *polar front*. The different densities of the air masses prevent mixing, and the cold, dense mass may slide under the warmer, lighter one. This generates a kind of twist between the air masses, appearing as the typical spiral pattern. In America, the most famous and most violent extratropical cyclones are the *northeasters* that sweep the Eastern seaboard up to 30 times each year.

Scientists continue to study, describe, and document these weather phenomena in hope of better understanding and more accurate forecasting, but there is still much to learn. New technology has helped, including continuously-operating Doppler radar, weather satellites, and advanced computer modeling. Despite all this, meteorologists still feel that the biggest challenges for the future are the need for a more thorough understanding of the factors that affect Earth's weather and climate, the unknown effects of global warming on Earth's climate and weather, and how all of this will affect Earth's human population.

Focus Your Learning

Learning Objectives

After completing this unit you should be able to do the following:

1. Understand the molecular composition of Earth's lower atmosphere, the fluctuating role of water vapor, and how all of this influences air's density.
2. Describe the phenomenon of convection and how it affects the movement of air over Earth's surface.
3. Explain the concept of Earth's "climate system" as it relates to the whole Earth system—ocean, atmosphere, and land.
4. Understand the basic sequence of events involved in Hurricane Mitch, and why it is used as an example of how the power of Earth's climate system affects the human population.
5. Explain the role of the sun's energy in atmospheric circulation, recognizing the absorption/radiation/re-radiation concept and the varying amounts of Earth's heat budget involved.
6. List and characterize the four major types of atmospheric convection cells—Hadley, Ferrel, Polar, and Walker.
7. Understand the Coriolis effect and how it influences atmospheric conditions.
8. Describe the various common global wind and storm patterns—where they form, their basic characteristics, and how they are named.
9. Define the concept of an air mass, how different air masses move and behave, and how they interact at boundaries and fronts.

Assignments

This lesson is based on information in the following text and video assignments. The key terms, focus points, and practice test are intended to help ensure mastery of the material presented in this unit of study.

Text: Chapter 8, "Circulation of the Atmosphere," pages 200–231

Video: Episode 10, "Something in the Air"

Video Focus Points

Review these points before watching the video assignment for this lesson to help you focus on key issues.

— Earth's climate system is based on interactions between land, ocean, and atmosphere.

— Hurricane Mitch (1998) is an excellent example of how the components of Earth's climate system interact in sequence, building into a dramatic and devastating event.

- — Atmospheric circulation is directly affected by the ocean, and vice-versa, as air heats, cools, rises, and falls.
- — All atmospheric motion is driven by solar heating and modified by regional shifts in Earth's heat budget.
- — The circuits of moving air that comprise Earth's atmosphere are known as cells, with different types being located above different surface regions.
- — The Coriolis effect is used to explain how Earth's rotation affects both atmospheric and oceanic movements.
- — The Intertropical Convergence Zone (ITCZ) is of special interest to meteorologists due to the extraordinary atmospheric turbulence and activities associated with it, particularly the monsoons.
- — As air masses move, they often meet at a front where, depending on their characteristics, they can progress into a variety of storms (cyclones, tornadoes, typhoons).
- — Modern technology (Doppler radar, weather satellites, computer modeling) is enabling scientists to learn more about the building blocks of weather, leading to better analysis and forecasting. However, there is still much to learn about how all of this will be affected by global warming and how it relates to Earth's human population.

Text Focus Points

Review these points before reading the text assignment for this lesson to help you focus on key issues.

- — There is a constant and dynamic interaction between the ocean and the atmosphere.
- — Earth's lower atmosphere is composed of nitrogen, oxygen, some minor gases, and water vapor.
- — Air has mass, and therefore density, which varies according to its temperature, water content, and proximity to Earth's surface.
- — Warm air rises and expands while cold air descends and condenses, setting up a circuit called a convection current.
- — A small amount of the sun's total energy strikes Earth. Some is re-radiated back into space; the rest is absorbed by Earth's atmosphere, land, and water, to be converted into heat and eventually re-radiated. This sequence of reception/absorption/re-radiation establishes a kind of thermal equilibrium called Earth's heat budget.
- — Solar heating is uneven at Earth's surface, depending on latitude.
- — Convection currents (warm air rising/cold air sinking) should, theoretically, move from equator to pole to equator powered by solar energy. But the rotation of Earth affects this movement as the Coriolis effect.
- — The Coriolis effect gives any moving mass a lateral clockwise deflection in the Northern Hemisphere and a counterclockwise deflection in the Southern Hemisphere. This interrupts the equator/pole/equator flow and establishes several large circuits called atmospheric circulation cells.
- — Circulation cells are described and named by location and basic characteristics—Hadley cells near the equator, Ferrel cells at mid-latitudes, and polar cells. Air movement between and within these cells is powered by solar energy and influenced by the Coriolis effect.
- — Depending on latitude and location (over land or sea) different types of air movements (wind patterns) are noted (doldrums, horse latitudes, trades, westerlies).
- — Monsoons are seasonal wind patterns formed where the Hadley cells converge near the equator at the Intertropical Convergence Zone.
- — An air mass is a large body of air with nearly uniform temperature and humidity. When two different air masses meet and cannot mix, they may form a front where storms often originate.
- — Storms are regional atmospheric disturbances. Cyclones (either tropical or extratropical) are the most common types of storms, and do the most damage.

Key Terms and Concepts

A thorough understanding of these terms and concepts will help you to master this lesson.

Air mass. A large body of air with nearly uniform temperature, humidity, and density throughout.

Atmospheric circulation cell. Large circuit of air driven by uneven solar heating and the Coriolis effect. Three circulation cells form in each hemisphere.

Convection current. A single closed-flow circuit of rising warm material and falling cool material.

Coriolis effect. The apparent deflection of a moving object from its initial course when its speed and direction are measured in reference to the surface of the rotating Earth. The object is deflected to the right of its anticipated course in the Northern Hemisphere and to the left in the Southern Hemisphere. The deflection occurs for any horizontal movement of objects with mass, and has no effect at the equator.

Cyclone. A weather system with a low-pressure area in the center, around which winds blow counterclockwise in the Northern Hemisphere and clockwise in the Southern Hemisphere. Not to be confused with a tornado, which is a much smaller weather phenomenon associated with severe thunderstorms.

Density (of air). A column of air one centimeter square, measured from sea level to the top of the atmosphere, weighs about 1.4 kg (2.3 lbs). Add to that the amount of water vapor it carries and the temperature of the mass to calculate density. This figure must then be modified to account for altitude (distance above sea level).

Doldrums. The zone of rising air near the equator known for sultry air and variable breezes. Also known as the Intertropical Convergence Zone (ITCZ).

Extratropical cyclone. A low-pressure mid-latitude weather system characterized by converging winds and ascending air rotating counterclockwise in the Northern Hemisphere and clockwise in the Southern Hemisphere. An extratropical cyclone at the front between the polar and Ferrel cells.

Ferrel cell. The middle atmospheric circulation cell in each hemisphere. Air in these cells rises at 60° latitude and falls at 30° latitude (see westerlies).

Front. The boundary between two air masses of different densities. The density difference can be caused by differences in temperature and/or humidity.

Frontal storm. Precipitation and wind caused by the meeting of two air masses, associated with an extratropical cyclone. Generally, one air mass will slide over or under the other, and the resulting expansion of air will cause cooling and, consequently, rain or snow.

Geographical equator. Zero degrees latitude, an imaginary line equidistant from the geographical poles.

Hadley cell. The atmospheric circulation cell nearest the equator in each hemisphere. Air in these cells rises near the equator due to strong solar heating there, and falls due to cooling at about 30° latitude.

Heat budget. An expression of the total solar energy received on Earth during some period of time, and the total heat lost from Earth by reflection and radiation into space during the same period.

Horse latitudes. Zones of erratic horizontal surface air circulation near 30°N and 30°S latitudes. Over land, dry air falling from high altitudes produces deserts at these latitudes (the Sahara).

Hurricane. A large tropical cyclone in the North Atlantic or eastern Pacific, with winds exceeding 74 miles per hour.

Intertropical convergence zone (ITCZ). The equatorial area at which the trade winds converge. The ITCZ usually lies at or near the meteorological equator. Also called the doldrums.

Land breeze. Movement of air offshore as marine air heats and rises.

Meteorological equator. Also called the thermal equator. An irregular imaginary line of thermal equilibrium between hemispheres. It is situated about 5° north of the geographical equator, and its position changes with the seasons, moving slightly north in northern summer.

Monsoon. A pattern of wind circulation that changes seasonally. Also, the rainy season in areas with monsoon wind patterns.

Nor'easter (northeaster). Any energetic extratropical cyclone that sweeps the eastern seaboard of North America in winter.

Polar cell. The atmospheric circulation cell centered over each pole.

Polar front. A boundary between the polar cell and the Ferrel cell in each hemisphere.

Precipitation. Liquid or solid water that falls from the air and reaches Earth's surface as rain, hail, or snow.

Sea breeze. Onshore movement of air as inland air heats and rises.

Storm. A local or regional atmospheric disturbance characterized by strong winds and often accompanied by precipitation.

Storm surge. An unusual rise in sea level resulting from low atmospheric pressure and strong winds; associated with a tropical cyclone. Onrushing sea water precedes landfall of a cyclone and causes most of the damage to life and property.

Thermal equilibrium. The condition in which the total heat coming into a system is balanced by the total heat exiting the system.

Tornado. Localized, violent funnel of fast-spinning wind, usually generated when two air masses collide. Not to be confused with a cyclone. The oceanic equivalent of a tornado is a waterspout.

Trade winds. Surface winds within the Hadley cells, centered at about 15° latitude, which approach from the northeast in the Northern Hemisphere and from the southeast in the Southern Hemisphere.

Tropical cyclone. A weather system of low atmospheric pressure around which winds blow counterclockwise in the Northern Hemisphere and clockwise in the Southern Hemisphere. Originates in the tropics within a single air mass, but may move into temperate waters if water temperature is high enough to sustain it. Small tropical cyclones are called tropical depressions, larger ones are called tropical storms.

Uneven solar heating. Differences in the amount of solar energy that is received/absorbed/reflected by Earth's surface, depending on latitude.

Walker cell. Cells operating in an east-west direction, relating to the activity along the equator between the warm sea surface temperatures in the western Pacific and the cooler eastern Pacific.

Water vapor. Water's invisible, gaseous form.

Westerlies. Surface winds within the Ferrel cells centered around 45° latitude, which approach from the southwest in the Northern Hemisphere and from the northwest in the Southern Hemisphere.

Test Your Learning

After working through these questions, check your answers against the key at the end of this book. If you answered any of the questions incorrectly, review the relevant sections of the text and video episode.

Multiple Choice

1. Which of these factors has the least effect on air's density?
   a. water content
   b. temperature
   c. argon content
   d. distance from Earth's surface

2. The solar energy that strikes Earth's surface is converted to heat and transferred back into the atmosphere by:
   a. conduction
   b. radiation
   c. evaporation
   d. all of these

3. Moving air accounts for what proportion of heat moving poleward?
   a. $\frac{1}{3}$
   b. $\frac{1}{3}$
   c. $\frac{3}{3}$
   d. $\frac{3}{3}$

4. The circular pattern set up as warm air rises and cool air sinks is called a ____________ current.
   a. circular
   b. convection
   c. conduction
   d. radiation

5. The Coriolis effect is related to:
   a. thermal equilibrium
   b. the tilt of Earth on its axis
   c. the water content of atmospheric air
   d. uneven solar heating and Earth's rotation

6. A Hadley circulation cell is located:
   a. on each side of the equator
   b. only on the equator in the Northern Hemisphere
   c. only on the equator in the Southern Hemisphere
   d. between the Ferrel and polar cells

7. When air masses collide, they will mix only in the presence of:
   a. high humidity
   b. energy
   c. Coriolis effect
   d. none of these

8. The location of the ITCZ determines the location and intensity of:
   a. extratropical cyclones
   b. monsoons
   c. tropical cyclones
   d. hurricanes

9. Regional atmospheric disturbances characterized by strong winds and rain are called:
   a. monsoons
   b. storms
   c. polar fronts
   d. westerlies

10. The trade winds are bands of "dependable" surface winds centered at:
   a. between 15°N and 15°S
   b. 45°N
   c. 45°S
   d. all of these

11. As air rises it tends to:
   a. compress and heat
   b. compress and cool
   c. expand and heat
   d. expand and cool

12. Water in its gaseous form (vapor) can occupy up to ________ percent of air's volume.
   a. 1
   b. 2
   c. 4
   d. 8

Short Answer

1. Much of the discussion and explanation of the global climate system is based on an understanding of a basic air mass. Define and discuss the air mass—its composition, characteristics, behavior, and interaction with Earth's land and water.

2. Meteorologists agree that there is still a great deal to be learned about weather, climate, and atmospheric circulation, with an eye toward more accurate forecasting. In the video, one scientist discusses several of the ongoing research projects—tornado detection, tracking, and warning; computer modeling using weather satellite data; hurricane detection and tracking. Briefly summarize the discussions of those research areas.

3. Discuss some of the roles of water in atmospheric activity.

Supplemental Activities

1. As your school's resident oceanographer, you have been asked to explain the Coriolis effect to a class of non-science majors.

2. Although your text names a number of the more common types of global wind and weather systems (trades, westerlies), the local/regional names for these phenomena can be much more colorful and descriptive. Local media weathermen often use these terms for the benefit and understanding of their audiences (Santa Ana, blue northern, knik, kona, to name a few from the United States). Names from other countries can also be interesting ("sirocco" from North Africa, "dahatoe" from Sumatra). Do some research (Web, weather manuals and almanacs, encyclopedias, local weather channels) and compile a list of the more colorful and interesting wind/weather

terms, including their definitions/descriptions, literal translations, and geographic origins.

3. Become an informed citizen about the weather where you live. Thanks to Doppler radar, weather satellite data, and computer modeling, weather and wind patterns are now observed and analyzed on a 24-hour basis at both local and national levels, and reported via newspapers, radio, television, and Websites. Document the jargon and symbols used to report and forecast the weather in your area, including how and when particular weather phenomena (fronts, storms) may form offshore and then modify (or not) as they approach your region. Keep a record of prediction success for a week or so. Then, using the various sources and resources mentioned, write a short report on the similarities and differences in reporting language, procedures, and formats (newspaper, radio, television, Web), eventually deciding which one you consider to be the most reliable as a weather information source.

4. Until the Atlantic Hurricane season of 2005, Hurricane Mitch (1998) had been called the most lethal Atlantic hurricane of modern times, and is used in the video as an example of how weather events interact. Using your text and any appropriate Websites, document the sequence of events leading up to Mitch's emergence as a category-5 storm. Begin with its entrance as a chain of thunderheads over the Central African coast. Use all of the applicable meteorological terminology in your description, and summarize the damage done in each geographic region. Conclude your report with a detailed glossary of the scientific and descriptive terms used.

Lesson 11

Going with the Flow

Overview

When the early Greek seafarers began to venture out past Gibraltar and into the Atlantic Ocean, they encountered a wide, strong, persistent southwest flow of water. It is now understood that the entire world ocean contains these large movements of water called *currents*, and that they are found both at the surface and at depths. Winds also form characteristic patterns at different latitudes, with most of Earth's surface wind energy being concentrated in the easterlies, or trade winds, and the westerlies. Energy from this moving air interacts with the sea surface, causing it to move, or flow, forming a *surface current*.

About 10 percent of the water in the world ocean is involved in wind-driven surface currents—water that flows horizontally in the uppermost 400 meters (1,300 feet). This boundary should not be confused with the *pycnocline*, which is a middle zone within which density increases rapidly with depth. Wind-driven currents affect waters both above and below the pycnocline, transferring water and heat from tropical to polar regions, and influencing weather, climate, and the distribution of oceanic nutrients and organisms.

Following the basic laws of physics, and pushed by the wind, these surface water masses tend to deform or pile up in the windward direction, elevating the sea surface on that side and establishing a *pressure gradient*. Gravity then pulls at the water pile, trying to move it back down its slope, against that pressure gradient, in the direction from which it came. The Coriolis effect enters the picture, deflecting the Northern Hemisphere surface currents in a clockwise direction and those in the Southern Hemisphere counterclockwise. This potentially continuous circumglobal flow is interrupted by the continents and the configuration of the ocean basins causing a circular flow called a gyre, the edges of which are at the boundaries of the continents. The water at the center of these gyres may be a meter or so higher than at the edges, eventually establishing a balance, or *geostrophic gyre*, between the pressure gradient and the Coriolis effect producing geostrophic currents. Due to the unique configuration of each of the major ocean basins and its continental margins, these gyres are essentially independent of each other in each hemisphere. There are six great currents in the world ocean—five are geostrophic gyres, while the other, the *Antarctic Circumpolar Current*, is not considered a true gyre because it does not flow around the periphery of an ocean basin. It is the greatest of all surface currents, moving endlessly around Antarctica, powered by ceaseless westerly winds, and never deflected by continental margins.

While the interaction between the wind-pushed waters, the Coriolis effect, and the ocean basin boundaries is responsible for establishing the gyres, there is another factor to be considered—the movement of water below the topmost layer, within a gyre. This is a complicated mechanism called an Ekman circuit, named for the Swedish oceanographer Vagn Ekman, who studied the interactions between wind energy and the Coriolis effect. The Coriolis effect pushes surface water sideways, while the winds move it windward, resulting in a compromise of movement of about 45° to the right of the wind in the Northern Hemisphere and to the left in the Southern Hemisphere. The next deepest layer then moves at about 45° to the one above, and on down the column, resulting in a cumulative spiral movement down to a depth of about 100 meters called the *Ekman spiral*. This is not a single, whirlpool-like spiral, but rather a pattern set up as the various thin layers of water in the column continue to move at 45° angles; the net result being called the *Ekman transport*. Putting all of this together, we see that most of the major ocean currents move as large circular patterns, or gyres, within boundaries established by the margins of the continents. The fastest and deepest are those on the western basin boundaries. All of these currents move great masses of water and heat poleward within the gyre. The largest of these *western boundary currents* is the *Gulf Stream* in the North Atlantic.

Each of these large currents is unique to its surrounding geography. In some areas a current may flow directly opposite to the prevailing winds. In a surface current this is called a *countercurrent*, while a subsurface countercurrent is an *undercurrent*. Both of these can influence sea surface conditions, thus affecting atmosphere/ocean dynamics. Surface currents and winds distribute tropical heat worldwide. As warm currents move poleward, the water transfers its heat to the air. The cooled water then returns to lower latitudes, absorbs more heat, and repeats the cycle.

While most wind-driven water movement tends to be horizontal, some vertical movement, called *wind-induced vertical circulation*, can occur. Upward movement, or *upwelling*, occurs when winds blown from onshore cause the near-coastal waters to move offshore and be replaced via the Ekman transport mechanism. Upwelling is an important factor in the distribution of gases and nutrients, especially in coastal regions. It can also affect the weather, as the temperature of coastal water masses change. *Downwelling* occurs when water driven toward a coastline is forced downward and back out to sea, helping to supply the deeper ocean with dissolved gases and nutrients, but having no direct impact on coastal climates or marine productivity.

Any alteration in the normal flow of the major surface currents causes changes in water temperature and nutrient content, and can drastically affect weather patterns and the nearshore biological productivity. This is best illustrated by the *El Niño* phenomenon. Normally, the "pile" of warm surface water at the western boundary of the North Pacific basin is higher than it is at the eastern boundary, off the coast of South America. Evaporation from this warm water establishes a characteristic cloud cover over Indonesia, the Philippines, and Australia resulting in a normally humid climate and a lot of rain. Every few years, however, the prevailing winds weaken, upsetting the east-west balance within the basin and allowing warm water to move back into the Central and Eastern Pacific, bringing the moist air and cloud cover with it and causing droughts in Indonesia and Australia, and coastal flooding and heavy rains in Peru and Chile. This may not be just a regional effect, however. Depending on the circumstances it can, and usually does, affect weather patterns over much of North America as well. In earlier history, the effects of an El Niño event (named after the Christ child because it usually appeared around Christmas) were generally noted along the coastal areas of Peru and Chile because the unnaturally warm waters caused fish to die or disappear, devastating the local economy.

At first considered to be primarily an oceanic event, El Niño is now known to have a major impact on the atmosphere and, thus, the weather. High pressure builds in the western Pacific, with corresponding low pressure in the eastern Pacific. Tropical Pacific winds (westerlies, trade winds) weaken or may totally reverse direction and blow from west to east. This change in atmospheric pressure and its accompanying change in wind direction is called the *Southern Oscillation*. It is directly coupled with the classic El Niño phenomenon, and called the *El Niño/Southern Oscillation* (*ENSO*). While effects are strongest in the Pacific, all oceans at the trade-winds latitudes can be affected. During major ENSO activity, sea level rises in the eastern Pacific and water temperatures increase significantly—as much as 7° higher than normal. This warming causes more evaporation and humid air, resulting in storms and high precipitation in normally dry areas like the Galapagos, Peru, and Chile. Natural marine and terrestrial habitats and their inhabitants can also be dramatically impacted by these sudden temperature and rainfall changes.

ENSO events, while not completely predictable, have a history of lasting about a year with some persisting for more than three years. Normal circulation and weather patterns are often reestablished very suddenly, bringing strong currents, powerful upwelling, and cold stormy conditions along the west coast of South America. This colder-than-normal "return" is called *La Niña* (the girl child).

Science is gradually learning how to gather data to reaffirm and better understand the importance of the atmosphere/sea-surface interactions, and the roles that ocean currents play in those dynamics. But reliable current measurements are difficult to make because they require that either data-gathering instruments be at sea for very long periods, or measurements be made from instrumented vessels that have a built-in inaccuracy component because they normally move faster than the currents they are tracking. The basic current meter, which measures current speed from the rotation of a propeller, has been the most reliable for many years, setting the standard for mechanical devices. It is now being supplemented or replaced by acoustic technology, which is able to measure a large column of water from one location. Satellite altimeters orbiting Earth can accurately report variations in sea surface height, indicating changes in surface currents. A complex moored-buoy array coupled with a satellite, extending from the western Pacific to the coast of South America—a system called *TAO-TRITON*—measures and relays data on temperature, humidity, wind pressure and direction, and other factors. More satellite-based technology, using the JASON-I satellite and the older TOPEX-POSEIDON is also used. Many of these resources are currently being focused on the successful forecasting/prediction of El Niño events. Some scientists consider this an impossible task because each El Niño recorded so far has been different, and their activity patterns are typically and historically chaotic. But El Niño remains a challenge to scientists and meteorologists.

Focus Your Learning

Learning Objectives

After completing this unit you should be able to do the following:

1. List and briefly discuss the physical factors that cause surface currents.
2. Understand the characteristics and dynamics of the geostrophic gyre, including the effects of wind, gravity, and the ocean basin boundary.
3. List and be able to locate on a map the six great currents of the world ocean.
4. Compare and contrast the characteristics of the eastern and western boundary currents.
5. Understand the mechanisms involved in the poleward transfer of water and heat by surface currents.
6. List and describe, briefly and in order, the events leading to an ENSO event.
7. Describe the La Niña phenomenon.
8. Describe the Ekman transport mechanism.
9. Become familiar with the Gulf Stream and the research done there.
10. Discuss the difficulties involved in studying surface currents, and the technology now being used to overcome them.

Assignments

This lesson is based on information in the following text and video assignments. The key terms, focus points, and practice test are intended to help ensure mastery of the material presented in this unit of study.

Text: Chapter 9, "Circulation of the Ocean," pages 232–263

Video: Episode 11, "Going with the Flow"

Video Focus Points

Review these points before watching the video assignment for this lesson to help you focus on key issues.

— The El Niño phenomenon tends to be a prime focus for oceanographic and meteorological research, particularly as it affects the countries on the eastern and western boundaries of the North Pacific basin.

— Both atmospheric and oceanic dynamics are involved in El Niño activity, with wind-driven surface currents of primary importance.

— The Coriolis effect and prevailing winds must be factored into surface current dynamics.

— Vagn Ekman studied the wind/Coriolis effect on surface waters, and described the spiral pattern of this interaction, called the Ekman flow.

— Ocean currents move in large circular patterns called gyres, within which there can be both horizontal and vertical water movement. All of this tends to move water and heat poleward.

— Within the ocean basins the wind and the Coriolis effect tend to deform the circulating water mass, causing it to "pile up" on one side and form a pressure gradient, against which gravity works trying to level it out.

— Within the basin, the horizontal surface currents can move against the prevailing winds, causing countercurrents and undercurrents.

— Knowledge of surface currents is critical to understanding ocean/atmosphere dynamics, but obtaining accurate surface current data is difficult.

— In addition to the standard current meters, modern research involves acoustic sensing, satellite altimeters, and complex moored buoy/satellite arrays.

— Prediction and study of the El Niño phenomenon is a top priority for oceanographers and meteorologists, as well as being of interest to the media and the general public.

— While most surface water movement is horizontal, there can also be vertical movement—upwelling and downwelling. These affect the weather and distribution of gases and nutrients to the near-coast biological communities.

— At the beginning of an El Niño event, the trade winds suddenly weaken, allowing the "pile" of warm water at the western boundary of the North Pacific basin to move eastward toward the west coast of South America, bringing warm water and humid cloud cover. This change in wind direction is termed the Southern Oscillation.

— The Southern Oscillation is directly coupled with the classic El Niño sequence, together termed the El Niño/Southern Oscillation (ENSO).

— This phenomenon is devastating to the countries at both the eastern and western North Pacific basin boundaries. The normally humid air follows the eastward-flowing warm water mass, leaving western boundary countries in drought and eastern boundary countries in flood.

— When the ENSO concludes, the return to normal can also be violent, causing storms, major upwelling, and cold weather. This is called the La Niña phenomenon.

Text Focus Points

Review these points before reading the text assignment for this lesson to help you focus on key issues.

— A surface current forms as a water mass moves horizontally in the uppermost 400 meters of the ocean, driven mainly by winds.

— A combination of winds and the Coriolis effect influence surface currents and direct their flow.

— Most of the major currents are located in ocean basins bounded by continental margin, causing the water mass to move in a circular pattern, or gyre.

— Within a gyre there are also certain patterns of water movement—Ekman transport, countercurrents, and undercurrents.

— Surface currents affect climate and weather patterns as they transfer heat and water worldwide.

Key Terms and Concepts

A thorough understanding of these terms and concepts will help you to master this lesson.

Acoustic Doppler Current Profiler. A device for measuring currents using sound waves.

Antarctic Circumpolar Current (West Wind Drift). The largest of all ocean currents, driven by powerful westerly winds north of Antarctica. Unrestricted by continental masses, it moves permanently eastward without changing direction.

Coastal upwelling. Upwelling adjacent to a coast, usually induced by wind.

Countercurrent. Surface current flowing in the direction opposite that of an adjacent surface current.

Current. Mass flow of water. This term is usually reserved for horizontal movement.

Downwelling. A circulation pattern in which the surface water moves vertically downward.

Eastern boundary current. A weak, cold, diffuse, slow-moving current at the eastern boundary of an ocean basin, off the west coast of a continent. Examples: Canary Current, Humboldt Current.

Eddy. A circular movement of water, usually formed where currents encounter obstructions, or between two adjacent currents flowing in opposite directions, or along the edge of a permanent current.

Ekmán spiral. A theoretical model of the effect on water of wind blowing over the ocean. Due to the Coriolis effect, the surface layer is expected to drift at an angle of 45° to the right or left of the wind (depending on the hemisphere). Water at successively lower layers then drifts progressively to the right (or left), though not as swiftly as the layer above it, due to friction.

Ekman transport. Net water transport; the sum of water layer movement due to the Ekman spiral pattern. Theoretical Ekman transport in the Northern Hemisphere is 90° to the right of the wind direction.

El Niño. A southward-flowing, nutrient-poor current of warm water off the coast of South America, caused by a breakdown of trade wind circulation.

ENSO. An acronym for the coupled phenomena of El Niño and the Southern Oscillation.

Equatorial upwelling. An upwelling in which water moving westward on either side of the geographical equator tends to be deflected slightly poleward and replaced by deep water, which is often rich in nutrients.

Geostrophic. Term describing a gyre or current in balance between the Coriolis effect and gravity, meaning literally "turned by Earth."

Gulf Stream. The strongest western boundary current of the North Atlantic, found off the east coast of the United States.

JASON-TOPEX/POSEIDON. An oceanographic satellite complex, using very accurate positioning to measure sea surface winds, water vapor, and sea surface height.

Kelvin wave. An eastward-moving equatorial surface wave, formed when easterly winds weaken and release warm water from the western Pacific.

La Niña. An event during which normal tropical Pacific atmospheric and oceanic circulation strengthens, and surface temperatures of the eastern South Pacific drop below average values. Usually follows an ENSO event.

Langmuir circulation. Winds that blow steadily across the ocean, and the small waves that they generate. Can induce long sets of counter-rotating cells in the surface waters.

Pressure gradient. A difference in water pressures caused by wind "piling up" water against the pull of gravity.

Pycnocline. A middle zone of the ocean, where density increases rapidly with depth, as temperatures fall and salinity rises.

Rossby wave. A westward-moving, wind-driven equatorial surface wave.

Southern Oscillation. Reversal of airflow between normally low atmospheric pressure over the western Pacific and normally high pressure over the eastern Pacific. Associated with El Niño.

Surface current. Horizontal flow of water at the ocean surface.

TAO-TRITON. An array of satellite-linked data buoys extending from the western Pacific to the coast of South America.

Transverse current. An east-to-west or west-to-east current linking the eastern and western boundary currents. Example: North Equatorial Current.

Undercurrent. A current flowing beneath a surface current, usually in the opposite direction.

Upwelling. A circulation pattern in which deep, cold, usually nutrient-laden water moves toward the surface. Can be caused by winds blowing parallel to shore or offshore.

Western boundary current. A strong, warm, concentrated, fast-moving current at the western boundary of an ocean, off the east coast of a continent. Examples: Gulf Stream, Kuroshio Current.

Westward intensification. The increase in speed of geostrophic currents as they pass along the western boundary of an ocean basin.

Wind-induced vertical circulation. Vertical movement in surface water—upwelling or downwelling—caused by wind.

Test Your Learning

After working through these questions, check your answers against the key at the end of this book. If you answered any of the questions incorrectly, review the relevant sections of the text and video episode.

Multiple Choice

1. The prime movers of oceanic surface currents are the:
   a. polar cells
   b. easterlies and westerlies
   c. Coriolis effect and water temperature
   d. basin boundary configurations
2. The fastest and deepest surface currents are along the _______ boundaries of the ocean basins.
   a. eastern
   b. western
   c. northern
   d. southern
3. Tropical heat is distributed world-wide by means of:
   a. upwelling
   b. the Ekman transport
   c. surface currents
   d. Longmuir circulation
4. The California current is well-known for supplying the North American west coast with cool, nutrient-rich water in the form of:
   a. Langmuir circulation
   b. upwelling
   c. downwelling
   d. "monsoon" currents
5. One of the first events in an El Niño sequence is:
   a. Langmuir circulation
   b. coastal upwelling
   c. strengthening of westerly winds
   d. weakening of trade winds
6. The only major global current that is not in a gyre-configuration is the:
   a. Gulf Stream
   b. West Wind Drift
   c. Canary Current
   d. Langmuir Current
7. The circulation pattern set up as layers of water that affect those immediately below them is called:
   a. Coriolis effect
   b. Ekman spiral
   c. undercurrent
   d. countercurrent
8. The force of gravity tends to act _______ a pressure gradient.
   a. with
   b. parallel to
   c. against
   d. none of these
9. In the Northern Hemisphere the Coriolis deflection is:
   a. about 90° to the right
   b. about 45° to the right
   c. negligible
   d. unmeasurable
10. Ocean current speed is difficult to measure because:
    a. instruments cannot operate in seawater
    b. they move so fast
    c. they move so slowly
    d. they are too sporadic
11. The TAO-TRITON buoy array reaches from:
    a. western Brazil to South Africa
    b. Australia to Falkland Islands
    c. Japan to California
    d. Western Pacific to South America
12. In the Northern Hemisphere, surface currents flow to the right of wind direction, due to:
    a. conflicting westerlies
    b. friction
    c. pressure gradients
    d. none of the above.

Short Answer

1. Ocean current measurements are extremely important in studying atmosphere/ocean dynamics, but obtaining accurate, long-term data remains a formidable problem. Trace the history of ocean current measurements from the early seafarers to the present.

2. Describe and discuss the Ekman transport system and how it affects the configuration of a gyre in a Northern Hemisphere ocean basin.

3. While the El Niño phenomenon remains mysterious and unpredictable, the sequence of events, once it starts, is fairly well-known. However, the accompanying return to normal patterns, known as La Niña, is not as well explained or documented. List, in sequence, the events involved in La Niña, and discuss how this also can affect circulation, temperatures, and coastal populations.

Supplemental Activities

1. The Sargasso Sea was mentioned several times in this lesson. Your text mentions it as the location of a "hill" of water in the North Atlantic, built by the combined forces of wind, Ekman transport, and Coriolis deflection. It is also noted as the place where Irving Langmuir observed, studied, and described the "windrows" of debris caused by the peculiar circulation patterns that now bear his name. But these brief mentions have not told you exactly where the Sargasso Sea is located, why it is important in surface circulation, or why it is called a sea in the midst of an ocean.

   Do some research and write a "Guide to the Sargasso Sea" for marine scientists. Include the latitudes and longitudes that comprise its boundaries. Expand on the phenomenon of the "hill" and the complexities of its formation and maintenance as they involve surface circulation. Expand on the phenomenon of Langmuir circulation and on Langmuir's observations and research. Using this information as background, research and describe the characteristics of the Sargasso Sea that entitle it to be called a "sea," and that support the masses of seaweed and their bizarre fauna.

   Your university or public library is a good place to start. Look through several world atlases to find the exact location and boundaries of what is called the "Sargasso Sea." The major encyclopedias (*Britannica, Americana, World Book*) contain material. A variety of sources can be found by conducting a search for "Sargasso Sea" on the Internet. One specific site is: http://www.bermuda-triangle.org/html/sargasso_sea.html.

2. Read about the historic 1969 journey of the research submarine *Ben Franklin* at http://seawifs.gsfc.nasa.gov/FRANKLIN/HTML/ben_franklin.html. The *Franklin* drifted in the Gulf Stream for a month conducting scientific studies and investigating the secrets of the Gulf Stream. Imagine that the submarine, which currently resides at the Vancouver Maritime Museum, is being re-activated for another drift, and that you have been asked to be one of the onboard oceanographers. But you need to know more about the vessel and its capabilities, what it might be like to be sealed up in it for a month, its several missions in 1969, and other facts which will help you decide if you wish to go. On the Internet site http://seawifs.gsfc.nasa.gov/FRANKLIN/HTML/ben_franklin.html, you will find information and a list of other Internet resources. Write a report covering all of the information you will need to decide whether or not you wish to make this voyage. (Note: do not confuse the *Ben Franklin* with the Navy's USS *Benjamin Franklin*).

3. El Niño/ENSO events are usually considered to be a Pacific Basin phenomenon, mainly at the eastern and western boundaries of that basin. Other regions, however, have also been affected (decreased seal births in Antarctica, flooding in Louisiana, poor harvests in Zimbabwe). Consult a global map and select one or more areas that are not in the Pacific Basin and subject to the direct effects of an ENSO event. Document the effects of an El Niño event in that area (climate change, natural disasters, effect on the economy). Most of your information will be found on Websites, although some other sources might be *National Geographic*, later editions of encyclopedias, NOAA publications. Two good Websites to begin with are:

   http://www.usatoday.com/weather/tg/wninogrf/wninogrf.htm

   http://www.elnino.noaa.gov

Lesson 12

Deep Connections

Overview

Currents are not a phenomenon of surface waters alone. They exist in the deepest reaches of the world ocean. Surface water is moved primarily by wind, but the currents below the *pycnocline* are density-driven. Termed *thermohaline circulation* or flow, this relatively slow-moving mechanism is responsible for most of the vertical movement of water in the ocean, and for both vertical and horizontal movement below the pycnocline layer.

Water is the least dense at the surface of the ocean and increases in density with depth. The world ocean contains three major density layers. Additional stratification exists in the tropical and temperate latitudes because the difference in temperature from surface water to deep water is greater at these latitudes than at the poles. While each of these layers has general depth boundaries, the characteristics of each layer are determined by the conditions under which it was initially formed.

Oceanographers identify the layers, or masses, of water according to their location in the water column. In the temperate and tropical latitudes the layers, from shallow to deep, are the surface water, central water, intermediate water, deep water, and bottom water. Water masses usually originate at the surface as a result of the heating, cooling, evaporation, and dilution that occurs there. Although they may move after their initial formation, these masses or layers tend to retain a "history" of their formation for long periods of time and over great distances before mixing with surrounding waters.

One of the more reliable methods for determining the "age" of a water mass is to compare its dissolved oxygen content with what was potentially available to it from the atmosphere when it was formed. While each of the named layers is defined by general depth boundaries within the water column and by its temperature and salinity characteristics, different combinations of temperature and salinity can result in the same density. Oceanographers represent this graphically as a T-S (temperature/salinity) curve, as illustrated in Figure 9.26 on page 254 in your text.

The deepest and most distinctive of the deep waters is the Antarctic Bottom Water. It is notable for being the densest water in the ocean, for being produced in great volume in the Antarctic, and for its massive, global, northward migration along the sea floor. When ice forms from sea water, it incorporates only about 15 percent of the salt, leaving a highly saline and very dense unfrozen *brine* beneath the ice. In the Antarctic, up to 50 million cubic meters of this brine can form every second, to sink along the Antarctic continental shelf, where it mixes with the Antarctic Circumpolar Current. From there it moves down the continental slope to the deep-sea floor, and creeps slowly northward into the Atlantic and Pacific basins—a journey that may take over a thousand years to complete. This same mechanism occurs in the Arctic Ocean, but the seafloor topography there prevents the dense bottom water from migrating freely.

Deep water masses can form in other areas, and their migration is always based on their relative density. North Atlantic Deep Water and Pacific Deep Water can both encounter Antarctic Bottom Water as it moves northward. The Mediterranean can also form a deep water mass, but its increased salinity is caused by excess evaporation rather than by the formation of sea ice. The densities, therefore the depth, of these water masses can vary, causing them to move either over or under the masses they encounter, rather than to mix.

The density-driven deep migrations of these masses of water create *thermohaline* circulation patterns because simple physics demands that these large bodies of dense, sinking water be replaced by equal volumes of less-dense, rising water. The cold deep water from the poles migrates toward the equator where it warms, rises, and slowly returns poleward, establishing a more-or-less permanent thermocline. This thermohaline circulation is constant, moving masses of water around and into potential contact with each other, at more-or-less predictable *convergence zones*—Antarctic, Arctic, and Subtropical. When the densities of these water masses are different, the heavy ones usually slide beneath the lighter ones, as happens with air masses. When water masses of the same density but different temperature and salinity components meet, they can combine to form a new mass of greater density than the two original masses, in a process called *caballing*. This newly-formed mass sinks because of its increased density.

Not all thermohaline activity occurs over long periods of time. The existence of strong, localized, fast-moving bottom currents, or *contour currents*, is evidenced by scour marks on the rocks of the sea bed, and by ripple marks in bottom sediments, made as these currents flow around the projections and over the terrains. While most bottom currents move toward the equator or around the western basin boundaries much more slowly than do the wind-driven surface currents, the Coriolis effect maintains its influence.

Thermohaline circulation patterns are also affected by the configuration of the continental land masses. In the distant past when the land masses were arranged differently, the thermohaline patterns were also different, implying the possibility of continuous reconfiguration over time affecting the ocean and its climatic influence. Present concern about global warming includes the possibility that as the slow, poleward thermohaline currents continue to release heat there could be a cumulative warming effect. This concern emphasizes the complex connection between deep ocean circulation and global climate, and the need for continued study of ocean currents.

Both surface and bottom currents are involved in poleward heat transfer in slow circuits that straddle Earth's hemispheres in a sort of global conveyor belt. Recent technology has enabled oceanographers to identify the major water masses and currents and document their movements. In addition to transfering heat, this slow three-dimensional flow of water mixes and distributes gases, nutrients, and marine organisms.

The technology for measuring current speed and direction has greatly increased in versatility and accuracy, but traditional methods of current study are still divided into two basic categories—float and flow. The *float method* uses data gathered from an object or instrument as it is carried along by the current; the *flow method* measures currents flowing past a fixed instrument. The simplest floaters or drifters can be drift bottles or cards, which are released at a point and retrieved somewhere else at some future time, telling how long they traveled but with no specifics about their path. Drifters can also be modified floaters that can be tracked via radio signals or radar. Two famous examples of accidental drifters involve, in one case, the release of thousands of tennis shoes from a ship in a storm. At another time, under essentially the same conditions, thousands of rubber bathtub toys escaped from a transport ship. In both cases, many of the escapees eventually drifted ashore, and their release/travel/recovery times were noted. Deeper currents can also be tracked using sophisticated, free-floating sonar pingers. Current meters are fixed devices that gather data, including speed and direction, as the water flows past them.

In addition to these two basic approaches to current measurement, advanced electronic buoys can now sense and record the electromagnetic forces generated by currents as they move through the Earth's magnetic field. A self-propelled instrument, powered by an interaction of gravity and buoyancy forces, can move through the water column and gather current and water-mass data, transmitting it to satellites. Satellites can also be equipped to sense sea-surface temperatures and topography, and even surface chlorophyll content. One of the newest techniques is *acoustical tomography*, which uses pulses of low-frequency sound to sense variations in water temperature, salinity, and movement. These instruments can follow mid- and deep-ocean water mass formation and circulation patterns, gathering types of data that were never before available. Another innovation is the profiler float concept, where several thousand instruments are released to gather and transmit data from the surface down to 2,000 meters.

Measuring bottom currents may no longer be a challenge thanks to a free-fall profiling instrument called a *High Resolution Profiler*, or *HRP*, capable of being programmed to stop at any predetermined depth and relay data to shipboard computers.

Man-made chemicals, such as the notorious *chlorofluorocarbons* (*CFCs*), were produced for industrial use and subsequently released into the atmosphere from the 1930s to the 1970s, before they were banned for causing damage to the ozone layer. These chemicals are soluble in seawater, and their residues can act as tracers for studying oceanic circulation. Their value lies in knowing when and where they were released, thereby enabling scientists to study atmospheric changes and to calculate when that water mass was last in contact with the atmosphere. Some of the nuclear weapons tests done in the 1950s and 1960s released radioactive tracers into the atmosphere, which made their way to the ocean. These can be tracked and observed as they are moved about by the currents. There are also a number of natural phenomena that can be used as tracers and indicators of current activity. Organisms that make calcite skeletons and shells use calcium carbonate molecules from the seawater, and the growth rings and other chemical factors associated with their biology can be studied.

The ultimate goal of the study of ocean circulation is to know as much about it as is known about the atmosphere and to more accurately and intimately connect and analyze the dynamics and interactions of the atmosphere and the sea surface.

Focus Your Learning

Learning Objectives

After completing this unit you should be able to do the following:

1. Identify and characterize the basic (named) oceanic water masses—three for the world ocean and an additional two in temperate and tropical latitudes.
2. Understand where, and under what conditions, the ocean's water masses are formed, how they retain a history of that formation over time, and how they eventually lose it.
3. Describe the T-S (temperature/salinity) curve, and how it is used to explain the layering of oceanic waters.
4. Describe and discuss the Antarctic Bottom Water—its formation, characteristics, migration, and mixing patterns.
5. Explain the basic thermohaline circulation patterns and mechanisms, including what happens when different water masses encounter each other.
6. Compare and contrast the slow-moving currents with the faster-moving contour currents, and tell how we know of the existence of the contour currents.
7. Recognize the roles played by both thermohaline flow and surface flow in the global heat budget.
8. Compare and contrast the two basic methods for studying ocean currents—flow and float—and the types of devices used in each.
9. List and briefly discuss chemical tracers as they are used in studying currents.
10. List and describe some of the satellite-involved technology used for ocean current studies.

Assignments

This lesson is based on information in the following text and video assignments. The key terms, focus points, and practice test are intended to help ensure mastery of the material presented in this unit of study.

Text: Review pages 250–263 in Chapter 9, "Circulation of the Ocean"

Video: Lesson 12, "Deep Connections"

Video Focus Points

Review these points before watching the video assignment for this lesson to help you focus on key issues.

- Ocean surface currents are driven by the wind, deeper currents are driven by the density/gravity interaction; both types are influenced by the Coriolis effect.
- The movement of water below the pycnocline is called thermohaline circulation, involving water masses with different densities and depth boundaries.
- Surface and deep circulation work together to transfer global heat.
- While surface waters retain a fairly uniform density due to constant mixing, deeper waters tend to retain a long-term memory of the conditions under which they were formed.
- Most deep water masses move slowly but constantly, entering and leaving the various ocean basins and frequently colliding with other masses.
- Considerable attention is being paid to the study of ocean currents to learn more about the ocean/atmosphere link and its impact on global weather and climate.
- Study methods include the use of floating devices, flow meters, satellite technology, chemical tracers and isotopes, and the structural biology of certain marine organisms.
- Midwater current study techniques are fairly well-established, but movement and mixing on the actual sea floor is more difficult and is requiring the development of very special instrumentation.

Text Focus Points

Review these points before reading the text assignment for this lesson to help you focus on key issues.

- There are three fundamental water mass stratifications comprising the world ocean, with an additional two in the temperate and tropical regions.
- Most water masses are formed at the surface, and are characterized by the conditions present at the surface during that formation. Once formed, these water masses tend to move, some staying at the surface and the others sinking. Once below the pycnocline they are powered by the density/gravity interaction.
- This density-driven movement is called thermohaline circulation, and is responsible for both vertical and horizontal movement in the so-called deep water, which comprises about 80 percent of Earth's ocean water.
- Seawater density is a function of temperature and salinity, and many different temperature/salinity combinations form water masses of the same density. This results in a layering, or stratification, of the deeper waters, and is explained by oceanographers through a T-S (temperature/salinity) diagram depicting stratification in the water column.
- The largest, densest, and most distinctive of all the deep-water masses is the Antarctic Bottom Water. It forms from the highly saline brine resulting from sea ice formation, creating massive amounts of bottom water that migrates northward toward the equator and into various ocean basins.
- Cold, sinking water is replaced by warm, rising water, contributing to thermohaline movement.
- Large, density-distinct water masses, propelled by gravity, can encounter each other in areas called convergence zones. The heavier masses tend to move under the less-dense ones.
- When two water masses of the same density but different temperatures and salinities meet, they may mix in a process called caballing, forming a new, denser mass that begins to sink.
- Most deep-water currents move slowly, but in some areas fast-moving localized bottom currents may form. These are called contour currents due to their particular type of motion along the sea floor.
- Surface and deep currents both contribute to the global heat conveyor belt and the global heat budget. This mechanism also moves and distributes dissolved gases, nutrients, and the planktonic forms of marine organisms.
- Traditional methods used to study currents are the float method and the flow method, with technological modifications that include advanced electronic data-gathering and satellite connections.
- Data on currents can also be gathered using chemical and isotopic tracers (usually unintentionally released) and acoustic tomography, which uses low-frequency sound to artificially

stratify water masses through differences in temperature, salinity, and movement.

Key Terms and Concepts

A thorough understanding of these terms and concepts will help you to master this lesson.

Acoustical tomography. A technique for studying ocean structure that depends on pulses of low-frequency sound to sense differences in water temperature, salinity, and movement beneath the surface.

Antarctic Bottom Water. The densest ocean water, formed primarily in Antarctica's Weddell Sea during Southern Hemisphere winters.

Artery. A term used in the video to describe the path of motion of a marine water mass (as in an "artery" of a road or highway).

Brine. Seawater with higher-than-normal salinity, such as that forming under Arctic and Antarctic sea ice.

Caballing. Mixing of two water masses of identical densities but different temperatures and salinities, such that the resulting mixture is denser than its components.

Chlorofluorocarbons (CFC). Chemicals developed for use as propellants and refrigerants in the 1930s to 1970s, then banned. Their secondary release into the atmosphere eventually brought them into contact with the sea surface, and their residuals are now used as tracers.

Contour current. A bottom current made up of dense water that flows around, rather than over, seabed projections.

Convergence zone. The line along which waters of different densities converge. Convergence zones form the boundaries of tropical, subtropical, temperate, and polar areas.

Float method. A method of current study that depends on the movement of a drift bottle or other free-floating object.

Flow method. A method of current study that measures the current as it flows past a fixed object.

Halocline. A subcategory of the pycnocline, referring to the rapid increase of salinity with depth.

High-Resolution Profiler (HRP). A free-fall profiler that is used to measure deep ocean mixing, it is equipped with weights that it releases when it reaches a pre-programmed depth and creates a density profile from the surface to the bottom.

Pycnocline. The middle zone of the ocean in which density increases rapidly with depth. Temperature falls and salinity rises in this zone.

T-S curve. A graph showing the relationship of temperature and salinity with depth.

Thermocline. A subcategory of the pycnocline, referring to the rapid decrease in temperature with depth.

Thermohaline circulation. Water circulation produced by differences in temperature and/or salinity, and therefore density.

Test Your Learning

After working through these questions, check your answers against the key at the end of this book. If you answered any of the questions incorrectly, review the relevant sections of the text and video episode.

Multiple Choice

1. Surface currents affect the uppermost _____ % of the world ocean.
   a. 2
   b. 5
   c. 8
   d. 10

2. A new global ocean observing system is being developed, in which several thousand floating sensors called _________ will be distributed throughout the world ocean.
   a. Slocums
   b. seabed drifters
   c. laser drogues
   d. profiling floats

3. How many common water masses are noted in temperate and tropical latitudes?
   a. 3
   b. 5
   c. an indeterminate number
   d. 7

4. Identify the water mass outstanding for its extreme density, the large quantities of it produced near Antarctic coasts, and its ability to migrate north along the seafloor.
   a. North Atlantic Deep Water
   b. Mediterranean Deep Water
   c. Antarctic Bottom Water
   d. Arctic Deep Water

5. The primary driving mechanism for deep water currents is:
   a. density
   b. the pressure gradient by depth
   c. the Coriolis effect
   d. the friction against the sea floor

6. Antarctic Bottom Water has been known to retain its character for up to _______ years.
   a. 375
   b. 1,600
   c. 2,500
   d. 800

7. A primary factor in the formation of Mediterranean Deep Water is:
   a. condensation
   b. caballing
   c. evaporation
   d. rapid heating and cooling

8. Deep water formation does not occur as extensively in the Pacific Ocean as it does in the Atlantic because:
   a. the Atlantic is less saline
   b. the Atlantic is more saline
   c. the Atlantic has a higher surface temperature
   d. none of the above

9. What drives thermohaline circulation?
   a. surface winds
   b. density
   c. currents
   d. heating of water at lower latitudes and cooling at higher latitudes

10. The continuous upwelling within the thermohaline circuit as the water returns poleward results in the formation of a permanent ______ at all low- and mid-latitudes.
   a. halocline
   b. pycnocline
   c. mixed layer
   d. thermocline

11. Caballing is a term used to describe what happens when:
   a. water from the pycnocline encounters surface water
   b. water masses with different salinities and temperatures but the same density collide
   c. water masses with different densities collide
   d. none of the above

12. Corals and foraminiferans are used to study the chemical history of seawater by analyzing their:
   a. calcium carbonate deposition rates
   b. intake of dissolved oxygen when alive
   c. their reactions to CFCs
   d. all of the above

Short Answer

1. The Global Heat Circuit has been analogized to a vast conveyor belt that carries surface water to the depths and back again. Describe/discuss that circuit, and the water masses and basins it encounters and affects, beginning with the formation of the North Atlantic Deep Water.

2. Some of the global water masses move very slowly, taking many hundreds of years to complete a circuit. These water masses tend to retain a memory of their initial formation at the sea surface. Because of this historical aspect to thermohaline circulation, scientists are trying to reconstruct some of the possible differences in circulation, and thus in climate patterns, as continental drift has rearranged the land masses over time and, consequently, the ocean boundaries and basins. From the video presentation, recall some of the thoughts and present concerns of those scientists as they

consider the consequences of long-term continental drift patterns.

3. Discuss the use of chemical tracers (specifically, CFCs) to study ocean circulation.

Supplemental Activities

1. Both the text and video for this lesson stress the importance of seawater density, and how it affects and controls thermohaline waters. This exercise will allow you to observe some of the density-dependent behavior of seawater.

   The text tells us that when sea ice (both Arctic and Antarctic) is forming, it incorporates only about 15 percent of the salt present in surface water, and that the salt remaining in the unfrozen water beneath the ice forms a frigid brine. The increased density of this brine causes it to sink. But how can the brine be denser than the original seawater if it has lost 15 percent of its salt? Because it also has lost a lot of its water in the ice-making process, and the overall result is a much saltier, but unfrozen, brine.

   a. LET'S TEST IT! The only piece of scientific equipment you will need is a hydrometer (simple device used for measuring the density of a liquid). Borrow one from your school laboratory, or purchase one for a few dollars at a store that deals in aquarium supplies. They are used to maintain the correct salinity in saltwater aquariums. The hydrometer is used by placing it in a "column" of water (usually a drinking glass) that is deep enough that it will not hit the bottom. Place it in the water, push it down a little and let it rise and stop moving. Read it where the surface of the water meets the skinny column of the hydrometer, noting the numbers on the column. Test it in a glass of fresh water (tap water is OK)—it should read about 1.00 (1 gram per cubic centimeter, which is the way density is measured and recorded).
   b. Now, obtain about 2 quarts of seawater to test. If this is not possible, make some by adding about 2 teaspoons of table salt to 2 cups of water (for starters), stir it well, and test it with the hydrometer. The density of seawater is about 1.020 to 1.030 gm/cubic centimeter. It is not necessary to be exact; the idea is to have a water sample that is considerably more saline than the fresh water. But try to get as close to seawater density as you can (1.020)—you may have to try a few times!
   c. Next, place the container of salt/seawater (about 2 cups) in the freezer for about 2 to 2.5 hours, checking it after about the first hour to see if ice crystals are forming. When there appears to be more ice than water in the cup, pour some of the water from the partly-frozen seawater into a glass, let it warm up, and test the density. Be aware that very small density increases are significant—your text calls the Antarctic Deep Water, at 1.0279, extremely dense. Record your findings and journal your experimental procedures.

   For your report on both of these, include the purpose of the experiment, the materials and methods used, the results, and your interpretation/conclusions about the density characteristics of seawater vs. fresh ("fresher") water.

2. Research and document the famous accidental release of the Nike tennis shoes and the rubber bathtub toys from the storm-tossed cargo ships in the 1990s. Document the cargo of the ship, its destination, the conditions under which the floaters were released, and their retrieval by coastal inhabitants. Explain how these escapees were used to study and document current patterns.

   For reference, start with the short account on page 258 of your text, then go to the Internet for more details. A couple of good sites are:

   http://www.islandnet.com/ ~ see/weather/elements/shoes.htm

   http://seawifs.gsfc.nasa.gov/OCEAN_PLANET/HTML/oceanography_currents_2.html

   For a broader search, just call up "rubber duckies January 1992" and "Hansa Carrier May 1990."

Lesson 13

Surf's Up

Overview

Do waves conjure an image of lazy sunsets and the gentle slap of water on the sand? Or maybe the roar of water littered with surfers? Waves can range from soothing to destructive. In its broadest definition, the term wave describes a disturbance caused by the movement of energy from a source, through a medium—solid, liquid, or gas. As that energy, or disturbing force, travels, it affects the medium in specific ways. The focus of this lesson is how ocean water is affected by a particular type of energy; namely, the wind.

Ocean waves seem like a moving ridge of water that can be ridden to shore. By definition, however, a *wave* is a moving ridge of energy, not water. Look at Figure 10.1 on page 266 in your text, in which a seagull sitting on the water's surface stays in essentially the same lateral position but is moved vertically as the forward-moving energy changes the wave form, or shape, from crests to troughs to crests. When energy contacts the sea surface as wind, some of it is transferred to the water beneath the surface, causing the water molecules comprising the wave form to move. These tend to form a circular, or *orbital*, pattern as the energy passes, but the water itself does not progress forward. This *orbital wave* concept applies to most of the more common types of waves.

The orbital motion of the water continues beneath the surface, but the diameter of the orbit is controlled by the depth of the water, rapidly diminishing when the depth of the water is less than half of the wavelength. Most ocean waves have moderate wavelengths, so the circular disturbance that propagates the wave affects only the uppermost water mass. It is important to have a clear mental image of this orbital activity to set the stage for understanding the forms and behaviors of the various types of wind waves. The diagrams found on pages 266 and 267 of your text will help.

Ocean waves are classified and named according to the disturbing forces that cause them, and the *restoring forces*, or surface tension and gravity, that attempt to restore the surface to smoothness. Wind is a major disturbing force, but earthquakes, landslides, erupting volcanoes, and tides can also cause waves. Waves can occur in any ocean, but are most common around the equator and in the Southern Hemisphere, where most storm energy originates. Large waves originating in these areas tend to move northward, affecting the ocean and shorelines in the Northern Hemisphere.

In order to discuss and understand wind wave formation and behavior, there are a number of descriptive terms to be learned. Wave shape consists of the *crest*, or high point, and *trough*, or low point. The *period* of a wave is its travel time expressed in seconds and measured based on the time it takes successive wave crests to pass a fixed point. Wave *size* refers to its height and length.

Wavelength is measured by the distance between adjacent crests or adjacent troughs. The *speed* of a wave is based on the ratio of the length of the wave to its period. It is ultimately controlled by gravity, wavelength, and water depth. As a general rule, the longer the wavelength is, the faster the wave energy will move through the water.

Height is an important characteristic of wind waves, and is measured from the trough to a line connecting the two adjacent crests. Moderately sized waves usually have about a 1:7 ratio of wave height to length. Waves with a ratio higher than this will often break at sea, dissipating their energy as *whitecaps* or *combers*. The highest wave accurately measured and recorded to date was 34 meters (112 feet).

Wind waves begin as small *capillary waves*, when some wind energy is transferred from the atmosphere to the sea surface. If the wind continues and grows stronger, the orbital motion gains momentum, while the surface tension and gravity—the restoring forces—try to flatten them out, creating a larger wave called a *gravity wave*.

Deep-water waves are those that move through water that is deeper than one-half their wavelength; *shallow-water waves* are found where the water is less than one-twentieth of the wavelength. Deep-water waves can become larger as the wind continues to blow, forming into a chaotic mix of waves of different lengths, called a *sea*. These storm waves tend to sort themselves out into groups with similar lengths and speeds, settling into a smooth surface undulation called *swell*, which may develop into progressive groups called *wave trains*. Wave trains coming from different directions sometimes meet in an interaction known as *interference*. It is called *destructive interference* if the waves counteract each other and *constructive interference* if they reinforce each other. If high crests from different wave trains meet, a *rogue wave*—one that is higher than is theoretically possible—may form. Rogue waves can also form when a wind wave meets a swift current.

As a deep-water wave approaches a shoreline it will eventually "feel" the bottom, which initiates a series of events that change the form and speed of the wave, eventually forming *surf* as the wave dissipates its water and energy on the beach. When gravity moves that water back down the beach, it can form swift, outward-moving *rip currents*, which can be dangerous to swimmers.

Most waves fall into the category of *progressive waves*; so named because the wave forms move forward. When a wave line approaches land, its configuration is affected by the geography of the local shoreline. Most waves move toward shore at an angle, causing the wave line to bend, or *refract*. If the wave line encounters an obstacle such as a rock or breakwater, it is interrupted or *diffracted*. When a wave series enters a bay or harbor the energy is *reflected* away from the land, causing a *standing wave*, or water mass that oscillates vertically without progressing.

Not all waves are surface waves; subsurface *internal waves* may form where water of different density masses meet. You may recall that most surface winds in the tropical Pacific, such as trade winds, move from east to west, but that at irregular intervals they can reverse direction and move from west to east. This is called the Southern Oscillation, and it causes Ekman transport in the area to change direction, too. Instead of flowing up and away from the equator it moves down and toward it. The warm water of the upper layer of the Pacific downwells into the lower layer, resulting in an oscillation where those two density-different layers meet. This oscillation forms an internal wave at the base of the pycnocline, especially at the bottom of a steep thermocline.

Internal waves are also important in transporting and mixing nutrients with surface waters, affecting both midwater and coastal planktonic organisms. Although deep-ocean internal waves are less well known than those along the coasts, space shuttle and TOPEX/Poseidon satellite data have recently documented their existence and basic behaviors.

Since they are such a fundamental component of the oceanic system, waves have been observed and studied since the early voyagers. Wave study became even more important in the mid-1940s when amphibious landings on the coasts of Europe were being planned during World War II. Since then significant progress has been made in the effort to describe, monitor, and predict wave formation and propagation using a number of highly innovative techniques and devices. One such device, called a *pressure sensor*, is mounted on the sea floor and actually records wave height as indicated by the weight of the water passing over it. Fixed electric gauges are used to record changing wave heights; and an array of wave buoys is in place along the West Coast of the United States measuring wave height and direction and sending the data to computers at the Scripps Institution of Oceanography in California. Satellite altimeters sense changes in the sea surface and transmit the data to oceanographic institutions or computers on ships at sea.

While adding to knowledge of the basic science and physics of wave formation and propagation, these instruments also track the dispersal of natural nutrients, pollutants such as oil and other chemicals, and the biology of phytoplankton as it affects the global carbon budget.

Focus Your Learning

Learning Objectives

After completing this unit you should be able to do the following:

1. Recognize that wind waves are formed and propagated by transmission of wind energy to the sea surface, and that it is the energy that moves as a wave, not the water.
2. Understand that this energy causes water to move in closed circular paths (orbits), which carry the energy forward as a progressive wave form.
3. Explain how waves are classified and named based on the strength and character of the forces that affect them.
4. Describe the relationship between wave length and wave speed.
5. Discuss the terminology used in describing wind waves and their effects—size and shape, crest-and-trough, rogue waves, surf, sea, and period.
6. Describe the chronological sequence of events that occur from the time a deep-water wave begins to approach a shore until it breaks as surf.
7. List and briefly discuss the behaviors—refraction, diffraction, reflection—exhibited by waves as they encounter different types of shorelines.
8. Recognize the role of wave study in oceanography, its importance in World War II, and the technologies now in use to study waves.

Assignments

This lesson is based on information in the following text and video assignments. The key terms, focus points, and practice test are intended to help ensure mastery of the material presented in this unit of study.

Text: Pages 264–283 in Chapter 10, "Waves"

Video: Episode 13, "Surf's Up"

Video Focus Points

Review these points before watching the video assignment for this lesson to help you focus on key issues.

— Wind waves are best known for their effect on shorelines and ships at sea.

— Wind waves are characterized by their height, length, and period.

— A wind wave is formed when moving air contacts the ocean surface. Water molecules begin an orbital motion and the wind energy is moved forward, but the water stays in essentially the same place.

— When waves encounter a shoreline they change shape and transfer their energy onto the land. Rip currents can form as the water retreats back down the beach to the sea.

— Waves impacting continental margins can cause erosion and major relocation of beach components such as sand.

— At sea, large wind waves can impact offshore structures, such as oil rigs and ships, in the form of very large, isolated rogue waves. Waves of different periods and lengths can also come together and reinforce each other into larger waves.

— Internal waves are formed at the subsurface boundary of two density-different subsurface water masses, and behave much like surface waves.

— The study of waves became important in the 1940s, during the planning of amphibious landings in World War II.

— A number of innovative techniques and devices are now used to study wave formation and propagation—pressure sensors, electric gauges, buoy arrays, and satellite altimeters.

— Wave activity and its effect on ocean circulation is important in studying and tracking nutrients, pollutants, plankton distribution, and Earth's carbon budget.

Text Focus Points

Review these points before reading the text assignment for this lesson to help you focus on key issues.

- Wind waves form as the energy of moving air is transferred to the sea surface, forming small capillary waves in which the water begins to move in an orbital pattern. Energy is moved forward, but the water particles remain essentially in the same place. At the same time, surface tension and gravity (restoring forces) act to return the sea surface back to smoothness.
- Ocean waves are characterized and described according to their height, length, and period.
- Shallow-water waves move in depths shallower than one-twentieth of their wavelength; deep-water waves move in water deeper than half of their wavelength. Only wind can form deep-water waves; shallow-water waves can also form from earthquakes, volcanoes, landslides, and tides.
- Wavelength determines the speed with which a wave can move. The longer the wavelength the greater the speed.
- If the wind continues to blow over water deeper than half the wavelength, waves can become larger and more erratic, forming a chaotic surface called sea.
- Storm waves sort themselves out into groups with similar lengths and speeds, resulting in the smooth, undulating surface called swell, which can organize further into wave trains.
- The growth and movement of wind waves is affected by wind strength and duration, and the uninterrupted distance over which it can move, called *fetch*.
- The ratio of wave height to length, about 1:7, is called the wave *steepness*. If this ratio is exceeded and the wave grows high, the excess energy will cause it to break at the crest, forming whitecaps or combers.
- When wave trains of different speeds meet, they may either cancel each other out (destructive interference) or reinforce each other to become larger waves (constructive interference). Huge, isolated rogue waves may form in this way. Rogue-type waves may also form when wind waves meet strong, swift currents.
- When wind waves encounter shallow water as they approach shore, a sequence of events begins that leads to their eventual breaking on shore, forming surf.
- Wave lines usually approach a shoreline at an angle, refracting from their original direction. Diffraction can occur when a wave line moves around an obstacle, such as a rock or breakwater; wave reflection happens when a wave train enters a harbor or bay and is reformed into a non-progressive standing wave.
- Internal waves may form at the boundaries of density-different subsurface water masses, usually at the steep thermocline seen at the bottom of the pycnocline. Internal waves may move rapidly, and have many of the same characteristics and behaviors as surface waves.

Key Terms and Concepts

A thorough understanding of these terms and concepts will help you to master this lesson.

$C = \sqrt{gd}$. Calculation for the speed of shallow-water waves. C represents speed (celerity), g is acceleration due to gravity, d is the depth of water in meters.

C = L/T. Calculation for the speed of deep-water waves. C represents speed (celerity), L is the wavelength, T is time.

Capillary wave. A tiny wave with a wavelength of less than 1.73 centimeters (0.68 inch), whose restoring force is surface tension; the first type of wave to form when wind blows.

CDip program. Coastal Data Information Program monitored by the Scripps Institution of Oceanography in California and consisting of a wave buoy array located on the West coast of the United States that records wave height and direction.

Constructive interference. The addition of wave energy as waves interact, producing larger waves.

Deep-water wave. A wave in water deeper than one-half its wavelength.

Destructive interference. The reduction of wave energy as waves interact, producing smaller waves.

Fetch. The uninterrupted distance over which wind blows without a significant change in direction, a factor in wind wave development.

Free wave. A progressive wave free of the forces that formed it.

Gravity wave. A wave with wavelength greater than 1.73 centimeters (0.68 inch), whose restoring forces are gravity and momentum.

Interference. Addition or subtraction of wave energy as waves interact.

Internal wave. Progressive wave occurring at the boundary between liquids of different densities.

Orbit. In ocean waves, the circular pattern of water molecule movement at the air-sea interface. Orbital motion contrasts with the side-to-side or back-and-forth motion of pure transverse or longitudinal waves.

Orbital wave. A progressive wave in which particles of the medium move in closed circles.

Pressure sensor. A device that is mounted on the sea floor, designed to weigh the water that passes over it, as a measure of wave height.

Progressive wave. A wave of moving energy in which the wave form moves in one direction along the surface of the transmission medium.

Restoring force. The dominant force acting to return water to flatness after formation of a wave.

Rip current. A strong, fast current formed by the seaward return flow of surf. Sometimes called, erroneously, rip tide.

Rogue wave. A single wave crest much higher than usual, caused by constructive interference.

Sea. A term used to describe simultaneous wind waves of many wavelengths, forming a chaotic ocean surface. Sea is common in an area of wind wave origin.

Shallow-water wave. A wave in water shallower than one-twentieth its wavelength.

Standing wave. A wave in which water oscillates without causing progressive wave forward movement. There is no net transmission of energy in a standing wave.

Surf. A term used to define the confused mass of agitated water rushing shoreward during and after a wind wave breaks.

Swell. Mature wind waves of one wavelength that form orderly undulations of the ocean surface.

Transitional wave. A wave moving through water deeper than one-twentieth but shallower than half its wavelength.

Wave. Disturbance caused by the movement of energy through a medium.

Wave crest. Highest part of a progressive wave above the average water level.

Wave diffraction. Movement of waves around obstacles.

Wave frequency. The number of waves passing a fixed point per second.

Wave height. Vertical distance between a wave trough and adjacent wave crests.

Wave period. The time in seconds that it takes for successive wave crests to pass a fixed point.

Wave reflection. The reflection of progressive waves by a vertical barrier. Reflection occurs with little loss of energy.

Wave refraction. Slowing and bending of progressive waves in shallow water.

Wave size. The height and length of a wave.

Wave speed. How fast the energy of a progressive wave is moving forward. Calculation differs for deep-water and shallow-water waves; see formulas at the beginning of this list, and in the text.

Wave steepness. Height-to-wavelength ratio of a wave. Theoretical maximum steepness of deep-water waves is 1:7.

Wave train. A group of waves of similar wavelength and period moving in the same direction across the ocean surface.

Wave trough. The valley between wave crests below the average water level in progressive waves.

Wavelength. The horizontal distance between two successive wave crests (or troughs) in a progressive wave.

Whitecaps/combers. A wind wave breaking at sea and releasing its energy at the crest.

Wind duration. The length of time the wind blows over the ocean surface, a factor in wind wave development.

Wind strength. Average speed of the wind, a factor in wind wave development.

Wind wave. Gravity wave formed by transfer of wind energy into water.

Test Your Learning

After working through these questions, check your answers against the key at the end of this book. If you answered any of the questions incorrectly, review the relevant sections of the text and video episode.

Multiple Choice

1. Wavelength is defined as:
   a. the horizontal distance a wave travels in one second
   b. the horizontal distance between successive crests (or troughs)
   c. the horizontal distance a wave particle travels during a 30-second interval
   d. none of the above
2. The basic concept of a wave is that of a _______ caused by the movement of _______ through a _______.
   a. crest / water / sea
   b. disturbance / energy / medium
   c. particle / gravity / liquid
   d. surface / wind / water basin
3. Rip currents are the result of:
   a. breaking waves returning to the sea
   b. longshore currents caused by wave diffraction
   c. energy expended against steep vertical shorelines
   d. all of the above
4. Surface tension/cohesion is the main _______ for capillary waves.
   a. disturbing force
   b. waveform
   c. restoring force
   d. dispersing force
5. Waves of different periods and wavelengths, combining and moving in different directions, is called
   a. dispersion
   b. wave train
   c. swell
   d. sea
6. Three factors directly affect the growth of wind waves—wind strength, wind duration, and:
   a. cohesion
   b. dispersion
   c. fetch
   d. group velocity
7. The growth of wind waves is affected by all but one of the following wind conditions:
   a. strength
   b. temperature
   c. fetch
   d. duration
8. Rogue waves are usually the result of:
   a. destructive interference
   b. surf break
   c. constructive interference
   d. wave diffraction
9. A deep-water wave is defined as a wave moving through water:
   a. more than 100 km deep
   b. in a fully-developed sea
   c. deeper than half of its height
   d. deeper than half of its length
10. Waves forming between layers of fluids of different densities are called:
    a. pycnocline rhythms
    b. transitional waves
    c. internal waves
    d. all of the above
11. When the maximum wave size theoretically possible for a wind of a specific strength, duration, and fetch is reached, this is called a:
    a. seiche
    b. rogue wave
    c. fully-developed sea
    d. wave train
12. When a shoreward-bound wave is propagated around an obstacle such as a rock or breakwater, it is said to be:
    a. reconstructed
    b. diffracted
    c. refracted
    d. reflected

Short Answer

1. Discuss the so-called rogue wave—what is it, how is it formed, how is it different from other wave forms and patterns, and what are some famous "rogue wave encounters."
2. Imagine a large progressive wave train approaching a wide harbor mouth straight-on, entering the harbor and moving through it, assuming there are no major obstructions to the rear of the harbor and impacting the sea-wall at the back. If conditions were right, the progressive wave train could be re-formed into a standing wave. Describe the sequence of events.
3. Describe and document the sequence of events occurring as a deep-water wave approaches a shoreline, from the first encounter with the bottom until it breaks as surf, and how the contours and characteristics of the nearshore bottom determine the nature of the breaking wave.

Supplemental Activities

1. Use the Internet and local libraries to document and expand upon Walter Munk's opening statement in the video – "It became clear early in World War II that amphibious landings would have to be concerned about the intensity of wave action on the beaches."
2. Contact some surfers (body or board) and/or look in surfing publications, and learn the basic jargon/terminology they use in describing wave activity and how they interact with it to pursue their sport. Then assemble a "guide" for the non-surfer, combining the descriptive jargon with the actual scientific explanations of the wave phenomena.
3. Page 282 of your text mentions "Desktop-ocean" devices, which illustrate some of the characteristics of internal waves. If you already own one of these devices, or can purchase one, use it to simulate what happens at the bottom of the pycnocline when liquids of different densities meet. Following the explanation in the text under "Internal Waves," document/describe/explain what happens as these fluids interact, as it relates to the behavior of internal waves.

If you wish to construct such a device, it is best done using a clear bottle, either round or rectangular. A round bottle, such as a liter wine bottle, will work, provided it has a screw-top that can be tightened. If you can find a small, square bottle, such as the type used for some types of cooking sauces, you will get a little better representation of the interacting fluids. Rinse out the bottle and soak off the label, then fill it with water and lay it on its side, to see if it is leak-tight. If it leaks, you can seal it with tape when it is filled. Half-fill the bottle with plain water, add a few drops of food coloring, and shake to mix. Blue is the most appropriate color, but any color will do to show the water mass boundary. Then, very slowly and carefully (so as not to create air bubbles), fill the bottle to the very top (almost overflowing) with mineral oil (available at any pharmacy). Screw the lid on tight, tape it if necessary, and tilt the bottle slowly and gently. Note, describe, and document the various ways that the layers interact. If you do this well, and the bottle is tightly sealed, it may make a nice desk ornament.

Lesson 14

Look Out Below

Overview

Large waves have captured humankind's attention since the times of the early voyagers, not so much as an oceanographic event but because of the damage and loss of life that usually accompanies them. The large wind waves and rogue waves you studied in the previous lesson tend to be open ocean phenomena. Disturbing forces in the form of earthquakes, cyclones, and tidal phenomena can also cause large waves, but these are more likely to affect shorelines than the open ocean. A category of waves that your text calls "large" waves are unique in their formation and potential for violence and destruction.

Although technically not a wave, *storm surge* is an aspect of the tropical cyclone that can be very destructive if it comes ashore. The low atmospheric pressure at the storm's center produces a "dome" of water that can be up to a meter higher than sea level in the open ocean. Waves and continuing strong winds add to the size of that dome as it moves landward, where it can suddenly rush ashore—not as a cresting wave but as a high, fast-moving mass of water looking very much like a tide-related event. If this happens during a normal high tide, accompanied by strong onshore winds or rain, the effect may become even more destructive. Even though storm surge is a short-lived event, the combination of factors involved can cause major catastrophes such as happened in 1992 when Hurricane Andrew's arrival on the Florida coast was made even more violent by its storm surge. Frontal storms, with their high winds and rains, can also be accompanied by storm surge.

Another potentially "large" wave is the seiche, which is a type of standing wave (see text page 282) as opposed to a progressive wave. A *standing wave* is caused by the straight-on encounter of a progressive wave with a vertical barrier such as a seawall, large ship hull, or jetty. The wave is reflected back from the obstruction and forms a vertically-oscillating standing wave. If this encounter occurs in a small, semi-enclosed space such as a bay, harbor, lake, or ocean basin, the standing wave can assume a rhythmic back-and-forth rocking pattern, called a *seiche.* Depending on the size and configuration of the confining area, this oscillation can continue for just a few minutes or for more than a day. The wave oscillates around a central point called a *node*, generating alternating crests and troughs, confined and controlled by the boundaries of the basin. There is a good illustration of this in Figure 10.23 on text page 282. As this oscillation continues, however, it does not necessarily lose energy or momentum but can actually increase energy with each cycle, through constructive interference between its crests and troughs. Along open coasts the impact of this wave is typically minimal, because although the wavelength may be huge the wave height rarely exceeds a few inches. In large but semi-confined areas, the wavelength may reach that of a tsunami with corresponding heights. In even more confined areas, such as harbors, bays, or lakes, the seiche phenomenon may even further compound itself, forming waves large enough to interrupt shipping and cause some shoreline damage—especially if it happens to be accompanied by large wind waves or a tsunami.

By far the most feared and legendary wave is the tsunami. The text uses the word tsunami as both singular and plural, while the video calls a single event a tsunami but uses the term "tsunamis" for the plural. Either form is acceptable. However, it is not correct to call a tsunami a tidal wave. The only true tidal wave is one associated with an actual tidal event. *Tsunami* can result from undersea earthquakes, undersea and shoreline landslides, calving glaciers, volcanic eruptions, or any event that will cause direct displacement of the sea surface. Tsunami caused by undersea earthquakes are appropriately named *seismic sea waves*, and tend to be the most common, the largest, and the most destructive of all tsunami.

Seismic sea waves are common in the Pacific due to the seismic activity along the *subduction zones* around the Pacific Rim plates. Movement and rupture along a submerged fault lifts the sea surface directly above it, while gravity immediately pulls the crest back down. This momentum causes the crest to overshoot and form a trough, and the crest/trough oscillation forms progressive waves that radiate out from the earthquake epicenter in all directions. Structurally, tsunami are considered shallow-water waves because their lengths are so great that they would never be found in water deeper than half the wavelength. Tsunami also differ from wind waves in that they energize the entire water column instead of just the upper region. As with wind-generated waves, however, tsunami are affected by bottom type and configuration as they approach shore, and their behavior is modified accordingly.

Tsunami can also be generated by the *calving* of glaciers, especially in a confined area like a fjord. The slab of ice breaks free from the glacier and hits the water below forming a tsunami, which propagates to the other side of the channel and may set up a seiche-like oscillation. The huge volcanic eruption of Krakatoa in Indonesia (1863) caused massive waves to form (up to 35 meters high), destroying over 100 villages and killing thousands of people. Geologic research has even uncovered evidence of a wave estimated to have been 300 feet high that impacted what is now central Texas over 60 million years ago, depositing sand, gravel, and even fossil sharks teeth. It is thought to have originated when the Chixculub meteorite struck the Yucatan at that time.

There are some structural and behavioral differences between wind-waves and tsunami. Tsunami tend to be less steep than wind waves but have a longer period, which is usually measured in minutes rather than seconds. This accounts for the fact that tsunami can pass essentially unnoticed under ships at sea, especially with all the surface-disturbing wind waves around. The speed of a tsunami is calculated with the same formula used for other shallow-water waves—by multiplying the acceleration of gravity times the water depth and taking the square root of the answer. Compared to wind waves, tsunami move very rapidly in open water—up to 472 m.p.h. (212 meters per second), making them capable of crossing the Pacific Basin in about a day!

Although they may be essentially unnoticeable at sea, tsunami behave much differently as they approach shore. Friction with the seabed causes the bottom of the wave to slow down, from several hundred miles per hour to about 50 or 60 miles per hour. This makes the height increase dramatically (up to 100 feet high in some areas) and to move ashore as a fast, onrushing flood of water, not the huge, plunging breaker that is often depicted. Depending on the configuration of the shoreline, this mass of water can then move inland from several hundred to several thousand feet. Tsunami often come ashore as a train of waves, reinforcing each other over a period of up to 15 minutes and causing a temporary rise in sea level. Depending again on the characteristics of the impacted shoreline, there can be spectacular damage and much loss of life. Destructive tsunami, usually generated by undersea earthquakes, are thought to strike somewhere on the planet about once a year. Some notable events include the 1960 Peru-Chile Trench earthquake, which killed over 400 people in South America then formed a tsunami that reached Japan (over 9,000 miles away), killing over 100 people and causing millions of dollars in damage.

One of the recurring stories about tsunami is the way people react to them. The curious, wishing to see the wave impact land, may first notice the water level begin to recede and reveal a large expanse of beach and dying fish, shells, and living marine organisms. These onlookers may then move out onto that beach, not considering that what goes out must come back in. This, and the need for better prediction and warning systems in tsunami-prone areas, has resulted in some better education programs and in a number of tsunami warning systems. Hilo, Hawaii, is one of the coastal areas where large wave fronts are common, A tsunami that killed over 150 people in 1948 stimulated the formation of the Pacific Tsunami Warning System. That tsunami was actually generated by an Alaskan undersea earthquake, calling attention to the submarine geology of that area and resulting in the establishment of the Alaska Tsunami Warning Center. In the mid-1990s, in response to a small tsunami in Northern California, the *National Oceanic and Atmospheric Administration* (*NOAA*) established a National Tsunami Hazard Mitigation Program, which is currently in place but has yet to be tested with a real event. This program has a Hazard Assessment component that addresses the vulnerability of an area; a Hazard Mitigation section, which is basically a preparation/evacuation plan; and a Warning System based on earthquake detection using seismometers and deep-water pressure sensors that send their data to satellites.

A tsunami in action can be viewed as a temporary change of sea level in a very localized area. There is, however, a potential for long-term sea level rise that is far more ominous, and that is currently occurring on a global scale. Research has documented about a two millimeter rise in sea level per year, planet-wide. This is more rapid than has been experienced in the last couple of centuries, and has been associated with the global warming phenomenon. While changes in sea level are natural events occurring over geologic time, such as has occurred during the ice ages, the present rise is of particular concern because it is happening more rapidly than it did in the past and because Earth's population is much larger, with many people living in the coastal regions that are in obvious jeopardy as water levels rise.

Heating of the atmosphere will melt glaciers and ice sheets, adding more liquid water to the oceans and eventually inundating shorelines. Another concern is that a significant change in sea level will affect tectonic processes world-wide, particularly on geologically-active coastlines. Scientists are trying to sort out patterns in present and past sea level fluctuations using computer modeling and arrays of tide-level gauges and satellites among other technologies.

Focus Your Learning

Learning Objectives

After completing this unit you should be able to do the following:

1. Describe the three types of immense waves—storm surge, seiche, and tsunami—and the disturbing forces that cause them.
2. Explain the sequence of events that can cause storm surge to form from a hurricane or frontal storm, and the type of impact storm surge can have as it comes ashore.
3. Understand the transition of a shoreward-bound progressive wave into a seiche, and its potential effect on the shores of enclosed or semi-enclosed bodies of water.
4. Recognize the uniqueness of the progressive wave called tsunami, and the types of disturbing forces that can cause them.
5. Describe the seismic sea wave—its formation, behavior, and impact on a shoreline.
6. Recognize the characteristics and behavior of tsunami generated by volcanic activity, landslides, and calving glaciers, as compared to seismic sea waves.
7. Identify some of the more notable historic events resulting from the impacts of storm surge, seiches, and tsunami, particularly the locations, damage done, and lives lost.
8. Identify the various educational and warning programs now in place to address tsunami and other large-wave threats.
9. Understand present concerns about rising sea levels and their association with global warming, the disciplines in which research is being done, and the technologies being used.

Assignments

This lesson is based on information in the following text and video assignments. The key terms, focus points, and practice test are intended to help ensure mastery of the material presented in this unit of study.

Text: Pages 283–295 in Chapter 10, "Waves,

Video: Episode 14, "Look Out Below"

Video Focus Points

Review these points before watching the video assignment for this lesson to help you focus on key issues.

— Tsunami are fast-moving, shallow-water waves that are barely noticeable at sea, but can become massive as they approach shore.

— When tsunami-type waves encounter shallow water they decrease in speed but increase in height as they approach shore, often in the pattern of sets or trains.

— Tsunami move onto shore as a mass of moving water, often destroying buildings and taking lives.

— People tend to gather when an approaching tsunami is announced, often putting themselves in harm's way.

— Tsunami are often misnamed tidal waves because of how they impact shores, but they have no direct relation to true tidal phenomena.

— Tsunami can be generated by undersea landslides, volcanic eruptions, and calving glaciers, but most are the result of undersea earthquakes at subduction zones; often around the Pacific Rim.

— As the result of several devastating tsunami, a number of Tsunami Warning Systems are now in place. The National Oceanic and Atmospheric Administration (NOAA) has a three-part system, but it has not yet been put to the test under real conditions.

— Planet-wide, sea level has been rising more rapidly than in the past, causing concern about the effect on populated shores and on nearshore tectonically-active areas. Global warming has been associated with this rapid change in sea level, and scientists are monitoring deep-sea depth fluctuations using bottom-mounted pressure sensors and satellite-based electronics.

Text Focus Points

Review these points before reading the text assignment for this lesson to help you focus on key issues.

— Using knowledge gained about wind waves, study progresses on immense waves, which are those waves generated by storms, reflection of progressive waves, or resulting from a major and rapid displacement of the sea surface.

— Storm surge is a short-lived but potentially violent wave caused by the formation of an area of atmospheric low pressure at the center of a hurricane. This dome can be added to by strong winds as it moves shoreward, and can impact shorelines as a fast-moving, wind-blown water mass, often causing significant destruction.

— Seiches are formed when progressive waves encounter a barrier and are reflected back. On open shorelines this has little effect, but if it happens in an enclosed or semi-enclosed body of water it can form into a very large oscillating standing wave, causing major shoreline damage and interruption of shipping.

— Tsunami are long-wavelength, shallow-water progressive waves generated by sudden and rapid displacement of surface sea water. They can be caused by undersea volcanic eruptions, undersea landslides, or calving glaciers, but are most often seismic sea waves resulting from undersea earthquakes.

— Rupture along seafloor faults can lift the sea surface above forming a crest, while gravity attempts to level it back down, forming a trough. The resulting crest/trough interaction forms a progressive wave that then radiates out in all directions.

— Tsunami are considered shallow water waves because their very long wavelengths assure that they will never be in water depths exceeding half their wavelength.

— Tsunami lack the steepness of a wind wave, but have a very long period (minutes instead of seconds) so they go unnoticed at sea.

— As tsunami approach shore, friction causes their lower parts to slow down but the tops keep moving and building, eventually coming onto shore and inland for up to one-half mile, not as the often-depicted breaking wave but as a series of fast-moving water masses resembling a tidal event.

— The configuration and characteristics of the impacted shoreline determine the effect of the wave and its potential for violence.

— Major, destructive tsunami occur somewhere on Earth about once a year. These events are well-documented as to the estimated size of the wave or waves, the damage done, and the lives lost.

— An international tsunami warning network has been in place around the seismically-active Pacific Rim since 1948, and has been credited with preventing major loss of life in several locations.

Key Terms and Concepts

A thorough understanding of these terms and concepts will help you to master this lesson. (since this lesson relies on knowledge of wind-waves, the student is reminded to refer back to those definitions in the previous lesson as well as the new ones pertinent to the present lesson)

DART program. An array of deep-ocean water-pressure sensors that gather data on wave height as a function of changing water column pressures. Data are relayed to the GO satellite and to a center on Wallop's Island in Virginia.

FEMA. Federal Emergency Management Agency.

Fjord. Deep, steep-sided inlets of the sea, usually located in the higher latitudes where mountains are adjacent to the ocean.

Glacial calving. A process whereby large pieces of ice break off the front of a glacier and impact the sea surface below.

NOAA. National Oceanic and Atmospheric Administration.

Node. The line or point of no wave action in a standing pattern.

Seiche. Pendulum-like rocking of water in an enclosed area; a form of standing wave that can be caused by meteorological or seismic forces or may result from normal resonance excited by tides.

Seismic sea wave. Tsunami caused by displacement of geologic material along a submarine fault.

Standing wave. A wave in which water oscillates without causing progressive wave forward movement. There is no transmission of energy in a standing wave.

Storm surge. An unusual rise in sea level caused by the low atmospheric pressure and strong winds associated with a hurricane. Onrushing seawater precedes landfall of the hurricane and causes most of the damage.

Subduction zone. An area at which a lithospheric plate is moving downward into the asthenosphere. Characterized by trenches in the sea floor and strong, deep-focus earthquakes.

Tsunami. Long-wavelength shallow-water wave caused by rapid displacement of surface water.

Test Your Learning

After working through these questions, check your answers against the key at the end of this book. If you answered any of the questions incorrectly, review the relevant sections of the text and video episode.

Multiple Choice

1. The dome of water formed by the low-atmospheric pressures at the center of a storm can be moved ashore to encounter land as a:
   a. cyclone
   b. tidal wave
   c. tsunami
   d. storm surge

2. Node, resonant frequency, and standing wave are all terms used in discussing a:
   a. seiche
   b. storm surge
   c. seismic sea wave
   d. all of the above

3. The seiche phenomenon is most dangerous when it:
   a. impacts an internal wave
   b. occurs in an enclosed or semi-enclosed body of water
   c. occurs at an extreme low tide
   d. impacts an open coast

4. Which of these phenomena cannot result in a tsunami?
   a. rupture at a subduction zone
   b. glacier calving
   c. storm surge
   d. undersea volcanic eruption
5. The DART program for tsunami monitoring is based upon:
   a. a fleet of self-propelled underwater vehicles
   b. satellite monitoring of sea surface height
   c. seismometer readings
   d. bottom-oriented pressure sensors
6. Once a tsunami is generated, its ratio of height-to-length, or steepness:
   a. immediately begins to increase
   b. slowly begins to increase
   c. is very low
   d. is incalculable
7. When a tsunami comes ashore, its typical appearance is as a:
   a. large, plunging breaker
   b. a fast-moving flood of water
   c. a huge, but slow-moving flood of water
   d. none of the above
8. When describing the energy distribution in a tsunami's water column, you would say that:
   a. like a wind wave, it only affects the top layers
   b. it extends below the surface but not to the full depth of the column
   c. it extends upward from its origin but does not impact the surface
   d. it extends the full length of the column
9. When calculating the speed, or celerity, of a tsunami:
   a. use the same formula as for that of a shallow-water wave
   b. use a different figure for the "acceleration due to gravity"
   c. determine water depth and modify it according to bottom type
   d. none of the above
10. As the tsunami crest approaches shore, the wave height ___________, the velocity ___________, and the period ___________.
   a. remains constant / increases / remains constant
   b. increases / drops / remains constant
   c. drops / increases / decreases
   d. increases / increases / increases
11. Most seismic sea waves are generated __________, due to the seismic activity there.
   a. along the Mid-Atlantic Ridge
   b. in the African Rift Zone
   c. around the Pacific Rim
   d. along the equator
12. In addition to the prospects that rising sea levels will inundate shorelines, scientists are also concerned about the global effect on:
   a. pollutant distribution
   b. presently active seismic areas
   c. the Mid-Atlantic Ridge
   d. nearshore fisheries

Short Answer

1. Describe, in some detail, NOAA's National Tsunami Hazard Mitigation Program.
2. Describe the sequence of events that led to the famous tsunami disaster at Hilo Bay, Hawaii, in April, 1948.
3. Describe and discuss the attraction of people to areas of seismic sea wave activity, and the characteristics of those waves that can, in turn, injure or kill those onlookers.

Supplemental Activities

1. There continues to be a great deal of controversy about global warming. Is it in fact happening? If so, is it entirely a natural, cyclical, historical phenomenon, or is it something different caused by recent human activities? What are some of the predictable consequences to Earth's ecosystems? Research these, and other questions that are relevant, to educate yourself about what is actually happening, what is opinion based partly on fact, and what is conjecture or opinion based on the emotional diatribe from radical, but uniformed, groups. Information can come from the Internet, from biology and ecology texts, scientific

journals, recent encyclopedias, and actual conversations with academics, governmental agencies, NOAA, and others. First, try to determine if there is a logical-sounding, more-or-less neutral explanation for the phenomenon, then use that to construct a list of questions to pose to your various sources. From that, write your own "Guide to Global Warming," covering the points you would use to explain it to someone else.

2. As your town's resident oceanographer, you have been asked by the local science teacher to make a presentation to an eighth grade Honors Science Class explaining the formation, behavior, and destructive potential of the immense waves they have read about in books and seen on television documentaries. Some, but not all, of those students have actually seen the ocean, but all are basically aware of waves and storms at sea. To your advantage, the school is well-equipped with computer technology, and the students are familiar with using graphic illustrations to supplement the written and oral word. To prepare for this, you must now write yourself a script, noting the main points you will cover, in the order in which you will cover them. Your text is a good place to start getting organized, using the early chapters as background, then leading into the three basic types of wave described in this lesson—storm surge, seiche, tsunami. The Web links for this chapter are a potential resource for you, as are many of the terms in the Key Words section. While you are assembling your "script," try to write it so it can serve as an outline to be photocopied and passed out to the students, so they can follow along with your presentation.

3. On page 291 of your text, reference is made to evidence suggesting that a huge wave, as high as 300 feet, may have formed when a comet or asteroid struck Earth on the Yucatan Peninsula some 66 million years ago, and that that wave may have carried pieces of the seafloor inland and deposited them in what is now central Texas. Using your text (chapter 13), as well as available geology texts, recent encyclopedias, scientific journals such as *Scientific American* and *American Scientist* and, of course, the Internet, summarize and document what is known about that event and whether or not you think it could have produced a wave that large that could travel that far!

4. The Thames Tidal Barrier is the largest of several movable defenses against the threat of steadily increasing tide levels, due to higher mean sea levels, more frequent storms, and other factors. If sea level continues to rise, other countries (the Thames Barrier is in England), may look to that type of system for protection. The Thames Barrier is mentioned and depicted in your text (page 284), but a better and more complete description of all aspects of that structure/concept can be found on the Internet. From the home page, select "Regional Information," then "Thames," and then look in "Key Issues" or do a search on "Thames Barrier."

   www.environment-agency.gov.uk

   Find this site, enjoy its pictures and presentations, and write a summary description of the structure and its operation.

Lesson 15

Ebb and Flow

Overview

In the previous lesson you encountered the term "tides," either as a misnomer for a disturbing force phenomenon or as an event that could enhance the violence brought about by such a phenomenon. *Tides* are periodic short-term changes in sea level in response to the combined gravitational forces of the moon and sun and Earth's motion. Tides are categorized as shallow-water forced waves because the forces that cause them are constant, unlike other types of waves caused by disturbing forces that come and go. This rhythmic and predictable change in water level on a daily basis was noted by seafarers and coast-dwellers of even the earliest civilizations.

It was Isaac Newton who calculated and explained the motions of the planetary bodies and their response to gravitational forces. One of his central findings, upon which such explanations and calculations are still based, concerns the gravitational relationship between any two bodies with mass—the gravitational attraction between two bodies is directly proportional to the masses of those bodies and inversely proportional to the square of the distance between them (see page 298 in your text). This general explanation applies to all bodies with mass, but it must be modified to address the sun/moon/tide relationships more accurately. While the main cause of the tides is the gravity of the sun and moon acting on the ocean, the forces that generate those tides vary inversely with the cube of the distance between Earth's center and the center of the force-generating object (the sun or moon). The basic concept is the same as Newton's more general theory, however, as it relates the masses of the objects to the distance between them. Thus, although the sun is millions of times more massive than the moon, it is also nearly 400 times farther from Earth, so its influence on the tides is only about half that of the moon.

To obtain an accurate understanding of Earth's tides, two theories must be considered—the 1687 explanation by Newton, which generalized the basics of gravitational attraction and related them to their effect on Earth's ocean, and the later theory of Pierre-Simon Laplace (1775), which added the factors of continental interference, wave behavior in shallow water, and the oscillation effect in ocean basins.

Newton's description of gravitational attraction as it affects tides is a more theoretical explanation, called the *equilibrium theory* because it does not include such effects as ocean depth and continents. It also assumes an ocean completely and evenly covering the planet, and tides that are governed by the same force that enables Earth to maintain a stable orbit around the sun, and the moon around Earth. It further assumes that Earth's ocean responds instantly and invariably to those forces, and thus is always in equilibrium with them.

Before this explanation can be assembled, there are some other players to introduce. While gravity works to pull bodies together, inertia (sometimes called centrifugal force) prevents them from crashing into each other. The basic definition of *inertia* is described in Newton's First Law of Motion: bodies at rest tend to stay at rest, while bodies in motion tend to remain in motion until an external force is introduced. Earth and the moon don't collide or fly apart because their mutual gravitational attraction is exactly offset by inertia, resulting in a stable orbit. Remember, too, that the moon does not revolve around Earth's center of mass, but that the whole Earth/moon system revolves, once a month, around the system's center of mass. This entire complex is illustrated on pages 299 and 300 of your text. Note that the outward-thrusting force of inertia acts in essentially a straight line, while the inward-pulling force of gravity does not, so the forces are not in exact balance at all spots on the moon and Earth. At some points (those closer to the moon) the gravitational attraction of the moon slightly exceeds the outward-moving tendency of inertia, so the ocean water under those points is pulled toward the moon. At other points inertia exceeds gravitational attraction and the water is moved to a spot opposite the moon. This constant imbalance is caused by tractive forces, which are responsible for the ocean's *tidal bulges* (see text figure 11.5), one on the side of Earth closest to the moon, the other on the opposite side of Earth. The only point on Earth where the forces of gravity and inertia are exactly equal and opposite is at the Earth's core.

While Earth does not respond significantly to these tractive forces, the oceans and the atmosphere do. Atmospheric reactions are essentially unnoticeable, but the oceanic reaction takes the form of tides. Remember that although Newton's equilibrium theory is mainly generalized, it still offers a basic explanation for why water levels can be seen to rise and fall with some regularity (see Figure 11.7 on page 301 of your text). Also remember that Earth does not move while the ocean remains stable in the clutches of the moon's gravity; the ocean moves as Earth moves! Think of the tidal bulges as the crests of planet-sized waves, and the spaces between them as the troughs. The island in Figure 11.7 will move as Earth moves, into and out of the bulges. Under the crests, or bulges, the island is submerged at high tide, then it moves out from under the bulge and into the trough at low tide. In reality, the moon does not stay exactly over the equator; it moves each month from about 28 degrees above it to 28 degrees below it, and the actual bulges are offset accordingly (see Figures 11.8–11.10 on page 302 of the text).

While most tidal behavior is explained by the position of the moon, the sun also has an effect, depending on its location relative to the moon and Earth. This involves the same tractive forces of gravity and inertia but with much less influence because of its distance. There are predictable solar bulges and solar tides, but since Earth revolves around the sun only once a year, these occur much more slowly. The effects of the sun and moon act together to create the tidal phenomenon, and Earth responds simultaneously to the tractive forces of both. This interaction is part of the descriptive jargon of the tidal phenomenon—spring tides and neap tides. Literally translated, they mean "to move quickly" (*springen*) and "hardly disturbed" (*neapa*). These phenomena alternate in two-week cycles, depending on the positions of the sun and moon relative to Earth. During *spring tides* there is considerable difference between the high and low tides during each cycle; during *neap tides* there is less difference between highs and lows. As can be seen in text Figure 11.11, spring tides occur when the sun, moon, and Earth are aligned, causing the lunar and solar effects to be cumulative. During neap tides the moon, sun, and Earth are at right angles and the tidal fluctuations diminish due to the opposing influences of the sun and moon, but with the moon always having the greater effect.

Although this is a theoretical explanation of the tides, the fact that the moon and sun orbit in ellipses rather than in perfect circles, and that their tidal forces are influenced by their relative distances from Earth at certain times must also be accounted for. Newton knew that his equilibrium model was not complete, since it was based mainly on theoretical calculations of mass physics and celestial mechanics, and the generalized configurations of Earth, sun, and moon.

In 1775 Pierre-Simon Laplace added the concepts of fluid dynamics, shallow-water wave behavior, the interference of the continents, and basin-oscillation to create the *dynamic theory of tides*. Predictions of tidal motion based on Newton's model could now be compared with the actual observed behavior of the tides and its deficiencies confirmed. As noted earlier, tides are a form of wave with crests, troughs, length, and speed. They are the longest of all waves because their crests are separated by half of Earth's circumference. Tides are forced waves since the disturbing forces of gravity and inertia are always acting, and their speed, or velocity, is determined by ocean depth. As Earth turns, land masses interrupt tidal crests and modify their theoretical movement, causing the observed high and low tides seen at continental coastlines. Some coastal areas have two high and two low tides each lunar day called *semidiurnal tides*; others have one high and low each day called *diurnal tides*. If the highs and lows differ significantly during the cycle they are said to be *mixed tides*. Different coastal areas have different tidal patterns. Tide prediction charts take that into account and are only made for very localized areas.

These differences in tidal patterns can also be clarified by remembering the concept of rhythmic oscillations set up in large ocean basins. The water level at the center of a basin tends to remain stable at the node, or *amphidromic point*. In large basins, the Coriolis effect also influences these oscillations. As can be seen in text Figure 11.15 on page 307, there are about a dozen amphidromic points in the world ocean. Because the effects of these points are so regional, reference points must be set up in each area from which to measure and predict water levels. These points are called the *tidal datum*, designated as the zero point, below which water level is considered to be a minus tide, and above which it is a plus tide. These points are based on the historical means, or averages, of water levels, depending on the types of tidal patterns in the area.

Tidal range is also considered, based on the differences between the historical high and low water levels for the area. In small- to medium-sized enclosed areas, tidal range is usually small to moderate, but in large oceanic basins it varies greatly. The greatest ranges usually occur at the edges of ocean basins, and their effect may be even further enhanced if their energy is directed into bays or inlets. One of the most famous, and spectacular, examples of extreme tidal ranges is the Bay of Fundy in Canada, where the low-high vertical distance can reach 15 meters, or 50 feet!

When such a tidal fluctuation occurs in certain coastal inlets associated with rivers, a tidal bore can form. A *tidal bore* is a true tidal wave; it is a steep wave that moves upstream under the energy of the tidal crest. If the crest of a tidal wave encounters, but does not enter, a coastal bay or harbor, the resulting high or low tide may effect the water level there, even though is does not actually enter the area. The rise or fall in sea level as a tidal crest approaches can form *tidal currents*, which behave differently depending on the characteristics of the given tidal wave. The resulting change in water level can cause the bay/harbor water to rush back into it causing high water, or to flow back out of it causing low water.

Tidal movement involves a lot of energy, which is mostly dissipated as heat. Much of this energy comes from the interaction, or friction, of Earth with the ocean. It has been calculated that this tidal friction is actually slowing Earth's rotation by a few hundredths of a second each century. It has also been suggested that the heat dissipated from tidal activity may affect global warming.

As with weather prediction, prior knowledge of the periodic rise and fall of sea level can be important to human activity. Fortunately, these changes in water level are somewhat more predictable than are changes in the atmosphere, and are based mainly on the study of past records. Prediction tables and graphs are published annually, and most are specific in both cycles and heights for specific regions. Naturally, other factors must be considered on an "at-the-moment" basis—weather changes, floods, and storm surges. One of the more interesting and best-studied effects of daily water-level changes deals with how those changes effect the organisms that live on the continental coastlines. The physical configurations vary from flat, sandy beaches to steep, rocky cliffs, and the physical conditions at any one place can determine the types of plants and animals that are found there. These organisms tend to occur in zones populated by the particular species best adapted to live there, based on their ability to thrive under the rather harsh conditions of constantly changing water levels, temperature, and salinity, and the number of predators in the area.

As noted earlier, the energy associated with tidal friction is dissipated primarily as heat. Recently, serious consideration has been given to harnessing some of that energy as a power source. Tidal power is the only marine energy source that has been successfully exploited on a large scale, mainly in oceans with large and predictable tidal fluctuations. The advantages of tidal power are its ready availability, low operating costs, and essentially non-polluting nature. There are, however, some potential disadvantages, including the potential for storm damage to coastal facilities, modification of natural tidal patterns, effects on marine life, and even the chance that the friction needed to turn the turbines could cause a tiny decrease in the rate of Earth's rotation.

Focus Your Learning

Learning Objectives

On successful completion of this unit of study, you should be able to:

1. Recognize that the periodic rise and fall of sea level, called tides, has been recorded and studied since the early explorers and coast-dwellers, particularly in the Middle East.
2. Understand the characteristics of the true tidal wave and related tidal phenomena, as compared to disturbing force waves.
3. Understand the contributions of Newton and Laplace in describing and explaining the tidal phenomenon.
4. Explain the roles of gravity and inertia as they affect the dynamics of the sun/moon/Earth system.
5. Understand the basic mathematical formulas for calculating the gravitational attraction between two bodies with mass, as part of the mechanics of the tide generating process.
6. Compare and contrast Newton's equilibrium model of the tidal phenomenon with Laplace's dynamic theory.
7. Understand the formation of the crest/trough configuration of the tidal "wave," the concept of tidal bulges, and how this explains the daily high-tide/low-tide fluctuations.
8. Define the terms low tide and high tide, neap tide and spring tide, tidal datum, and tidal range.
9. Explain the interest in accurate tidal prediction, and the methods used to accomplish it.
10. List the pros and cons of using the tides as a source of power.
11. Characterize the basic intertidal zone that is inhabited by marine organisms.

Assignments

This lesson is based on information in the following text and video assignments. The key terms, focus points, and practice test are intended to help ensure mastery of the material presented in this unit of study.

Text: Chapter 11, "Tides," pages 296–313

Video: Episode 15, "Ebb and Flow"

Video Focus Points

Review these points before watching the video assignment for this lesson to help you focus on key issues.

— The daily fluctuations in seawater level along coastlines has been noted since early history, mainly as related to commerce and coastal civilizations.

— The astronomer Newton related the roles of the sun and moon on tidal activity to his theory of gravitation.

— Based on Newton's explanation, the gravitational attraction of the sun and moon on Earth's surface is correlated with their relative distances from Earth; the moon is smaller but closer, so it has more effect.

— Newton's equilibrium theory of gravitational attraction assumed Earth to be a sphere uniformly covered with water, and that the sun, moon, and Earth would interact accordingly. His theory did not account for continental boundary interruption or water depth.

— The later dynamic theory of Laplace adds the laws of fluid dynamics and accounts for irregularities on Earth's surface

— Depending on the particular coastal location, tidal fluctuations can range from one per day (a high and low) to two per day (two highs and two lows).

— Local geography and climate conditions can also affect water level fluctuations.

— The influence of the tides has been well-studied in what is called the intertidal zone, especially as it affects the plants and animals living there.

— Tidal fluctuations can have varying degrees of impact on coastal areas, depending on their particular geographic locations.

— Some satellites are equipped to study sea level changes as a part of the global oceanographic data-gathering process, but to accurately assess global changes they must filter out the daily tidal fluctuations.

— Tidal friction—the interaction of the water with the sea floor or with density-different midwater layers—is an important factor in studying ocean circulation, and has been linked to long-term changes in the sun/moon/Earth system.

— Tidal power has been successfully harnessed by several countries, but has ecologists concerned because of its potential environmental impact.

Text Focus Points

Review these points before reading the text assignment for this lesson to help you focus on key issues.

— Tides are periodic, short-term changes in sea level caused by gravitational interactions and movements of the sun/moon/Earth system.

— "Waves" caused by tidal activity are the longest wave of any type, and are known as forced shallow-water waves because the forces that form them are always acting, and because the water depth controls their speed.

— Although early scientists had recognized the possible connection between the tides and the sun/moon positions, it was Isaac Newton who related the tides to the gravitational attraction of bodies with mass. However, his equilibrium theory was based on an "idealized" Earth, and did not account for water depths, continental interference, and other factors.

— Newton's theory assumed that Earth's ocean responds to the same forces that govern the orbits of the sun, moon, and Earth—the continuous struggle for equilibrium between the forces of gravity and inertia. It sets up theoretical tidal bulges in the ocean; one just below the position of the moon, the other on the opposite side of Earth. Earth and its ocean waters move under and out from under these bulges, creating high and low tides.

— The moon has a greater influence on the tides than does the sun simply because it is closer to Earth. When the sun, moon, and Earth are aligned, they reinforce each other's tractive forces and cause tidal fluctuations to be large spring tides. When they are at right angles to each other, the tidal height differences are less, causing neap tides.

— The orbits of the earth and moon are not circles but ellipses, causing the distances of the sun and moon from Earth to change as they move.

— In 1775, Simon-Pierre Laplace modified Newton's equilibrium theory to include some fluid dynamics and to recognize the importance of local coastal geography in tidal activity. Patterns may differ from region to region. This can be explained by the concept of rhythmic oscillation, which assumes a no-motion midpoint in the basin called a node—or, in this case, an amphidromic point—around which tidal crests move. There are about a dozen such points in the world ocean basins.

— When studying tidal fluctuations, one uses a reference point called a tidal datum, upon which to calculate high and low water movement.

— Tidal range is also important historically, recording the patterns of maximum high and low tides.

— A tidal bore can form if a tidal wave crest moves into a river mouth and breaks. A tidal current occurs when a tidal crest or trough encounters a bay or harbor, and the resulting high or low tide modifies the natural flow of water in and out of the bay or harbor.

— Tidal friction, the interaction of tidal waves with the sea bed, is associated with a great loss of energy, mainly dissipated as heat. This friction is also related to a perceptible slowing of Earth's rotation over geologic time.

— Tidal activity is important to marine commerce and coastal populations, and accurate prediction of periodic changes in water level are necessary. Most of this is done using the historical tidal data for the particular area, but current local factors such as storm surge or floods must also be considered.

— The ocean between the high and low tide marks on coastlines is called the intertidal zone and is of interest to scientists who study the organisms that live there and their adaptations for survival.

— The harnessing of the energy involved in tidal fluctuations has been successful in several countries, but is still under close study because of various environmental concerns.

Key Terms and Concepts

A thorough understanding of these terms and concepts will help you to master this lesson.

Amphidromic point. A "no-tide" point in an ocean caused by basin resonance, friction, and other factors, around which tide crests rotate. About a dozen amphidromic points exist in the world ocean. Sometimes called a node.

Diurnal tide. A tidal cycle of one high tide and one low tide per day.

Dynamic theory of tides. Model of tides that takes into account the effects of finite ocean depth, basin resonance, and the interference of continents on tide waves.

Ebb current. Water rushing out of an enclosed harbor or bay due to the fall in sea level as a tide trough approaches.

Equilibrium theory of tides. Idealized model of tides that considers Earth to be covered by an ocean of great and uniform depth capable of instantaneous response to the gravitational and inertial forces of the sun and moon.

Flood current. Water rushing into an enclosed harbor or bay because of the rise in sea level as the tide crest approaches.

High tide. The high water position corresponding to a tidal crest.

Inertia. The tendency of non-moving objects to remain at rest, and of moving objects to continue moving in a straight line, until affected by an external force.

Intertidal zone. The region on a coastline between the highest high and the lowest low tidal marks, especially noted for the types of marine organisms living there.

Low tide. The low water position corresponding to a tidal trough.

Lunar tide. Tide caused by gravitational and inertial interaction of moon and Earth.

Mean sea level. The height of the ocean surface averaged over a few years time.

Meteorological tide. A tide influenced by the weather. Arrival of a storm surge will alter the estimate of a tide's height or arrival time; as will a strong, steady onshore or offshore wind.

Mixed tide. A complex tidal cycle, usually with two high tides and two low tides of unequal height per day.

Neap tide. The time of smallest variation between high and low tides occurring when Earth, moon, and sun align at right angles. Neap tides alternate with spring tides, occurring at two-week intervals.

Semidiurnal tide. A tidal cycle of two high tides and two low tides each lunar day, with the high tides of nearly equal height.

Slack water. A time of no tide-induced currents that occurs when the current changes direction.

Solar tide. Tide caused by the gravitational and inertial interaction of the sun and Earth.

Spring tide. The time of greatest variation between high and low tides occurring when Earth, moon, and sun form a straight line. Spring tides alternate with neap tides throughout the year, occurring at two-week intervals.

Tidal bore. A high, often breaking wave generated by a tide crest that advances rapidly up an estuary or a river.

Tidal bulge. Areas of high water on Earth's surface—one under the location of the moon and the other on the opposite side of Earth.

Tidal current. Mass flow of water induced by the raising or lowering of sea level owing to passage of tidal crests or troughs.

Tidal datum. The reference level (0.0) from which tidal height is measured.

Tidal range. The difference in height between consecutive highs and lows.

Tidal wave. The crest of the wave causing tides; another name for a tidal bore. Has also been used as a general term for a tide-generated wave.

Tide. Periodic, short-term change in the height of the ocean surface at a particular place, generated by long-wavelength progressive waves that are caused by the interaction of gravitational force and inertia. Movement of Earth beneath tide crests results in rhythmic rising and falling of sea level.

Tractive forces. The combined interactions of inertia and gravity, to create the imbalance that causes the tidal bulges, and thus the crests and troughs of a tidal wave.

Test Your Learning

After working through these questions, check your answers against the key at the end of this book. If you answered any of the questions incorrectly, review the relevant sections of the text and video episode.

Multiple Choice

1. When calculating the gravitational attraction between the moon and Earth, we relate the masses of those two bodies to the:
   a. distances between the two closest surfaces
   b. distance from the center of the moon to the nearest amphidromic point on Earth
   c. distance from Earth's largest tidal bulge to the surface of the moon
   d. distance between the centers of both bodies

2. Which of these concepts was not added by Laplace in modifying Newton's equilibrium theory?
   a. observed tidal behavior
   b. continental interference
   c. effect of inertia
   d. rhythmic oscillation

3. Which of the following processes is/are affected by tides?
   a. coastal estuarine sedimentation
   b. pollutant transport
   c. invertebrate reproduction
   d. all of the above

4. The concept of a node in the resonant oscillation phenomenon is called ___________ in a large ocean basin.
   a. a tidal bulge
   b. an amphidromic point
   c. the tidal datum point
   d. mean sea level

5. Due to its closer proximity, the gravitational effect of the moon on Earth is about ______ that of the sun.
   a. twice
   b. half
   c. 81 times
   d. none of the above

6. When discussing tidal forces, centrifugal force is often used as a synonym for:
   a. gravity
   b. inertia
   c. tidal bulge movement
   d. solar tides

7. An ebb current is a type of:
   a. amphidromic wave
   b. slack water
   c. tidal current
   d. tidal friction

8. Solar tides are caused by the sun's ________ acting on Earth.
   a. perihelion
   b. tractive force
   c. amphidromic point
   d. equilibrium

9. Tidal friction plays a significant role in:
   a. deep ocean mixing
   b. the changing orbit of the moon
   c. global climate regulation
   d. all of the above

10. A tidal bore is:
    a. associated with river inlets
    b. considered to be a true tidal wave
    c. exposed to great tidal fluctuation
    d. all of the above

11. Which of these is not considered one of the three basic tidal patterns?
    a. diurnal tides
    b. amphidromic tides
    c. semidiurnal tides
    d. mixed tides

12. The difference between high tide and low tide levels is called the:
    a. tidal bore
    b. mean sea level
    c. tidal datum
    d. tidal range

Short Answer

1. Every continental coastline, whether a sandy beach, rocky shore, or a steep cliff, has an area that is covered by water at high tide and exposed to the air at low tide. This is called the intertidal zone, and is best known and studied for the marine organisms that live there. Describe basic physical aspects of the intertidal zone, and how the organisms that live there cope with the changing conditions.
2. Tides are defined in several ways—as periodic, short-term changes in sea level; as a form of wave. Expand upon this second description, listing and discussing the various descriptive components that justify their being called waves.
3. One of the concepts Newton did not consider in his equilibrium theory of the tides, but which was later used in Laplace's dynamic theory, is the rhythmic oscillation of water in an ocean basin. This standing wave phenomenon was encountered in the lesson on wind waves and again in the discussion of seiches, but we must now consider it in a larger venue as it relates to ocean basins. Explain the rhythmic oscillation phenomenon, how it relates especially to water movement in the world's ocean basins, and how it affects the tides.

Supplemental Activities

1. The energy associated with tidal fluctuations has been harnessed successfully by several countries to generate electrical power. This concept has many advantages, along with a corresponding number of disadvantages. Beginning with pages 311 and 312 in your text, research these pros and cons with the idea of becoming an advocate for one side or the other. As you know, a successful advocacy depends on a thorough knowledge of both sides. Using the Web resources suggested below (and others) prepare a preliminary list of pros and cons, or claims and counterclaims regarding the use of tidal power, then make a short summary statement recognizing your advocacy for one side or the other, and your reasoning behind this choice.

   Since this is a relatively new concept, it is unlikely that the standard encyclopedias will be of much help. Conducting a search of "tidal power" on the Web will yield a number of sites.

   A good explanation of tidal power can be found at:

   http://en.wikipedia.org/wiki/tidal_power

2. Knowledge of "at-the-moment" coastal sea-level fluctuations (tides) is of interest to maritime commerce, coastal dwellers, recreational boaters, swimmers, surfers, and beach-goers. Tidal information can be obtained from local newspapers, radio and television news, pamphlets in sporting-goods and fishing stores, online charts and graphs, just to name a few sources. Most of this prediction of tide cycles and levels is based on analysis of the historical tidal patterns of a given area, considering the six or seven most important factors that allow this prediction to be done mathematically. Begin your written report with a "History of Tidal Analysis and Prediction" (see the Website http://tidesandcurrents.noaa.gov/predhist.html), describing the early tide-predicting machines and their limitations. Then write a summary description of the types of computers and software used today, and list some specific sources for up-to-the-minute predictions.

   For software, try http://scilib.ucsd.edu/sio/tide. And, of course, for broader coverage, try "tidal prediction" on Google.

3. To many of us the term "maelstrom" conjures up images of confusion, turbulence, and chaos, usually associated with excessive forces of wind or water and often with a whirlpool. In Homer's *Odyssey*, Ulysses encountered a whirlpool named Charybdis, in the Strait of Messina. You may be surprised to learn that there is a real whirlpool, called the Maelstrom, in the Norwegian Sea. Poe describes it graphically in his story "Descent into the Maelstrom." Speaking oceanographically, what is a whirlpool, or a maelstrom, how does it form and perpetuate, and is it as potentially lethal as some may think?

Using your text as a starting point, following up with selections from your encyclopedias (*Britannica, World Book, Americana*), and the Web, research the whirlpool/maelstrom concept from both the scientific and the literature/mythology points of view. Write a short report covering both of these aspects to summarize and explain this phenomenon.

By calling up "Maelstrom, in the Norwegian Sea" you will see a number of sites. One of the better ones is:

www.math.uio.no/ ~ bjorng/tidevanns-modeller/tidemod.html

Another good site is:

www.math.uio.no/maelstrom

To look at the legendary Strait of Messina and the whirlpool Charybdis, try:

www.wikipedia.org/wiki/Charybdis

homepage.mac.com/cparada/GML/Charybdis.html

And, of course, broader coverage can come from calling up "Charybdis, Strait of Messina."

Note: be sure you use the full titles "Maelstrom, in the Norwegian Sea" and "Charybdis, Strait of Messina." By just entering "Maelstrom" will get sites for furniture design, computer programs, video games and an underground music magazine. If you just enter "Whirlpool" you will learn about Whirlpool washers and dryers!

Lesson 16

On the Coast

Overview

People on the west coast of the United States go to the "beach." In the Northeast summers are spent at the "shore." Before doing either, it's essential to tune in to the "coastal" weather report. A student of oceanography is immediately aware of the descriptive classification of the land/sea margin, which can be almost as varied as its actual physiography. This lesson examines that classification system, the short- and long-term effects of the natural coastal processes and the man-made forces that impact the coastline.

The *coast* is the place where the ocean meets the land. The activities at this junction can affect, and be affected by, features and processes of both land and sea. Over time, many of these limits and boundaries have been established and described differently, depending on their geographic and political circumstances. The fact that the physical constructions and profiles of the coastline are constantly changing further confuses attempts to establish standardized descriptive terminology.

Since the mid-1950s, the number of people moving toward the coasts—the Gulf, Pacific, and Atlantic—has been steadily increasing. It is now estimated that about 80 percent of the population of the United States lives in coastal counties. The impact of human intrusion on these environments is a major concern to coastal ecologists. Of particular interest at present is the increased possibility of significant rises in sea level, bringing the land/water boundary closer to populated areas.

The term *eustatic change* is used to describe variations in sea level that can be measured throughout the world ocean. These variations can be due to changes in the actual amount of seawater in the world ocean, changes in the size of the ocean basin due to seafloor spreading, and changes in the volume of the existing water mass due to fluctuating global temperatures. Other changes are primarily short term, including the constant adjustment and relocation of the actual coastal plate boundaries.

Until fairly recently, the descriptive terminology related to coastal areas has been based on the general appearance of the component features. With the acceptance of plate tectonics theory, the terms passive and active were applied to coasts, designating structures and processes related to the trailing or passive, and leading or active, edges of the continental plates. Other factors considered when classifying and describing coasts are the short-term and continuing erosion and sediment transport caused by waves, currents, and storms. Because of all these factors, there is no universal system for the descriptive classification of the coastal zones. But there is general agreement that they usually begin with the actual sea/land juncture and extend in both directions from there; those distances being determined by the people doing the measuring and describing.

The video episodes employ the classification scheme devised by Francis Shepard, who termed primary coasts as being "younger" coasts still forming and being affected by terrestrial processes, and secondary coasts as "older" coastlines being shaped and modified by marine processes and which retain few of their initial terrestrial characteristics. The text refers to primary coasts as erosional and secondary coasts as depositional. This guide will use the terms erosional and depositional.

Erosional coastlines tend to be rough and irregular, formed initially by terrestrial processes such as glaciation, volcanic activity, and stream and river deposition. They are still in the process of being modified by wave action, currents, and erosion. Drowned valleys, *fjords*, and large and small river *deltas* are typical features of erosional coasts. Upward or downward movement of the shoreline at fault zones, evident on the Alaskan and the Pacific coasts, also produce erosional coasts.

Depositional coastlines began as erosional coasts but have been changed over time, largely by marine-oriented processes such as wave action, currents, and sediment transport. In many depositional coastal areas, the marine processes are still accompanied by land-based activities, such as erosion by wind and water, extremes in temperature, and the effects of terrestrial vegetation. This constant assault from both land and sea continues to smooth and straighten shorelines, often resulting in such interesting features as *sea caves*, massive *sea cliffs*, and the remnants of old coastlines preserved as a series of *wave-cut platforms*.

Depositional coasts can be further subdivided into *high-energy coasts* where wave action and erosion is rapid and extreme, and *low-energy coasts* where wave action and erosion are slow and generally quiet, as along the protected shores of the Gulf Coast. The geographic location and physiography of an area determine whether a coastline is high- or low-energy. Characteristic of depositional coasts is the noticeable smoothing and straightening of its once-rough and irregular erosional structure. This is a regional phenomenon based on the comparative resistance of the shoreline geology to wave action, and to the strength and persistence of the waves and currents.

There are other systems of coastal classification. One such system relies on comparison of the relative rate of movement of the land to sea level. In this system, an emergent coastline is rising as a result of continents being relieved of the weight of recent glaciations. In contrast, submergent coastlines are steadily sinking, either actually or apparently, due primarily to flooding by rising sea levels.

The contour and shape of the sea floor at the land/sea junction determines the fate of materials being eroded and removed from the land. On a steep submerged slope the eroded sediments will quickly move into deeper water beyond the shoreline and become part of the continental shelf. In less-steep offshore areas this material will tend to remain near shore, to be transported by nearshore wave and current action in a process called *longshore drift*. Depending on the behavior of the local waves and currents, and the shape of the adjacent land, a beach may be formed.

The term *beach* has many different meanings: to oceanographers and geologists it is a feature of depositional coasts comprised of some type of sediment covering all or part of a shore. These particles can be of similar or different grain size, depending on their original source and the forces that carried and deposited them. The shape or slope of the adjacent land helps determine the distribution of the sediment. Usually the flatter the beach, the finer the material. Since the beach is constantly being impacted by waves, it is in a constant state of change based on its porosity, or the ability of water to move between the sediment grains. Larger grains permit water to percolate down between them as it returns seaward, while small, fine grains interlock and prevent percolation, causing the returning water to move quickly seaward over the beach surface.

There are certain features that tend to be common to most true beaches, based on the tendency of wave and tidal actions to deposit sediment in largely predictable patterns whether a coast is a high- or low-energy coast. However typical or predictable these patterns may be, they must be considered temporary, subject to total disassembly and reassembly according to seasons, tidal fluctuations, and storms.

A phenomenon peculiar to high-energy beaches is the *rip current*. This is a small but fast-moving current formed as large amounts of water pile up on a beach and return, out of balance with the ability of the *longshore current* to absorb them. These are erroneously called rip tides or undertow, and are safety concerns for coastal swimmers and surfers.

Coastal cells are related to the transport of sediment along a coastline. Most new coastal sand comes from streams and rivers that empty into the ocean, to be moved and deposited by *longshore drift*. Stable beaches are generally in balance between the incoming contributions of streams and rivers and the outgoing materials being washed off and carried out to adjacent deeper waters and submarine canyons. A natural section of coastline on which sand deposit and removal are in balance is called a *coastal cell*. Large cells are typically found at the passive trailing edges of continental margins and smaller cells are found on the more active leading edges.

Barrier coastal systems are a feature of depositional coasts related to offshore wave energy and sediment transport. They consist of spits, bars, and islands. *Sand spits* form from sediment deposited when a longshore current slows down as it meets a headland or bay mouth. If that spit actually closes off the bay it can form a *bay mouth bar*, which can eventually be broken through to re-establish the bay water/ocean water connection in the form of an inlet. Sandbars that form offshore and separate from land features are called *barrier islands*, and are a major feature of many coasts world-wide. Other near-shore, sediment-formed structures are *lagoons*, *sea islands*, and *tombolos*.

Along very active tectonic margins, such as the Pacific Coast of the United States, continental shelves tend to be narrow and their offshore features very steep, dramatically effecting nearshore wave and current action and sediment transport.

Coasts can be formed and modified by mainly biological processes as well as by physical processes. Nearshore structures made by millions of generations of reef-forming corals, such as the famous Great Barrier Reef of Australia, are formed in the warmer tropical waters world-wide. Charles Darwin described and classified three basic types of coral reefs: *fringing reefs*, which are connected to land; *barrier reefs*, which are close to land but separated by a lagoon; and *atolls*, which are ring-shaped offshore coral islands often built around sunken volcanoes.

Estuaries are bodies of water formed in shoreline inundations, where fresh water from the land mixes with ocean water. These are areas of extraordinary biological diversity and productivity, and are of increasing interest to the public as concern grows regarding human impact on the coastal ecosystems. Estuaries are classified and described according to their origin, their relationship to the adjacent land mass, and their sources of fresh and salt water. Other land-bordered seawater/freshwater entities are lagoons and wetlands. Although the term wetlands is often used interchangeably with estuary, the two have somewhat different characteristics.

As previously noted, there are three coasts associated with the continental United States—Pacific, Atlantic, and Gulf. The Pacific Coast is very active tectonically, while the Atlantic and Gulf Coasts are less active. The Gulf Coast has even greater descriptive differences, being more protected from major currents and wave activity—except for hurricanes—and having a smaller tidal range. The rate of subsidence of the Gulf Coast is also higher than that of either of the other coasts due to human activity, including sediment removal and compaction, dredging, and the mining of oil and natural gas.

Recent population migration to the coasts has the potential to impact natural coastal processes and ecosystems. Alteration of beaches and other natural coastline structures is done to make land available closer to the water for construction and recreational purposes. As sea level rises, those areas are in danger of erosion, flooding, and total destruction. Also of concern is the disruption of natural outlets for the dissipation of wave energy and sediment transport and deposition caused by the construction of harbors and *breakwaters*.

Coastal management and protection plans and strategies are being written in response to these intrusions. Most of these include input from national, state, and local governmental agencies, coastal engineers and scientists, sociologists, and coastal ecologists. Hopefully, this will eventually produce a coastal management strategy that balances the needs of the human population with those of the natural processes and ecosystems.

Focus Your Learning

Learning Objectives

After completing this unit you should be able to do the following:

1. Understand the basic concepts of the coast and coastline, as well as the existing variety of descriptive sub-categories under that general terminology.
2. Identify the various schemes and systems used for classifying coasts, and the concepts upon which they are based.
3. Understand the differences between the terrestrial and marine processes that affect coasts, and how and where they interact.

4. Characterize the basic differences in appearance, structure, and dynamics among the three main U.S. coasts—Pacific, Atlantic, Gulf.
5. Discuss the general concept of a beach (as opposed to other shoreline structures), and understand the terminology used to describe the various beach shapes, slopes, features, and material compositions.
6. Explain the formation of barrier spits, barrier islands, sea islands, bay mouth bars, and lagoons.
7. Describe the unique physical features of estuaries, lagoons, and wetlands; compare and contrast their ecological significance.
8. Describe the major points of controversy and discussion relating to human interaction with natural coastlines, and some of the possible management solutions and strategies.

Assignments

This lesson is based on information in the following text and video assignments. The key terms, focus points, and practice test are intended to help ensure mastery of the material presented in this unit of study.

Text: Chapter 12, "Coasts," pages 314–343

Video: Episode 16, "On the Coast"

Video Focus Points

Review these points before watching the video assignment for this lesson to help you focus on key issues.

— Since the 1950s there has been a major population shift to the coastal margins of the United States. It has been estimated that about 80 percent of the U.S. population now lives in coastal counties, impacting their resources for living space, water, fisheries, and recreation.

— While there is no universal definition of the term coast, most agree that it begins with the land/water interface. The differences in descriptions begin when attempting to define the landward and seaward boundaries of the coastal zone.

— The scientific approach to coastal classification is also variable and confusing, including schemes based on the effects of tectonic plate activity; whether or not the main constructional influences are from the land or the sea; if the actual coastline is being deposited/eroded or is emerging/submerging; and a descriptive system based on the energy of waves that impact the shore. The most useful (and most used) scheme for initial description appears to be based on the position and movement of the tectonic plates.

— An active coastal zone is associated with active plate boundaries; a passive coastal zone is at the trailing edge of a tectonic plate.

— Another system of coastal zone description/classification involves the combination of tectonic activity over geologic time, the effects of changing sea levels over time, and the short-term processes presently active, such as waves and currents.

— All coastlines tend to be unique, based on their geologic composition, their exposure to eroding processes such as waves and storms, and the fact that their shapes and compositions are part of a constantly-changing dynamic.

— The younger, more active coastlines at the leading edges of the tectonic plates are usually more rugged and irregular in shape, as they continue to be formed and reformed by natural forces. Older coastlines at the trailing plate edges have been subjected to erosive forces for much longer and are smoother and straighter. These are the primary explanations for the differences in our Pacific, Atlantic, and Gulf Coasts.

— The Atlantic and Gulf Coasts, classified as passive margins, tend to have wide continental shelves and substantial offshore wave action, resulting in significant amounts of eroded material being carried onshore and longshore, to be deposited as barrier islands or spits.

— If sand coming from offshore or longshore accumulates around a landside projection, it can form a barrier spit, as an extension of the land. This may cause a buildup of water in the resulting bay or estuary, which can then break through the spit and separate it from the land, forming a sand-rich barrier island.

— By way of contrast, the tectonically-active Pacific Coast characteristically has little or no continental shelf, and deep water is reached just offshore in many areas.

— Despite their basic structural and behavioral differences, all three types of U.S. coastlines are still affected, in both the long and short term, by the natural forces of wind, waves, erosion and deposition, and sea level changes. The changes caused by these forces and processes are a major concern to the coastal human population. Building and resource management strategies must be carefully planned and implemented, using both sociological and scientific input, to reach a balance between the ocean and the coastal population.

Text Focus Points

Review these points before reading the text assignment for this lesson to help you focus on key issues.

— Study of the coastal zone relies heavily on a system of descriptive classification and terminology. This chapter uses the erosional-depositional classification scheme.

— The term coast includes not only the shore, where the sea meets the land, but a wide zone on each side of that junction. A coast may have sandy beaches, salt marshes, and rocky cliffs on the land side, and submerged features like troughs and sandbars offshore.

— Active coasts are found at the leading edges of continental plates; passive coasts at trailing edges.

— Since the coastal zone includes the land/sea interface, long-term changes in sea level (eustatic changes) are an important consideration. These may be due to glaciation, volcanic activity, materials brought by comets, variations in the actual size of the world ocean basin due to seafloor spreading, and even changes in the total volume of seawater as a result of global warming.

— Erosional coasts are usually young coasts that are still essentially as they were at the end of the last ice age, and are formed primarily by terrestrial processes (land erosion, river deposition). Depositional coasts tend to be older, and to be formed and influenced more by marine processes (wave action, sediment transport).

— The young erosional coasts are characteristically rugged and irregular in shape, because there has not been time for the sea to smooth out the features caused by recent sea level changes, glacial scouring, deposition at river deltas, volcanic activity, and plate movement.

— The older depositional coasts, while having been re-shaped and smoothed by wave action and currents as sea level stabilized, still continue to be affected by terrestrial processes like stream erosion, rapid temperature changes, and rainfall. The most familiar feature on a depositional coast is the beach.

— The erosion of depositional coasts can produce a number of interesting coastline features and structures, depending on the material comprising the land and the energy of the waves. Steep sea cliffs are usually the result of the undercutting of a shoreline by wave action, sea caves result from continuous wave activity at a localized zone of structural weakness, and wave-cut platforms mark the history of shorelines resulting from a series of sea-level changes.

— As the rough, irregular, young erosional coasts continue to be affected by wave action, they tend to straighten out, again based on the geology of the shoreline and the intensity of the wave action. During this process, much sediment can be moved and re-positioned by wave action, longshore currents, and drift, often forming beaches of various configurations and compositions.

— Along with beaches and small localized features, depositional coasts can also produce a number of much larger structures. Sand spits and bars form when currents carrying sediments encounter shoreline irregularities. Sometimes these sand accumulations can become separated from the land and form either permanent or temporary islands.

— Coasts can also be formed and modified by the activities of both terrestrial and marine plants and animals. Colonies of coral-forming animals, continuously growing on each other over geologic time, can form massive coral reefs and atolls. Land-based but saltwater-tolerant mangrove trees can form and shape coastlines, as in southern Florida.

— Estuaries are formed in partially-enclosed coastal areas, where fresh water from the land mixes with ocean water. There are a number of types of estuaries, named after the processes and circumstances of their origins. These unique areas can be a biological wonderland, offering terrestrial and marine waters and substrates, and protection from the waves and currents of the open coast. This concept of a partially enclosed/protected coastal area also includes lagoons and wetlands.

— The continental borders of the United States support three basic coasts, each with different characteristics. The Pacific Coast is tectonically very active, while the Atlantic Coast lies along the passive plate margin and is tectonically calm; it has also been described as subsiding. The Gulf Coast is also a subsiding coast, but due to sediment compaction and the activities of humans, not tectonics.

— While natural forces continue to form and reshape the coastlines, these processes often collide with the way humans wish the coastline to be. Altering these natural longshore sand movements may set into motion unknown events and activities that will seriously impact the stability of the shorelines. It is important to find a balance between the needs of the environment and those of the coastal population.

Key Terms and Concepts

A thorough understanding of these terms and concepts will help you to master this lesson.

Active margin. Continental margin near an area of lithospheric plate convergence. Also called a Pacific-type margin.

Atoll. A ring-shaped island of coral reefs and coral debris enclosing, or almost enclosing, a shallow lagoon from which no land protrudes. Often form over sinking, inactive volcanoes.

Backshore. Sand on the shoreward side of the berm crest, sloping away from the ocean.

Backwash. Water returning to the ocean from waves washing up a beach.

Barrier island. A long, narrow, wave-built island lying parallel to the mainland and separated from it by a lagoon or bay. Also see sea island.

Barrier reef. A coral reef surrounding an island or lying parallel to the shore of a continent, separated from land by a deep lagoon. Coral debris islands may also form along the reef.

Bay mouth bar. An exposed sand bar attached to a headland adjacent to a bay and extending across the mouth of the bay.

Beach. A zone of unconsolidated sediment extending from below water level to the edge of the coastal zone.

Beach scarp. A vertical wall of variable height marking the landward limit of the most recent high tides. Corresponds with the at extreme high tides.

Berm. A nearly horizontal accumulation of sediment parallel to shore. Marks the normal limit of sand deposition by wave action.

Berm crest. The top of a berm; the highest point on most beaches. Corresponds to the shoreward limit of wave action during most high tides.

Breakwater. An artificial structure of durable material that interrupts the progress of waves to shore. Often used to shield harbors.

Coast. The zone extending from the ocean inland as far as the environment is immediately affected by marine processes, and seaward to a locally-determined distance based on measurement and terminology.

Coastal cell. The natural sector of a coastline in which sand input and sand outflow are balanced.

Delta. The deposit of sediments found at river or stream mouths, sometimes triangular in shape; hence the name after the Greek letter.

Depositional coast. A coast that is growing because its rate of sediment accumulation exceeds erosive processes. (Referred to in the video as *secondary coasts.*)

Dissolution. The dissolving by water of minerals in rocks.

Emerging coast. A coastline that is slowly rising in relation to sea level. Often the rebounding of a coastline as it loses the weight of glacial ice.

Erosional coast. Coasts on which terrestrial influences dominate.(Referred to in the video as *primary coasts.*)

Estuary. A body of water partially surrounded by land, where fresh water from a river or stream mixes with ocean water, creating an area of remarkable biological productivity.

Eustatic change. A worldwide change in sea level, as distinct from local changes.

Fjord. A deep, narrow estuary in a valley originally cut by a glacier.

Foreshore. Sand on the seaward side of a berm, sloping toward the ocean, to the low tide mark.

Fringing reef. A reef attached to the shore of a continent or island.

Groin. A short, artificial projection of durable material placed at a right angle to shore in an attempt to slow longshore transport of sand from a beach. Usually deployed in repeating units.

High-energy coast. A coast exposed to large waves.

Lagoon. A shallow body of seawater generally isolated from the ocean by a barrier island. Also the body of water enclosed within an atoll, or the water within a reverse estuary.

Longshore bar. A submerged or exposed line of sand lying parallel to shore and accumulated by wave action.

Longshore current. A current running parallel to shore in the surf zone, caused by the incomplete refraction of waves approaching the beach at an angle.

Longshore drift. Movement of sediments parallel to shore, driven by wave energy.

Low-energy coast. A coast only rarely exposed to large waves.

Moraine. A hill and/or ridge of sediment left by retreating glaciers. Examples include New York and Cape Cod, Massachusetts.

Partially-mixed estuary. An estuary in which an influx of seawater occurs beneath a surface layer of fresh water flowing seaward. Mixing occurs along the junction.

Passive margin. Continental margin near an area of lithospheric plate divergence. Also called an Atlantic-type margin.

Reverse estuary. An estuary along a coast in which salinity increases from the ocean to the estuary's upper reaches, due to evaporation of sea water and a lack of freshwater input.

Rip current. A strong, narrow surface current that flows seaward through the surf zone and is caused by the escape of excess water that has piled up in a longshore trough.

Salt wedge estuary. An estuary in which rapid river flow and small tidal change cause an inclined wedge of seawater to form at the mouth.

Sand spit. An accumulation of sand and gravel deposited down current from a headland. Sand spits often have a curl at the tip.

Sea cave. A cave near sea level in a sea cliff cut by processes of marine erosion.

Sea cliff. Cliff marking the landward limit of marine erosion on an erosional coast.

Sea island. Island whose central core was connected to the mainland when sea level was lower. Rising ocean separates these high points from land, and sedimentary processes surround them with beaches.

Shore. The place where ocean meets land. On nautical charts, the limit of high tides.

Submerging coast. Coastlines that are sinking in relation to sea level, usually as a result of rising sea level.

Swash. Water from waves washing onto a beach.

Tombolo. Above-water bridge of sand connecting an offshore feature to the mainland.

Watershed. A region or area bounded by water parting and draining ultimately to a particular watercourse or body of water.

Wave-cut platform. The smooth, level terrace sometimes found on erosional coasts, that marks the submerged limit of rapid marine erosion, and often indicates the history of ancient coastlines.

Well-mixed estuary. An estuary in which slow river flow and tidal turbulence mix fresh and salt water in a regular pattern through most of its length.

Test Your Learning

After working through these questions, check your answers against the key at the end of this book. If you answered any of the questions incorrectly, review the relevant sections of the text and video episode.

Multiple Choice

1. The Cape Hatteras lighthouse, as well as Atlantic City, Miami Beach, and Palm Beach, are actually built on:
   a. sea islands
   b. coastal cells
   c. barrier islands
   d. deltas

2. About _______% of the U.S. population is now estimated to live in coastal counties.
   a. 50
   b. 20
   c. 75
   d. 80

3. Which of the following is the officially recognized standard definition of the term coast?
   a. a strip of land of indefinite width that interacts with a major body of water
   b. any portion of the ocean or fresh water that interacts with the land
   c. where the beach ends
   d. none of the above

4. Eustatic change refers to:
   a. the mechanism by which fjords are formed
   b. globally-measured sea level variations
   c. water levels based on changes in river deltas
   d. dissolution of minerals from shoreside rock formations

5. Which type of estuary is particularly common along the Atlantic coast of the United States?
   a. bar-built
   b. drowned river mouth
   c. fjord
   d. tectonic

6. Common features of a depositional coast include:
   a. beaches
   b. sea caves
   c. natural arches
   d. all of the above

7. Where are deltas most common?
   a. at the mouths of most sediment-laden rivers
   b. where marine processes dominate
   c. on the low energy shores of enclosed seas and the tectonic trailing edges of some continents
   d. at converging plate margins

8. Drumlins and moraines are associated with:
   a. delta formation
   b. retreating glaciers
   c. sandy beaches
   d. fault coasts

9. Sea caves are formed:
   a. in cliffs at local zones of rock weakness
   b. in wave-cut platforms
   c. mainly on low-energy coasts
   d. mainly on depositional coastlines

10. The movement of sand along a coast by wave action is called:
    a. active margin transport
    b. backwash
    c. rip current
    d. longshore drift

11. Which of the following sediment grain sizes would most likely form the steepest beach?
    a. fine to very fine sand
    b. large particles, or cobbles
    c. very coarse sand
    d. pebbles

12. The Pacific Coast of North America is considered very active tectonically, resulting in which type of continental shelf?
    a. broad and slightly-sloping seaward
    b. essentially non-existent
    c. similar to the Atlantic Coast but with more fjords
    d. one which is constantly rising and falling

Short Answer

1. While oceanographers and other scientists are busy classifying and describing the world's coasts and their associated features (estuaries, coral reefs, beaches) sociologists and coastal ecologists are concerned that most of the U.S. population now lives in what are considered coastal counties, building towns and cities, tap-

ping into coastal fresh water supplies, and impacting other coastal resources. List those concerns and some of the proposed approaches to solving or addressing them.

2. Coral reefs are a major feature in many tropical oceans, and a significant attraction for both scientific and recreational scuba divers and snorkelers. During his historic voyage on the *Beagle*, Charles Darwin observed and described (from the surface) reefs at many of the areas he visited, and devised a three-part classification system that is still used today. Compare and contrast the formation and appearance of those three basic types of coral reefs.

3. Parents often caution their children to watch out for rip currents, which are usually called rip tides. Parents, lifeguards, and fellow swimmers tend to be concerned about this narrow, seaward-returning mass of swift-flowing water, and the basic escape routes from it. Characterize and discuss the rip current as an oceanographic phenomenon and as a concern to swimmers.

Supplemental Activities

1. To determine the physical composition of a beach, or of any sample of unconsolidated particles of geologic material (sediment), scientists will normally use a complex series of screens to separate the different grain sizes in the sample and determine their relative abundance. If your college/university/lab has access to this type of equipment, use it to complete this activity. If not, a somewhat less accurate but still instructive job can be done with the following tools:
   - a hand lens or low-power stereo microscope capable of magnifying at least 10X
   - a medium-sized sewing needle
   - a piece of poster-board or other sturdy material upon which to glue your display
   - some fast-drying liquid glue
   - a small sample (a pinch) of what you would generally call sand, from a beach, child's sandbox, stream or lake shore, construction site sandpile, etc.
   - a penny or dime, and a quarter, to use as guides for circles
   - a 4 x 6 index card
   - a toothpick with a flat end
   - a 6-inch ruler with metric scale
   - a small measuring spoon (1/4 teaspoon size)

   Step 1. Using the ruler, section off the index card into four 2 x 3 boxes. Referring to page 282 of the text, mark the boxes as follows:
   - fine sand – up to 5 mm
   - coarse sand – 0.5 to 1.0 mm
   - very coarse sand – 1.0 to 2.0 mm
   - granules – 2.0 to 4.0 mm

   Step 2. Using the penny or the dime as a template, draw a circle at the center of the card where the lines cross.

   Step 3. With the flat end of the toothpick, very carefully scrape some of the sand out of the measuring spoon you put it in (1/4 teaspoon or less), spreading it out into a single layer within the boundary of the circle.

   Step 4. Locate the ruler at a convenient place on the index card and, very slowly and carefully, using the eye end of the needle, begin to move sand grains to the edge of the metric scale on the ruler, to measure them. Grains up to 0.5 mm (fine to medium sand) may be done with the unaided eye, but keep the hand lens handy for confirmation.

   Step 5. Measure about 100 grains and move them to their appropriate-sized boxes on the card (this won't take as long as you may think).

   Step 6. Based on the number of grains of each size, estimate the percentage of each grain size in your sample and determine what kind of sand comprised this beach—very fine to medium, coarse, or very coarse. If the sand is predominantly larger, granule-sized grains, that should also be noted.

   Step 7. To preserve your work, use a quarter as a template to draw circles on the poster board. Spread a very thin layer of glue within each circle and, using a tiny paintbrush (slightly moistened) pick up the grains you have measured out and transfer them onto the glue. (Hint: do not touch the brush to the glue, but gently scrape the grains off the brush with the toothpick or the needle and let them drop onto the glue.) If your sample did not contain granules, pebbles, or cobbles, all of which are considered unconsolidated or loose sediments and potential beach material, you may wish to obtain those sizes from a stream bed or construction site where concrete is being made.

Step 8. Let the glue dry and you have an excellent teaching/demonstration display illustrating the various sediment grain sizes and how beach composition is classified.

2. The text chapter for this lesson (Chapter 12) begins with an account of the relocation of the Cape Hatteras lighthouse in response to the threat posed by the continuous erosion of the Hatteras Island coastline. Historically, lighthouses have been necessary features of exposed coastline to warn ships of the near shore dangers. Even with modern-day radar and other high-tech devices, shoreline lights of some kind are still in use on many U.S. coasts, headlands, and harbor entrances. Using the Websites suggested below, and the resources of your public library, write a report documenting the technological and structural history of the lighthouse, beginning with the so-called Lighthouse at Alexandria, one of the Seven Wonders of the World. In your report, pay particular attention to the basic structure and location, the various sources of the light, and how and by whom they were manned. Conclude with the modern-day lighthouse and how it works, and a little bit about what is happening to some of the older lighthouses that are no longer in use but are still standing. If you have access to a lighthouse on your coast, or in your travels, visit one for a personal interview.

   Here are some suggested Websites, all of which either give a great deal of information or refer you to other sites. Calling up "History of Lighthouses" in your search engine may also be productive.

   - http://www.uscg.mil/history/h_lhindex.asp
   - www.usalights.com
   - http://ce.eng.usf.edu/pharos/wonders/pharos.html
   - www.unmuseum.mus.pa.us/pharos.htm
   - http://en.wikipedia.org/wiki/history_of_lighthouses

3. There are a number of coastal features and processes related more to biological activity than to the physical effects of wave action, sediment transport, and so forth. Wetlands are one of the coastal features of interest, especially to coastal city planners and managers in their discussions about managing and protecting estuaries, and the biological activities and concerns about human impact on these fragile environments. As a member of a city Coastal Zone Management Advisory Committee, you have been asked to make a presentation specifically defining and clarifying the wetlands preservation issue in your state.

   Conduct some research using either the Internet or your local library on one particular wetland area in the United States to create a presentation for the Committee. Select one wetland for your presentation and answer the following questions. Be sure to include the following information in your presentation:

   - The generally accepted definition of a wetland, as compared to other estuary-related terms (slough, salt marsh, swamp).
   - What are the physical characteristics of wetlands?
   - What is the wetlands' biological importance and the descriptions of some of the organisms that call it home?
   - Why should this area be protected?
   - How have humans impacted the wetlands both negatively and positively?
   - Are the wetlands you selected being actively protected by the community, and if so what sorts of difficulties have arisen in the effort to protect them?

   Suggested Websites:

   - www.epa.gov/owow/wetlands
   - http://ceres.ca.gov/wetlands
   - http://www.floridaplants.com/eco-wet.htm

Lesson 17

Due West

Overview

Author's Note. This is a case study employing interviews with scientists involved in on-going studies at four coastal sites along the southern California coast, and does not directly involve a chapter in the text, except for reference and review of terms and concepts. The "text" for this episode will be from the video presentation, and you are encouraged to take notes while viewing that video.

"It's the story of the land and the ocean struggling to withstand the pressures of development and population shifts, coupled with the natural processes doing exactly what they've always done." This statement introduces the video episode and sets the stage for its focus and content: the combined effects of man's activities and the ongoing natural processes of coastal systems. In the last lesson you learned about the structures and characteristics of coasts, the processes that shape them, and the terms used to describe them. Armed with that background, you can now address the impact of human interaction on those natural processes and dynamics. While this interaction is probably occurring on all of the populated coasts of the civilized world, four sites have been chosen on the coast of southern California to use as representative case studies. These sites are located along a 120 mile-long stretch of coastline and while having many things in common, each also has concerns peculiar to its own location.

Ventura Marina. The northernmost site in this case study is around the city of Ventura, where geological records show significant changes in the natural coastline configuration within the past 100 years. In particular, the building and maintenance of the Ventura Marina, a recreational harbor, is often cited as a "poorly-conceived coastal modification" that continues to have many problems. Apparently, the initial planning for the marina did not consider natural longshore sediment transport processes and the probable need for frequent maintenance dredging to keep the harbor entrance open and navigable. Breakwaters and jetties were added sporadically over the years to solve immediate shoaling problems, but there was no long-term protocol in place. Sediments moved in their natural progress from up-coast to down-coast, causing the harbor entrance to *shoal*. Under natural conditions, the upcoast sediments would have replenished the beaches downcoast; but this natural transport was interrupted at the harbor mouth resulting in erosion of the downcoast beaches.

The Ventura city government began dredging the shoaled areas in the marina, mainly to keep the harbor entrance navigable. But the funding fell short of needs, so the U. S. Army Corps of Engineers assumed the responsibility and today removes about 40,000 cubic yards of sediment annually, transporting it to the sand-deprived downcoast beaches. In 1971 a sand trap was built to capture downcoast-moving sand and prevent it from shoaling the harbor entrance, and plans are currently being formulated to construct a bypass system of pumps to move that trapped sand past the harbor entrance and onto downcoast beaches on a continuous basis. This would, hopefully, increase the time between maintenance dredging from annually to every three or four years.

Point Mugu Naval Air Station. Downcoast a few miles is Point Mugu Naval Air Station, known for its extensive salt marshes. At present, the United States Navy is actively protecting those areas through an organized ecology program with the goal of fulfilling their military mission while being good stewards of their lands. The Mugu wetlands are considered to be the one of the largest and most diverse of the southern California wetlands, attracting both local and migratory waterfowl, several of which are either threatened or endangered species.

In contrast to the situation at the Ventura Marina, the Point Mugu wetlands are threatened primarily by urbanization and agriculture. This large watershed acts like a funnel, receiving runoff from many acres of surrounding farmland and rapidly-expanding urbanization. This runoff, while not carrying large amounts of sediment as is the case in Ventura, brings significant amounts of contamination such as grease and oil into the wetlands system. The activities of this large military base also create significant quantities of such contaminants, and the Navy has a major cleanup procedure and protocols to deal with that. In addition to maintenance cleanup, the Navy plans to remove previously contaminated soils and to restore the wetlands physiography to better serve the resident and migratory wildlife.

In the recent past, sewage from the military operation was treated on-site. It is now linked to a nearby city sewer system, and the Navy has plans to clean the oxidation ponds used with the previous disposal system. These plans include treating the sludge residue from the ponds to remove the heavy metals once dumped there, and to line the existing emergency overflow pond to prevent further ground contamination.

To monitor the success of these efforts, studies are being done to compare the resident plant and animal populations with those at nearby, uncontaminated sites, and the results have been encouraging. Of particular concern is the fate of some of the endangered and threatened birds, especially the clapper rail.

Malibu. Urbanization has been the main reason for ecologists to study the Malibu ecosystem for many years. Since the early 1920s the Malibu section of the southern California coastline has attracted the wealthy, who first settled there in small beach homes, and later in large mansions. This has created two concerns. The first concern is contamination of the Malibu Lagoon with sewage discharge. Each of the Malibu homes has its own septic tank system for sewage, but these tanks often overflow onto the beach or into the Lagoon and its surrounding wetlands.

An even more serious concern is coastal erosion and landslides, which is a much more widespread and potentially devastating problem. Rising and falling sea level coupled with periods of heavy rain makes the entire area fragile and unstable, and has triggered numerous major landslide events. While these are natural processes, exacerbated somewhat by road building, the main concern at Malibu is the danger to residents who build their homes on cliff-tops to get a better view, exposing themselves and their buildings to the dynamics of the landsliding process. A dewatering program is in effect, which actually pumps the water out of the soil to make it more stable. Other attempts to stabilize the coastal lands include grading and reforming the unstable areas and using fill to reconstruct them.

All of this notwithstanding, however, the major effort in the Malibu area is to educate the population about the instability of the coast and encourage careful selection of building sites. The residents of Malibu must accept the fact that natural erosion of the coastline by the ocean and weakening of the soil by heavy rains cannot be stopped; it must be anticipated and dealt with as a part of any land-use consideration.

Laguna Beach–Aliso Creek. The southernmost case study site is near Laguna Beach, particularly around Aliso Creek. The coastline here is essentially stable, and is of much less concern than the effects of urbanization on the watershed around Aliso Creek over the past 20 years. Major population increase, with its accompanying construction, have drastically altered the natural landscapes with extensively paved roads and parking areas that prevent the natural percolation and filtering of watershed waters. As a result, the runoff and effluents from those dwellings flow into Aliso Creek and are transported to the ocean, affecting not only the Creek ecosystems but those of the nearshore marine environments.

Since the Laguna Beach–Aliso Creek area is also attractive to beach-goers, the potential health threat from contaminated runoff is significant. Urban development is also thought to have resulted in creek bed erosion, causing the adjacent sewer lines to leak into the runoff waters rather than being absorbed into the soil.

Although each of these coastal sites has its own particular concerns, the common theme is how the areas are impacted by the increasing human presence. As you learned in the last lesson, it is now estimated that about 80 percent of the population of the United States lives in coastal areas. A number of activist groups and environmental organizations are trying to raise public and government awareness to those conditions, particularly pollution from urban runoff. Such concerns, if expressed by enough of the population and reinforced by reliable scientific data, must eventually be acknowledged at the appropriate levels of regulatory and governmental agencies, and changes made to ensure the future integrity of coastal areas.

Focus Your Learning

Learning Objectives

Note to the student: since this lesson is based on four specific case studies that are presently in progress, there is little or no reference to the text. Students will be getting information mainly from the video presentation and the appropriate components of the study guide.

1. Characterize each of the four southern California study sites, in terms of location, general physiography, and geology.
2. Describe and explain, for each case site, the human factors and their potential impacts and effects, including urbanization, contamination runoff, and road-building.
3. Discuss the plans or processes being used or considered to mitigate the negative effects of human intervention for each case study.
4. Understand the issues surrounding the concept of endangered species at Point Mugu, and how this concern is being addressed.
5. Understand the various natural factors involved in the landslides and shoreline erosion activities in Malibu, including sea level changes, heavy rains, and mountain building.
6. Identify and discuss the various levels of involvement of area residents, regulatory agencies, politicians, and scientists, in addressing the problems and concerns of each location studied.

Assignments

This lesson is based on information in the following text and video assignments. The key terms, focus points, and practice test are intended to help ensure mastery of the material presented in this unit of study.

Text: Review Chapter 12, "Coasts"

Video: Episode 17, "Due West"

Video Focus Points

Review these points before watching the video assignment for this lesson to help you focus on key issues.

— The increasing human impact on coastal areas and their adjacent terrains and wetlands is of national concern. Four sites along the southern California coast have been selected as case studies to represent those concerns.

— The construction of a small-boat marina at Ventura in the 1950s, interrupted the natural longshore transport of sand, causing the harbor entrance to shoal and the downcoast beaches to erode. Periodic maintenance dredging programs and sand-bypass systems are currently being carried out or considered.

— At Point Mugu Naval Air Station, the activities of both the military and the surrounding agricultural and urban communities are disrupting the natural ecosystems of one of the area's largest wetlands. Mitigation includes eventual elimination of contaminated runoff and reconfiguring of the wetland back to a more natural state.

— The Malibu coast has attracted home builders to construct on the cliffs overlooking the ocean, further weakening the already geologically-unstable coastline. Landslides and their potential threat to homes is of concern, as is the contamination of the Malibu Lagoon from local septic tank runoff.

— The Aliso Creek area, near the city of Laguna Beach, has been affected by urbanization for many years, especially where over-paving due to road construction has interrupted the natural percolation/filtering activities in the soil. This results in surface runoff down the Creek and into the ocean eroding the creek bed and carrying contaminated debris and pollution from the urbanized areas to the ocean.

— Each of the four case studies presented has somewhat different impacts resulting from human activities, yet they are all faced with the same basic concerns: identifying the problem areas and documenting them with good scientific data; informing and educating the affected populations about these concerns and involving them in the solution; and dealing with regulatory, financial, and political entities.

Text Focus Points

Since this lesson deals mainly with four ongoing case studies, there is little or no involvement with specific text sites. You are encouraged, however, to review the last episode (Episode 16 and its accompanying text Chapter 12) for some of the processes and terminology used to discuss these case studies.

Key Terms and Concepts

The terms included here are more or less specific to the cases in this lesson. However, as stated earlier, you are encouraged to review the applicable terms in the last lesson (Episode 16) and the text glossary.

Dredging. The removal, by mechanical means, of sediments from a site, usually to deepen a channel or harbor entrance.

Jetty. A kind of wall built out into the water to restrain currents and to protect a harbor or the end of a pier.

Landslide. A general term describing a wide variety of processes and landforms involving the downslope movement, under gravity, of soil and rock material.

Maintenance dredging. The periodic and scheduled removal of sediment from an area, usually to keep channels or harbors navigable.

Marina. A small harbor or boat basin providing dockage, supplies, and services for small pleasure craft.

Marine terrace. A wave-cut platform along a seacoast that has been exposed by uplift or by lowering of sea level.

Mitigation. Plans or programs designed to lessen the negative effects of a potentially disruptive process or activity.

Navigation channel. The entrance or mouth of a harbor or marina deep enough for vessels to navigate safely.

Oxidation pond. A man-made pond engineered to contain the water or sewage coming from a sewage treatment plant in order to dry out the effluent sludge.

Salt marsh. A shallow estuary inhabited by plants that can withstand only limited tidal submergence.

Septic tank system. A domestic waste-disposal system in which waste matter is allowed to putrefy and decompose through bacterial action. Defective systems may leak effluent into the surrounding soil.

Shoaling. Accumulation of sediment under water, decreasing the depth of a navigation channel.

Sludge. In this context, a term used to describe the more-or-less solid component of material being discharged from a sewer treatment plant.

Urban runoff. A general term describing the collection and movement of water, and usually contaminants, from populated areas into a watershed or into the ocean.

U.S. Army Corps of Engineers. An organization of military and civilian personnel, making up the world's largest public engineering, design, and construction management agency.

Wetland. Refers to land on which water dominates the soil development and the types of plant and animal life that live in that soil and on its surface.

Test Your Learning

After working through these questions, check your answers against the key at the end of this book. If you answered any of the questions incorrectly, review the relevant sections of the text and video episode. Note that all answers will come from the video presentation.

Multiple Choice

1. The accumulation of sediment at a harbor entrance is called:
   a. entrapment
   b. shoaling
   c. bypassing
   d. upcoast erosion
2. Which of these has not been a main concern at the Aliso Creek case study site?
   a. Landslides
   b. street and parking lot runoff
   c. impact on offshore kelp beds
   d. pathogenic bacteria in the nearshore waters
3. One of the largest wetland areas in southern California is located at:
   a. Aliso Creek
   b. Ventura Harbor
   c. Point Mugu
   d. Huntington Beach.
4. Geological evidence indicates that sea levels at the Malibu coastal study site have risen and fallen _______ times in the past ______ years.
   a. two, 100
   b. three, 200
   c. two, 150
   d. four, 150
5. Which of these activities is not associated with ecological issues at the Point Mugu study site?
   a. urbanization
   b. agricultural runoff
   c. extensive road building
   d. sewage disposal
6. At the Ventura Marina, the Army Corps of Engineers has used all the following techniques to keep the harbor open, including:
   a. annual removal of sediment through dredging
   b. construction of sand traps
   c. construction of breakwaters
   d. all of the above
7. At the Point Mugu site, three endangered bird species have been identified. Which of these is considered the most endangered?
   a. brown pelican
   b. least tern
   c. snowy plover
   d. clapper rail
8. According to the video, Malibu has had some success in controlling landslides by:
   a. paving the unstable areas
   b. dewatering the unstable hillsides
   c. sinking concrete pilings through the unstable areas
   d. creating retaining walls
9. Which of these concerns is common to the Point Mugu, Aliso Creek, and Malibu areas?
   a. unstable coastal geology
   b. sewage effluent discharge
   c. longshore transport interruption
   d. agricultural runoff
10. Groundwater pollution due to heavy metals from ______________ have been a cause for concern at the Point Mugu site.
    a. the Lag Four public works yard
    b. agricultural runoff
    c. urbanization of the surrounding area
    d. sewage sludge

Short Answer

1. The instability of the Malibu coastline, and the resulting landslide threat, has been attributed to a variety of factors. List those factors, and briefly discuss them.
2. Describe and discuss in some detail, the Point Mugu wetlands ecosystems and how they are being restored.
3. List, in approximate chronological order, the steps involved in the planning, construction, and maintenance of the Ventura Marina.

Supplemental Activities

1. The term dredge can be used as either a verb or a noun; the former referring to the act of removing material from the bottom of the ocean, a lake, or river, the latter to the actual machinery used to do that. The increasing popularity of recreational boating has resulted in the construction of numerous marinas along the coastlines, often interrupting the natural downcoast transport of sand, and causing it to shoal at the entrances to these harbors making them un-navigable until it is removed. Various kinds of machinery are used to do this, depending on the type of sediment, depth of water, and funding available.

   Write a report explaining dredges and dredging. If you live near a marina or harbor, contact the marina managers or the local Harbor Department and find out about their maintenance dredging program. You may even be able to visit and photograph their operation. But first, acquaint yourself with the basics of the process, and the equipment and terminology used. One of the best-documented dredging operations in the United States is the Santa Cruz Harbor Dredging Program in California. You can find information at http://www.taylorengineering.com/dmshome/DMSManual/littoral.htm.

   Another good reference is for the harbor in Portland, Maine: http://www.epa.gov/owow/estuaries/coastlines/98summer.pdf. It reports on some of the financial, jurisdictional, and political issues.

   For an even broader view of dredging, calling up "dredging process" on the Web may also be fruitful.

2. When one speaks of urbanization, particularly in coastal areas, the term urban runoff soon appears, usually in a negative context. The premise is that water from rain, agriculture, street cleaning, etc., move downhill collecting contaminants and debris that often end in a wetland, stream, or the ocean. Much of that runoff is funneled into street gutters and storm drains, but do we know where it goes from there? An additional concern is what happens to domestic sewage after it leaves the treatment facility. Choose one of the disposal or runoff activities in your city, and document the journey of waste from its source to its end point.

   A suggested beginning might be to choose the runoff/disposal activity that interests you, then go to the public records that deal with that activity. After you have done your homework, try contacting the local city official in charge of that activity and add to your knowledge, and to your report, from that interview. Often, local treatment facilities or regulatory agencies will share information as a public relations bonus for them, and some even offer tours for visitors. If practical, illustrate your report with photos of the journey of your runoff or effluent, and/or the facilities involved.

3. During our video visit to the Point Mugu Naval Air Station in southern California, reference was made to their being an ecology department devoted to looking after the natural systems.

   You have been asked by your City Council to represent the City on a newly-formed joint Advisory Board to consider environmental issues for your local military base. Before you attend your first meeting you would like to know some of the history of such activities at other military installations. If you live near such a base, try contacting their Public Relations Department to find out about their ecology programs and to meet their resident ecologists. If you are not near such a facility, you might begin your search with the Websites listed below. Realize that much of the effort now is being directed toward clean-up and reuse of recently-closed installations.

   Write detailed notes about the general philosophy of the U.S. military regarding such stewardship and assemble a list of questions you would like to ask and points you would like to make at the upcoming meeting. Remember that you are representing your City on the Joint Board, and will be expected to make a detailed report to them after the meeting.

   http://www.brad.ac.uk/university/newsandviews/96-01/university_helps_USNavy.html

   http://www.cinms.nos.noaa.gov

   http://ceres.ca.gov/wetlands

   Also search for "U.S. Military Base Ecology Programs" on the Internet.

Lesson 18

Building Blocks

Overview

This is the first of several lessons dealing with oceanic life. Biology—the scientific study of life—is dominated by three important concepts:

— Biological processes are governed by the same laws of physics and chemistry that govern non-living things.

— Living things are composed of cells.

— Living things evolve by natural selection, known as Darwinian evolution.

A useful definition of *life* developed by NASA is, "Life is a self-sustained chemical system that is capable of undergoing Darwinian evolution."

Twenty three hundred years ago Aristotle wondered if being alive was the same for all living things. Is life the same thing for a mushroom as it is for a dolphin, a jellyfish, or seaweed? Modern biologists agree that it is. Hidden within the great diversity of life is an even greater underlying unity of life. Living things, from the simplest bacterium to the most complex mammal, store their biological blueprints in genes containing DNA and use these instructions to construct their own uniquely functioning life form. *DNA* molecules can do two important things. First, they can replicate or copy themselves, which is the essence of reproduction. Second, DNA uses RNA to read its information and can then manufacture proteins according to that information. It is the proteins—and perhaps most importantly, the category of proteins called *enzymes*—that make life happen. The translation of DNA's genetic instructions into the life processes orchestrated by proteins,

(DNA → RNA → Proteins → Life)

is fundamental to all life on Earth. Living things are highly organized and self-regulating. They take in, transform, and use energy; they cycle matter; they reproduce; and as stated earlier, they evolve.

Evolution by natural selection, or Darwinian evolution, is the single most important unifying idea in biology. It has even been said that biology does not make sense without it. Proposed in 1859 by Charles Darwin and Alfred Wallace, evolution by natural selection is actually a rather simple idea. Although the word "gene" did not exist in 1859, evolution can be defined as a change in the genetic make-up, or gene pool, of a population.

Darwin made four important observations:

1. Natural populations reproduce excessively (a cod fish lays 28 million eggs, an oak tree has thousands of acorns), and yet
2. population sizes stay relatively constant over time;
3. individuals in a population exhibit variations in structure, function, and behavior; and
4. some of these variations can be passed from parents to their offspring through reproduction.

From these four observations Darwin concluded that:

— If organisms reproduce excessively yet populations stay relatively constant in number then there must be competition for food, mates, and other resources.

— If organisms are competing and they have variations in structure and function then there will be natural selection, sometimes called "survival of the fittest."

— If there is natural selection and certain characteristics can be passed from parents to offspring (blue eyes can, pierced ears can't), then there is evolution by natural selection.

In other words, better-adapted individuals tend to survive longer and reproduce more, while less well-adapted individuals don't survive as well, die younger, and reproduce less. The survivors contribute more of their genetic traits to the next generation and the population evolves to possess those genetic characteristics that favor survival. Because evolution is a change in the gene pool of a population it is only populations that can evolve, never individuals. Over short spans of time natural selection causes changes in the genetic makeup within a species. Over long periods an accumulation of many genetic changes can produce new species from old ones. Species that cannot adapt to changes in their environment will eventually die out or become extinct. Thus, life changes over time. But where did life come from to begin with and why is it possible, indeed necessary, for it to change?

The debate about how life began is a vigorous one. One popular hypothesis suggests that simple molecules present in the oceans approximately four billion years ago became more complex through natural chemical processes driven by lightning, volcanic heat, or ultraviolet energy. This is sometimes called the pre-biotic soup hypothesis. It has been demonstrated in the laboratory that simple molecules such as methane (CH_4), ammonia (NH_3), water (H_2O), cyanide (HCN), and others can combine to form the more complex molecules associated with life. These include molecules such as amino acids, sugars, and some of the components of DNA. Furthermore, under certain conditions these molecules will self-assemble into nucleic acids, proteins, and lipids. Four steps are necessary for the initial formation of life, or biosynthesis:

1. making the small biomolecules such as simple sugars, amino acids, and nucleotides;
2. hooking them together to make long chains of small molecules or polymers, such as RNA and proteins;
3. developing the capacity to store and copy genetic information; and
4. encapsulating it all within a membrane.

The details of how this occurred are murky and will probably remain so. One hypothesis focuses on RNA. Modern life is based on a combination of DNA and proteins (DNA → RNA → Proteins → Life). DNA can both replicate itself and carry the instructions for making the proteins that make life function. So, DNA is needed to construct proteins. But it is also true that special proteins, or enzymes, are needed to construct DNA. So, DNA is required to make proteins and proteins are required to make DNA! This is biology's chicken and egg question. Which came first, the DNA or the proteins? It seems possible that the answer is neither. RNA is a simpler version of DNA. It is an intermediate molecule between the DNA genes and the making of proteins. It is not only true that RNA has been created in a laboratory under conditions simulating the atmosphere of primitive Earth, but RNA also copies itself. It is a molecule that reproduces! Perhaps, within the ocean of primitive Earth, at the borderline of pre-life and life, there was an *RNA world*, a world of RNA molecules directing their own reproduction. Perhaps both DNA and proteins, the pillars of modern life, evolved from RNA. Since DNA and RNA can mutate and reproduce, if certain DNA or RNA molecules could do this better than others they would become more abundant through natural selection. Amplification (reproduction), *mutation*, and selection—the essence of evolution—would have existed in the chemical world that preceded the biological world, so NASA's definition of life as "a self-sustained chemical system capable of Darwinian evolution" could apply to RNA molecules. It is difficult to delineate where chemistry ended and biology began in the *pre-biotic soup* of an RNA world, but that boundary was apparently crossed about four billion years ago.

Life is a chemical system in which matter and energy interact. All living things utilize energy and cycle matter and there is no difference between the way matter and energy behave inside living things and outside them. The behavior of energy is described by physical laws called the laws of thermodynamics, the first law of which states that energy can be converted from one form to another. For example, sun's nuclear energy is converted to light energy, heat energy, and other forms. Through the process of photosynthesis, and less commonly *chemosynthesis*, some of the light energy is converted to chemical energy as sugar in a plant. The chemical energy might then be converted to mechanical energy in the muscle of an animal that eats the plant. But the second law of thermodynamics states that these conversions of energy are never 100 percent efficient. There is a continuous increase in the disorder, or *entropy*, of the universe; a perpetual march toward disorganization. As electrical energy enters your light bulb it changes to light. But heat energy is also produced. The first law says you may convert the chemical energy of your food into the mechanical energy of a muscle contraction but the second law says you must lose heat as you do so, and the entropy of the universe increases as your lunch fuels your walk across campus. The conversion of energy from one type to another is explained by the first law; the heat loss and increased disorganization is explained by the second. Because they are highly ordered, or low entropy systems, living things appear to disobey the second law. This, of course, is impossible; they maintain their low entropy or organization only through a continuous consumption, transformation, and utilization of energy.

While energy flows through living things, being disordered as it goes, matter cycles in and out. The same atom of carbon in a sugar molecule ($\mathbf{C}_6H_{12}O_6$) inside a blade of grass on Monday may be inside a cow on Tuesday, exhaled into the atmosphere as carbon dioxide ($\mathbf{C}O_2$) on Wednesday, taken into another grass plant on Thursday, and used in photosynthesis to make more sugar on Friday. Alternatively, the cow's $\mathbf{C}O_2$ may dissolve in the ocean on Wednesday and become part of a clam shell ($Ca\mathbf{C}O_3$) which falls into the sediment, gets carried by seafloor spreading to a subduction zone, where it is carried down into the mantle and spews out of a volcano as $\mathbf{C}O_2$ one hundred million years later. These examples illustrate the two loops of the *biogeochemical cycle* for carbon, also known as the carbon cycle. The text outlines the details of biogeochemical cycles for carbon, nitrogen, phosphorus and silicon, iron, and other trace metals.

The combination of flowing energy, mostly from the sun, the cycling of matter, such as the carbon atom in your liver that may have been in the first living thing on Earth, plus four billion years of reproduction, mutation, and selection has brought us dolphins, pine trees, bacteria, humans, and approximately five million other extant species and innumerable extinct ones lost forever in the history of our planet. This lesson presents a broad and general overview of marine life. Many of the rich details of these organisms will emerge in the lessons that follow.

Focus Your Learning

Learning Objectives

After completing this unit you should be able to do the following:

1. Define life and discuss scientific hypotheses about how it might have begun.
2. Explain the relationships among matter, energy, and life and explain how the second law of thermodynamics and entropy relate to life.
3. Outline the biogeochemical cycles for carbon, nitrogen, phosphorus, and iron and compare and contrast these cycles.
4. Outline the theory of evolution by natural selection and discuss its importance in biology in general, and biological oceanography in particular.

Assignments

This lesson is based on information in the following text and video assignments. The key terms, focus points, and practice test are intended to help ensure mastery of the material presented in this unit of study.

Text: Chapter 13, "Life in the Ocean," pages 344–375

Video: Episode 18, "Building Blocks"

Video Focus Points

Review these points before watching the video assignment for this lesson to help you focus on key issues.

— Life in the ocean exhibits great diversity but this masks an underlying unity. This unity is related to the central role of, and relationship between, DNA and proteins and can be traced back to, and perhaps before, the origin of life on Earth.

- The major steps in the origin of life were probably: the production of simple biomolecules like amino acids and sugars from even simpler molecules like ammonia, cyanide, and water; the polymerization, or stringing together, of these small biomolecules to make longer chain-like molecules such as RNA and proteins; the storing of genetic information in RNA or DNA; and the encapsulating of all these molecules within a lipid membrane thus producing a cell-like structure.
- Biology's most important unifying idea is Darwinian evolution. The roots of this process—amplification or reproduction, mutation or variation, and selection or choosing the fittest—were probably present in the molecules that gave rise to life.
- The essence of life is that DNA makes proteins via RNA and proteins make biology function. A popular hypothesis suggests that life began with RNA and both DNA-production and protein-production developed from that RNA.
- Life arose about four billion years ago. It is almost certain that it arose in the oceans because the conditions there were much more favorable than those on land.
- Extinction is as much a part of life as its origin and evolution. Earth's history is punctuated by occasional mass extinctions such as the one that led to the disappearance of the dinosaurs and many other groups 65 million years ago.

Text Focus Points

Review these points before reading the text assignment for this lesson to help you focus on key issues.

- Life is an organized biochemical system of low entropy. Living things take in and transform energy, utilize and cycle matter, and are self-regulated. They reproduce, mutate, and evolve by natural selection.
- The movement of matter between living things and the environment is accomplished via biogeochemical cycles such as the carbon cycle.
- Life on Earth is characterized by both diversity and unity. About five million different species of life are currently known. All of them function by essentially the same processes, which is centered around genetic information of DNA utilized to construct proteins.
- The central idea of biology is Darwinian evolution. Excessive reproduction, random genetic variation, competition for survival, and natural selection are the bases of this theory.

Key Terms and Concepts

A thorough understanding of these terms and concepts will help you to master this lesson.

Biogeochemical cycle. The processes that cycle a particular nutrient between living things and the environment.

Chemosynthesis. The biosynthesis of food molecules using energy released from the oxidation of simple chemicals such as sulfur and ammonia. Light energy is not required for chemosynthesis.

DNA. Deoxyribonucleic acid, the genetic molecule of life. In the simplest sense the genetic information stored in DNA is used to manufacture the particular proteins that make life function.

Entropy. The degree of disorganization of a system. According to the second law of thermodynamics, the entropy of the universe is always increasing. Being a low-entropy system, life must continuously utilize energy to maintain its organization.

Enzyme. A category of protein molecules, enzymes are biological catalysts. These molecules regulate and accelerate metabolic reactions and, to a large extent, permit the functioning of life.

Evolution. A change in the genetic make-up of a population. When certain individuals of a population have a greater probability of surviving because they have genetic variations that make them better adapted, they tend to pass their genetic traits to future generations.

Life. A universally accepted definition of life is problematic. NASA defines life as "a self-sustained chemical system that is capable of undergoing Darwinian evolution."

Mutation. A change in the genetic code. Both DNA and RNA can mutate, and these mutations are the origin of life's diversity and the raw material for evolution.

Pre-biotic soup. The oceans before the origin of life. It is hypothesized that lightning, heat, and other energy caused complex biomolecules to form from simpler molecules in Earth's atmosphere, oceans, and crust. These accumulated creating a molecular "soup" from which life arose.

RNA world. The chemical world just prior to the origin of life. It is hypothesized that life may have arisen from RNA, which later evolved the ability to make both proteins and DNA.

Test Your Learning

After working through these questions, check your answers against the key at the end of this book. If you answered any of the questions incorrectly, review the relevant sections of the text and video episode.

Multiple Choice

1. The level of disorder or randomness of a system is called its:
   a. enthalpy
   b. entropy
   c. thermodynamic quotient
   d. biogeochemical quotient
2. The oldest known fossils of primitive life on Earth come from:
   a. Antarctica
   b. The Canadian Shield
   c. Africa and Australia
   d. the seafloor
3. A group of organisms that is actually, or potentially, interbreeding and reproductively isolated from other such groups is a:
   a. cohort
   b. race
   c. stock
   d. species
4. Which of the following is not among the processes that are fundamental to Darwin's theory of evolution by natural selection?
   a. respiration
   b. reproduction
   c. mutation
   d. selection
5. Of the many possible conditions such as light, pH, temperature, and salinity, that might control the success of a species, the one(s) that do are called:
   a. minimal factors
   b. limiting factors
   c. homeostatic parameters
   d. lethal parameters
6. It seems highly likely that life arose in the oceans because water:
   a. contains the fatty acids needed to construct membranes
   b. does not contain excessive oxygen that would damage DNA
   c. blocks out ultraviolet radiation
   d. dissolves RNA which will crystallize if it dries
7. In the carbon cycle animals primarily:
   a. release carbon as carbon dioxide (CO_2)
   b. utilize carbon dioxide to manufacture sugars
   c. release carbon as methane (CH_4)
   d. utilize methane to manufacture sugars
8. Which of these is not critical to the process of (Darwinian) evolution?
   a. Individuals in populations compete for resources, space, mates, etc.
   b. Organisms typically produce more offspring than the environment can support.
   c. Mutations occur only when it is necessary to produce more "fit" individuals.
   d. Mutations can be passed on to offspring through reproduction.
9. According to the video, the molecule that may bridge the gap between the primordial soup and the origin of life as we know it today is:
   a. carbohydrate
   b. DNA
   c. protein
   d. RNA

10. The mass extinction which caused the demise of the dinosaurs occurred about:
    a. 220 million years ago at the end of the Triassic Period
    b. 120 million years ago at the end of the Jurassic Period
    c. 65 million years ago at the end of the Cretaceous Period
    d. 5 million years ago at the end of the Paleocene Period

11. Compared to the land, the oceans contain:
    a. more species but fewer phyla
    b. more phyla but fewer species
    c. both more phyla and more species
    d. about the same number of phyla and species

12. Molecules like starches, proteins, and DNA that are long chains of simpler building blocks are called:
    a. polymers
    b. enzymes
    c. bio-catalysts
    d. ester linkages

Short Answer

1. Outline the mechanism of evolution by natural selection (Darwinian evolution).

2. Explain the role of biogeochemical cycles in biology and describe the carbon cycle in the ocean.

3. The video states that there are three or four steps necessary for biogenesis. List and describe them.

Supplemental Activities

1. One of the most compelling ideas in this lesson is the idea of mass extinctions. Explore this phenomenon at the www.space.com Website. Type "Woodleigh Crater" in the search box and begin exploring there. Find out how asteroid impacts cause extinction. Examine how they may also stimulate explosions of evolutionary activity. Be sure to check out the hypotheses regarding impact craters and oil. This will turn your view of where oil comes from upside down.

2. When starting a home aquarium you must allow for a "running in" period. That is, you must place only one or two small inexpensive fish in the aquarium while the bacterial populations of the gravel become abundant. The bacteria in question are nitrifying bacteria that convert the toxic ammonium from the fish urine into less toxic nitrate. Study the nitrogen cycle in your textbook and explain exactly why
    a. the aquarium must be allowed to "run in" for several weeks before adding a large number of expensive fish,
    b. the gravel bed of an established healthy aquarium should never be thoroughly washed, and
    c. while denitrifying bacteria may be important in ocean sediments, they are probably not very important in your aquarium.

3. Enzymes are essential for regulating and accelerating the chemical reactions of life. They are polymers built of sub-units called amino acids and each enzyme is coiled and then folded in a very specific shape. The shape is necessary for it to function, and heat and pH changes can denature (unfold) and destroy enzymes. Pineapples contain enzymes that digest protein. Obtain a fresh pineapple and cut several small cubes from it. Place one cube on a small lunch-size cup of gelatin (a protein). Boil another cube for five minutes, cool it to room temperature and then place it on another cup of gelatin. Check both cups the next day and explain the results. Go to the Protein Data Bank Website (www.rcsb.org) and click on "Molecule of the Month." Several proteins are showcased here. The enzymes' names end in -ase, -zyme, or -sin. Choose 10 enzymes and write two sentences about each enzyme's function and importance.

Lesson 19

Water World

Overview

Lesson 18 explored life's unity, its requirement for energy and matter, and its ability to evolve. We turn now to the result of that evolution—life's diversity—and to specific aspects of how life deals with the physical, chemical, and biological conditions of the ocean world.

The marine environment is complex, and evolution has produced a bewildering variety of life forms. Systems of classification help in dealing with this complexity. Just as locating a movie in a video store is aided by a classification system, schemes of classification in biology create organization, foster unambiguous communication, and build understanding. Marine organisms are classified by their evolutionary and ecological relationships, and by their behavior. Ecologically, for example, organisms are classified as producers (autotrophs) or consumers (heterotrophs or decomposers). Behaviorally, marine organisms are classified as pelagic or benthic. Benthic organisms live on, in, or closely associated with the ocean bottom and include clams, worms, corals, seaweeds, and flounders. *Pelagic* organisms live in the water column and include drifters, or plankton, such as jellyfish or diatoms and the swimmers, or nekton, such as fish, dolphins, and squids.

The most familiar scheme of classification to anyone who has studied biology is taxonomic classification. This system categorizes organisms based on their evolutionary relationships. This system is often approached beginning with the six kingdoms—Archaea, Bacteria, Protista, Fungi, Plantae and Animalia—and working down through the smaller categories of phylum, class, order, family, genus, and species. In reality, this classification scheme is based on the only "real" category, the species, and then builds upward. A species is a group of similar organisms that are potentially interbreeding and reproductively isolated from other species. That is, the members of a species typically reproduce with each other and do not reproduce with different species. This is the only taxonomic category to which organisms actually pay attention because the species is the category that determines with whom they may reproduce. The higher categories of genus, family, phylum, and kingdom are all man-made. Unlike species, there is no biological definition of a genus, class, or phylum. A genus is a group of species that experts determine are closely related to each other. Thus, experts have decided that the tiger, lion, and jaguar are in the same genus, Panthera, while the Asiatic wildcat, the Chinese mountain cat, and the domestic house cat are in a different genus, Felis. A group of similar genera make a family, a group of similar families constitute a class, classes combine to make a phylum, and related phyla constitute a kingdom. Humans construct the higher taxonomic categories both to unravel evolutionary relationships among organisms and, quite simply, to create order and facilitate communication.

Naturalists have been classifying life for millennia, but Carolus Linnaeus, an eighteenth century Swedish biologist, was a zealous categorizer of life and might legitimately be called the father of taxonomy, although he did not believe in evolution. Linnaeus devised the system of assigning each species a scientific name such as *Panthera leo* for the lion. The first part of this two-part or binomial name is the genus name and the second part is the species name. Linnaeus also divided the living world into two supreme kingdoms—plants and animals. Contemporary opinion views life as being divided into six kingdoms. The kingdoms Archaea and Bacteria are the two most primitive groups. They are unicellular organisms and are prokaryotic. The other four kingdoms—Protista, Fungi, Plantae, and Animalia—are eukaryotic. All life is cellular, but prokaryotic and eukaryotic are the only two fundamentally different types of cells in the living world. Prokaryotic organisms have very simple cells that are most notably distinguished by their lack of a discrete nucleus. In contrast, eukaryotic cells enclose their chromosomes within a nucleus and are much more complex than prokaryotes in many other ways.

In addition to classifying marine life, scientists also classify the marine environment. Discussion of the classification of the marine environment began in Lesson 8. You may look back at Lesson 8 to review the classifications of the ocean environment by light—euphotic, disphotic, and aphotic—and by location—littoral, benthic, epipelagic. The taxonomic classification of the witch flounder, *Glyptocephalus cynoglossus*, is illustrated in your text on page 372. Employing additional modes and levels of classifying, it is also instructive to state that *Glyptocephalus cynoglossus*, is a benthic, eukaryotic member of the Kingdom Animalia and it is a heterotrophic carnivore. In other words it is a type of flounder that lives on the bottom and eats other animals. The giant kelp of coastal California is classified in the kingdom Protista. Furthermore, it is a benthic, sublittoral organism restricted to the photic zone.

It is widely held in biology that each species of living thing occupies a unique niche in the environment. The niche is a multidimensional description of where an organism lives. A niche is not just a place but includes all the factors that affect the organism's survival. If a particular factor is not present within an organism's range of tolerance it is called a *limiting factor*. Consider temperature and salinity for example. Every marine organism has an optimal range for temperature. It flourishes within its optimal range, is under stress at higher or lower temperatures, and perishes beyond the coldest and hottest tolerable temperature extremes. The same is true for salinity. An organism has an optimal range, suboptimal high and low ranges, and at some highest salinity and some lowest salinity point, it dies. Every species requires a proper mix of a wide variety of physical, chemical, and biological factors to survive and flourish.

As land creatures we are familiar with many of the limiting factors, but in a multitude of ways living in the ocean is strikingly different than living on land. The video and text discuss a dozen of these factors including such things as temperature, light, oxygen, pH, pressure, osmotic balance, and viscosity. Pay particular attention to each one and attempt to grasp the absolute importance of the factor to marine life, and the relative differences between how marine life and terrestrial life deal with it. For example, the availability of oxygen in air is about 30 times greater than that in water while the pressure of water can be 1,000 times greater than that of air. The thermal properties of water are quite different from those of land. Air temperature fluctuates more widely and changes more rapidly than does water temperature. *Endothermy*, or warm-bloodedness, is a characteristic of large organisms of the land—birds and mammals. Most sea creatures are *ectothermic*, or cold-blooded. They are not subjected to wide ranges of temperature and maintaining a constant body temperature seems to be relatively unimportant, or at least energetically unprofitable, to sea creatures. Warm-blooded animals of the sea such as whales, seals, and penguins came from land ancestors that were endothermic before they evolved their marine way of life. Yet a few large, fast fish such as the bluefin tuna have evolved the ability to maintain elevated body temperatures to meet their high metabolic demands and can be considered endothermic.

Marine life must deal with varying salinity. If the ambient water is *hypertonic*—that is, it has too much salt and not enough water—organisms will lose water by osmosis and their cells will shrivel. If the surrounding water is *hypotonic* because it has too little salt and therefore too much water, their cells will absorb water and swell or burst. Organisms must either have a mechanism to gain or lose salt and water, or they must confine themselves to water with acceptable salinity.

The interconnectedness of limiting factors is also important to an organism's survival. For example, salinity and osmotic balance are intimately related. Via photosynthesis, light affects oxygen levels while pH is closely related to carbon dioxide levels. The only creatures on land or sea that are warm blooded are large ones. Generally, small organisms and certainly tiny ones could not possibly be warm-blooded because of the relationship of their external surface area to their internal volume. Endothermy may be useful to birds and mammals, but it is impossible for insects and bacteria. With such small size, they would lose or gain heat at very high rates.

Both terrestrial and marine species are composed largely of slightly salty water, but marine species are also surrounded by it. Terrestrial organisms, of course, are surrounded by air. The striking differences between the marine and terrestrial environments has led to some very peculiar adaptations, and makes the ocean a very unusual and fascinating place to study.

Focus Your Learning

Learning Objectives

After completing this unit you should be able to do the following:

1. Discuss the behavioral and natural systems of classification of ocean life.
2. Compare and contrast the six kingdoms of life.
3. Discuss the physical limiting factors—light, temperature, salinity, dissolved nutrients and gasses, pH, and hydrostatic pressure—as they relate to marine life.
4. Define osmosis and discuss its importance to life in the sea.
5. Discuss the relationships between surface area and volume, gravity and buoyancy, and viscosity and movement. Explain the importance of each to marine life.
6. Describe how the marine environment is classified on the basis of light penetration and location.

Assignments

This lesson is based on information in the following text and video assignments. The key terms, focus points, and practice test are intended to help ensure mastery of the material presented in this unit of study.

Text: Review pages 356–364 and 369–372 in Chapter 13, "Life in the Ocean," and read pages 421–423 in Chapter 15, "Marine Animals"

Video: Episode 19, "Water World"

Video Focus Points

Review these points before watching the video assignment for this lesson to help you focus on key issues.

— As an environment for life, the ocean is quite different from the terrestrial world.

— The ocean is more three dimensional, has the distinct thermal properties of water, and is a liquid world.

— Water is more viscous, more dense, less well-oxygenated, and can have exceedingly greater pressure than air. Light in the water environment is quantitatively less and qualitatively different, and it is a salty place.

— As with all life on Earth, marine species have been challenged by the conditions of their environment and have adapted by natural selection to flourish in their particular niches.

— From the totally lightless, constantly near-freezing, oppressively high-pressured depths of the abyss to a shallow tropical tide pool flooded with sunlight and hotter than an average bath tub, marine life seems alien to us. But evolution produces logical, functional, and some say beautiful, solutions to the problems that confront it. Marine life and terrestrial life share in this fact but the different problems have yielded different solutions.

— Scientists find it useful to unravel the complexity of the marine environment and the organisms that inhabit it by searching for patterns and generalizations. These patterns create a framework of classification and foster greater understanding.

Text Focus Points

Review these points before reading the text assignment for this lesson to help you focus on key issues.

— A variety of physical, chemical, and biological factors affect life in the sea. When a particular factor is excessively present or absent, it is called a limiting factor. Examples of factors that can affect life and might be limiting factors are light, temperature, pH, salinity, pressure, buoyancy, and food availability.

— Ocean zones inhabited by marine life are classified according to physical factors. Location, depth, and availability of light define these zones.

— The interplay of physical, chemical, and biological factors affecting a particular type of organism within a particular type of environment determines the success of each species and regulates their abundance and distribution.

— There are several useful ways to classify life. The natural system of classification groups organisms into kingdoms, phyla, classes, species, and so forth, based or their evolutionary relationships while the behavioral system of classification groups organism as plankton, nekton, and benthos.

Key Terms and Concepts

A thorough understanding of these terms and concepts will help you to master this lesson.

Atmosphere (pressure). One atmosphere of pressure is equal to the air pressure at sea level. This is equal to 14.7 pounds per square inch and is caused by the weight of all of the thousands of feet of air above sea level. Water is much denser than air and hydrostatic (water) pressure increases by one atmosphere with only 33 feet of depth. Pressure exceeds 1,000 atmospheres in the deepest trenches of the oceans.

Counter-current exchange. The exchange of chemicals or heat between two fluids flowing in opposite directions. This greatly improves the rate of exchange. It can be seen in fish gills where oxygen exchange between seawater and blood is improved or in certain blood vessels in large tunas where warm blood flowing outward from the muscles retains body heat by exchanging heat into cool blood flowing inward from the skin.

Dissolved oxygen. Often called "D.O.," this the concentration of oxygen gas dissolved in seawater. It is measured in ml/L which is equivalent to parts per million. Where the D.O. is very low, less than 0.5 ppm, an oxygen minimum zone exists.

Ectotherms. Organisms that freely lose the heat they generate by metabolism to the surrounding environment. Because their body temperatures vary with and match the environment's temperature they are also called poikilotherms. Such organisms are often called cold-blooded.

Endotherms. Organisms that retain metabolic heat and consequently are usually warmer than the surrounding environment. Many can maintain a constant body temperature, and so are also called homeotherms. Often called warm-blooded.

Hemoglobin. A complex protein molecule found in the blood of many organisms. It greatly improves the ability of blood to pick up, transport, and release oxygen. It is red in color.

Hypertonic solution. A solution with a greater concentration of solutes than another solution to which it is being compared. Seawater is hypertonic to fresh water. Water moves into a hypertonic solution from a hypotonic one by osmosis.

Hypotonic solution. A solution with a lesser concentration of solutes than another solution to which it is being compared. Fresh water is hypotonic to seawater. Water moves from a hypotonic solution into a hypertonic one by osmosis.

Limiting factor. Any factor, such as light, temperature, and salinity, that limits the survival of an organism.

Natural system of classification. The system of bioclassification based on evolutionary interrelationships among organisms. Because they are both mammals, a whale and a bat are evolutionarily more closely related to each other than a whale is to a fish or a bat is to a bird.

Pelagic. Of the open ocean.

Salt-balance. The maintenance of a proper relationship between the salt content and water content of organisms' body fluids. Water will move in or out of organisms by osmosis. Some organisms can adjust salt and water using kidneys or other organs. Others can't adjust and are restricted to water whose salinity matches that of their cells.

Viscosity. The resistance of a fluid (gas or liquid) to flow. It is the "syrupiness" of the fluid. Pancake syrup is more viscous than water. Cold water is more viscous than hot.

Test Your Learning

After working through these questions, check your answers against the key at the end of this book. If you answered any of the questions incorrectly, review the relevant sections of the text and video episode.

Multiple Choice

1. The natural system of classifying organisms into kingdoms, phyla, species, etc. is based on:
   a. their evolutionary relationships
   b. their ecological relationships
   c. their anatomical structure
   d. their appearance
2. Two kingdoms that contain organisms that are prokaryotes are the Archaea and the:
   a. Bacteria
   b. Fungi
   c. Protista
   d. Plantae
3. Which of these environmental factors varies least with changing depth?
   a. light
   b. temperature
   c. pressure
   d. salinity
4. The two major organs in fish used to regulate salt-balance are the:
   a. kidneys and liver
   b. intestine and skin
   c. kidneys and gills
   d. gills and skin
5. Unlike humans, many shallow water fishes have eyes adapted to detect:
   a. light between red and green on the spectrum
   b. violet, indigo, blue and ultraviolet
   c. orange, yellow, red, and infrared
   d. the same spectrum as humans plus ultraviolet
6. Which of these is least likely to be a limiting factor in the ocean?
   a. nitrate
   b. phosphate
   c. carbon dioxide
   d. iron
7. Which of these organisms are endothermic?
   a. only mammals are
   b. only birds are
   c. mammals, birds, and a few fish are
   d. mammals and birds are, but fish are not
8. The cells of a fresh water algae placed in seawater would probably:
   a. swell
   b. shrivel
   c. burst
   d. remain unaltered
9. The scientific name of the green sea turtle is *Chelonia midas. Chelonia* denotes its:
   a. order
   b. family
   c. genus
   d. species
10. Which kingdom includes the organisms known as extremophiles?
    a. Archaea
    b. Bacteria
    c. Protista
    d. Protozoa
11. A fluid's resistance to flow is called:
    a. drag
    b. turbulence
    c. density
    d. viscosity
12. The littoral zone of the marine environment is:
    a. the zone between high and low tide
    b. the zone associated with the continental shelf
    c. the upper layers of the open ocean
    d. between the abyssal and the hadal

Short Answer

1. Tuna fish, dolphins, penguins, and torpedoes have teardrop shaped bodies. Discuss this and other adaptations exhibited by marine organisms that swim rapidly, efficiently, and for sustained periods.
2. List the six kingdoms of life, give examples of each, and briefly compare and contrast them.
3. Explain the limiting factor concept and discuss temperature as an example of a limiting factor.

Supplemental Activities

1. The ratio of surface area to volume (SA:V) of a cell or organism is a critical geometric relationship in biology. Observe the effects of this relationship by performing the following experiment:

   Take 3 glasses of water and sprinkle a pinch of flour in one, a pinch of sugar in another, and drop a sugar cube into the third. Observe their relative sinking rates and calculate the SA:V ratio of each. Assume the sugar and flour grains are spheres. The radius of the flour grains is 0.01 mm and the radius of the sugar grains is 0.3 mm. A side (s) of the cube is 12.0 mm. (SA of sphere $= 3\pi r^3$ and $V = 4/3\pi r^3$. SA of the cube is $3s^3$ and the $V = s^3$. Based on your observations, explain why it is helpful to phytoplankton, which require light and must absorb nutrients, to be small.

2. Visit the Websites below that provide on-line demonstrations and experiments for diffusion and osmosis and observe the animations at each site, then conduct the demonstrations of osmosis that follow.

   http://physioweb.med.uvm.edu/bodyfluids/osmosis.htm

   http://www.vivo.colostate.edu/hbooks/cmb/cells/pmemb/hydrosim.html

   Formulate a hypothesis as to what you predict will happen in each experiment and then allow each to run for several hours or overnight and describe and explain the results.

   a. Take two fresh celery stalks and place one in tap water and one in very salty water made by dissolving six tablespoons of salt in a quart of water.
   b. Take two unpeeled potatoes and cut them in half. Boil one half-potato to destroy its cell membranes and allow it to cool. Cut a small well (1x1x1 inch) in the rounded side of each half and place all four flat-side down in a pan of tap water. Place a teaspoonful of salt in one well, a teaspoonful of sugar in another, leave one empty and place another teaspoonful of salt in the boiled half.
   c. Take two hardboiled eggs and place them on a piece of paper and draw marks indicating their length and width. Decalcify them by placing in a cup of vinegar overnight. The shells should dissolve exposing the membrane below. Place one in tap water and the other in very salty water and wait several hours. Measure them again.

3. Choose ten marine organisms of interest to you with at least one from each kingdom except the fungi. Go to the Internet and attempt to classify each to kingdom, phylum, class, order, family genus, and species. Classify each one also by its ecology using all of the following terms that apply: autotroph, heterotroph, producer, consumer, and decomposer. Finally, designate each of the ten as plankton, nekton, or benthos.

Lesson 20

Food for Thought

Overview

Life requires energy, and glucose is the major fuel that supplies it. Glucose is the sugar manufactured by photosynthesis. It is also the sugar that is measured when blood sugar is tested. Glucose can be converted to lipids, starches, and a variety of other molecules used to supply or store energy. The equation below summarizes photosynthesis. It simply says that six molecules of carbon dioxide and six molecules of water are converted by light energy into one complex, energy-rich glucose molecule and six oxygen molecules. Memorize this equation. You will need to know it before you proceed with this lesson.

$$6CO_2 + 6H_2O \rightarrow \text{☼} \rightarrow C_6H_{12}O_6 + 6O_2$$

Through the process of photosynthesis, green plants produce most of the food and oxygen on Earth. In the oceans the major photosynthesizers are the phytoplankton. These microscopic, unicellular organisms produce 90 to 96 percent of the ocean's food. Seaweeds, sea grasses, and chemosynthetic bacteria also contribute to marine food production.

The biological production of food is called primary production and is accomplished by organisms called *autotrophs,* or producers. Phytoplankton and other producers manufacture food and then use it to fuel their own biological processes. Consumers, or heterotrophs such as animals, are unable to manufacture food so they must consume other organisms to get their fuel. Both autotrophs and heterotrophs release energy from glucose through cellular respiration. This process is essentially the opposite of photosynthesis—glucose and oxygen react to produce carbon dioxide and water. The critical difference is that photosynthesis takes in energy while respiration releases it. The energy taken in by photosynthesis is solar energy; the energy released by respiration is chemical energy suitable to power the processes of life. Both light and gasoline contain energy but only gasoline will power your car. Similarly, both light and glucose contain energy, but only glucose will power your cells.

The three main questions asked when studying primary productivity are:

- Who is doing it?
- How fast are they doing it?
- How much of them are doing it?

Who is doing it? The question of who is carrying out primary production has many answers. Overwhelmingly, it is the phytoplankton but this includes a wide diversity of organisms. Among these, the diatoms and dinoflagellates are larger and have long been known to be major ocean primary producers. Locally, the smaller coccolithophores and silicoflagellates can be abundant producers. Most diatoms and dinoflagellates are net-plankton. They are large enough to be strained from the ocean with a plankton net. Smaller species must be collected by allowing them to settle from a water sample, centrifuging the water, or filtering seawater through fine-pored membrane filters. These tiny species are called nanoplankton, picoplankton, and even ultraplankton. Each group includes individuals that are smaller than those in the group before it. These previously overlooked cells include certain cyanobacteria and other photosynthetic prokaryotes. Their discovery has revolutionized science's understanding of marine primary production.

Though tiny, their numbers are staggering and they may be responsible for as much as 70 percent of the primary production in some ocean regions.

More conspicuous, but much less important overall are the larger primary producers. These include the red, green, and brown seaweeds, all of which are simple algae classified in the kingdom Protista. They can be important in some regions and may dominate an ecological community, as is the case with the giant kelp communities of California. Most species are attached to the bottom and therefore restricted to shallow water where light penetrates to the seafloor.

The word plant is sometimes used loosely to describe any photosynthetic organism, but true members of the kingdom Plantae are rare in the ocean. The angiosperms of the kingdom Plantae are the familiar plants, trees, grasses, etc. They have flowers, seeds, roots, and leaves and are evolutionarily advanced organisms almost entirely restricted to the land. They generally cannot tolerate salt and because they are rooted in the substrate, they too are restricted to the shallow fringes of the ocean. The submerged sea grasses can be particularly important in certain shallow tropical and temperate regions. The mangrove trees of the tropics and the temperate salt marshes are emergent vegetation. They are never totally submerged. Sea grasses, salt marsh plants, and mangrove trees can provide food, shelter, and a substrate for attachment for a variety of organisms. They can also stabilize and protect shorelines.

How fast? Biomass and productivity rate can be related, but they are different things. The rate of production measures how rapidly glucose is being manufactured. An acre of one inch grass growing rapidly but being grazed down by cattle could have a higher rate of production than an acre of foot tall grass growing slowly in a field with little water and fertilizer.

Because photosynthesis uses carbon dioxide and water and produces sugar and oxygen, its rate can be measured by the rate of disappearance of CO_2 or H_2O or the rate of appearance of $C_6H_{12}O_6$ or O_2. The most widely-used method utilizes radioactive carbon-14 to measure how fast a form of radioactive CO_2 is converted to radioactive $C_6H_{12}O_6$. Production of oxygen by phytoplankton in a transparent bottle compared to its consumption in an opaque bottle can also be used to measure rates of photosynthesis and respiration. Both occur in the light bottle while only respiration occurs in the dark one. Adding the rate of loss of oxygen in the dark bottle to the rate of production of oxygen in the light bottle allows the calculation of total production for the enclosed phytoplankton sample.

The *light/dark bottle experiment* demonstrates the two key measurements of primary productivity—gross primary production and net primary production. Gross and net in photosynthesis have much the same meaning as they do in your paycheck. Gross is a larger value and net is a smaller one. Gross productivity is total productivity of the primary producers. But producers make food and then use it. The amount they use for their own respiration must be subtracted from the gross production to calculate the net production, which is what is left after the producers meet their own energy requirements. It is net primary production that feeds the entire food web above the primary producers.

As light penetrates the ocean it diminishes. The depth at which the light intensity permits phytoplankton to produce only enough sugar to meet their own needs is called the *compensation depth*. This is the break-even depth. Somewhat below the compensation depth they may still carry out photosynthesis but they will use sugars faster than they make them. At even greater depths the light is so diminished that photosynthesis is impossible. A phytoplankter's starvation is certain if it is not washed up into more sunlit water at or above the compensation depth. The compensation depth marks the bottom of the euphotic zone, which amounts to less than two percent of the ocean. Most of the remaining 98 percent of the ocean depends on a constant raining down of tiny organic food particles from this thin upper veneer of sunlit water.

Two major factors—light and nutrients—limit primary production in the ocean. Primary producers must have both to flourish. Light is only found near the surface while nutrients tend to be concentrated in the depths. Primary production is generally greatest where nutrients are brought into the photic zone. Upwelling regions and places where rivers or melting sea ice bring nutrients into the euphotic zone are highly productive. Coral reefs are an exception. They are highly productive because of the tight recycling of nutrients by their symbiotic zooxanthellae as discussed in Lesson 9.

Some general patterns of primary production distinguish tropical, temperate, and polar regions. Because of the intense sunlight, the tropics generally have a well established and permanent thermocline that traps nutrients below. Despite the plentiful sunlight, the absence of nutrients prevents high rates of productivity over much of the open tropics. Exceptions are discussed in the text. The polar regions are too cold for a permanent thermocline, so nutrient-rich water is not impeded from mixing upward. It is the low light of the polar regions that prevents productivity except in the summer season. High levels of upwelled nutrients coupled with summer runoff from ice edges results in very high productivity during the polar summer when the period of daylight can approach 24 hours.

The temperate zone is somewhat like the polar ocean in the winter and somewhat like the tropical ocean in the summer. A seasonal thermocline forms in the spring, strengthens throughout the summer, then weakens and finally disappears in the fall. The water is well mixed and nutrients are brought up in the winter. As sunlight increases in the spring and couples with the nutrients there is a bloom of phytoplankton. As the sun-drenched summer approaches, the thermocline also forms and nutrients tend to get trapped below it causing productivity to decline in the summer. In the fall the thermocline breaks down and wind and storms mix the nutrients upward again. As long as this fall upwelling occurs before the sunlight declines too much, there will be a fall bloom as well. While the spring bloom is triggered by increasing sunlight and terminated by decreasing nutrients, the fall bloom is caused by upwelling nutrients and ends with declining sunlight.

How much? Determining how much requires measuring biomass. This is the total mass of producers in a given region. An acre of grass one foot tall has a greater biomass than an acre of one-inch tall grass. There are various ways to collect phytoplankton, but once collected it can be dried and weighed to measure its biomass. More often biomass is estimated by measuring the amount of chlorophyll in a water sample. In addition, significant progress has been made in using satellites to measure chlorophyll in the ocean. Such large-scale remote sensing enables scientists to measure the phytoplankton biomass of entire oceans much more quickly, frequently, and ultimately cheaply, than would be possible with even an entire fleet of ocean research vessels.

Focus Your Learning

Learning Objectives

After completing this unit you should be able to do the following:

1. Define the term primary production, explain how it is measured, and describe its importance.
2. Compare and contrast diatoms and dinoflagellates and discuss the role of dinoflagellates in harmful algal blooms.
3. Discuss the importance of the picoplankton.
4. Compare how light and nutrients interact to produce the general seasonal patterns of primary production in tropical, temperate, and polar oceans. Discuss notable exceptions to these general patterns.
5. Compare and contrast foraminifera, copepods, and krill.
6. Outline the major groups of macroalgae and angiosperms and discuss their role in marine primary production.

Assignments

This lesson is based on information in the following text and video assignments. The key terms, focus points, and practice test are intended to help ensure mastery of the material presented in this unit of study.

Text: Review pages 346–356 in Chapter 13, "Life in the Ocean," and read Chapter 14, "Plankton, Algae, and Plants" pages 376–401

Video: Episode 20, "Food for Thought"

Video Focus Points

Review these points before watching the video assignment for this lesson to help you focus on key issues.

— Primary production is the production of organic food molecules by autotrophs and is accomplished by photosynthesis or chemosynthesis.

— The most important primary producers of the ocean are the phytoplankton including the diatoms and dinoflagellates. Seaweeds such as kelps and other attached algae are important producers in shallow coastal water.

— Advanced techniques for collecting phytoplankton have revealed an enormous quantity of previously unknown, very tiny prokaryotic phytoplankton. They may be the oceans' most important primary producers.

— Food manufactured by primary producers is passed to consumers and decomposers through the ocean's food webs. Most higher level consumers, including commercially important fisheries species, depend ultimately on the microscopic phytoplankton.

— Light and nutrients are the two most important factors regulating photosynthesis. Light availability changes on daily and seasonal cycles. Cloud-cover, water clarity, and surface conditions such as ripples also affect the photic zone.

— Nutrients such as N and P are also required for primary production. Their distribution in the photic zone is also variable, being abundant in many coastal, ice-edge, and upwelling zones and much less abundant in open waters above a stable thermocline.

— Excessive nutrient runoff from the land into coastal waters is called eutrophication.

Text Focus Points

Review these points before reading the text assignment for this lesson to help you focus on key issues.

— The photosynthetic light energy captured and stored in glucose molecules is later released in the process of cellular respiration. In an endless cycle, photosynthesis captures energy from the sun and respiration releases it in a biologically useful form.

— Organisms that produce the sugar that fuels biology are called primary producers. They employ photosynthesis, and less commonly, chemosynthesis.

— Primary producers, or autotrophs, both make and use food. Consumers, or heterotrophs, can not make food so they must consume other organisms.

— Phytoplankton are the most important marine primary producers. Diatoms and dinoflagellates are among the most important phytoplankton, but the surprising discovery of extremely tiny phytoplankton is revolutionizing our understanding of primary production in the oceans.

— The rate of photosynthesis is determined by measuring the production of oxygen or the uptake of radioactive carbon dioxide. The carbon method is more common.

— Sunlight and nutrients are the major factors that control primary production. Nutrients are abundant where they well up from the deep or where they runoff from land. Sunlight is affected by seasons, depth, and water clarity. The interaction of these factors produces generally distinct patterns of production in polar, temperate, and tropical seas.

— Zooplankton are the major consumers of the ocean's phytoplankton. Foraminifera, copepods, and krill are among the most important.

— Larger primary producers include the red, green, and brown seaweeds and the sea grasses and mangrove trees.

Key Terms and Concepts

A thorough understanding of these terms and concepts will help you to master this lesson.

Algae. A general term applied to any photosynthetic member of the kingdom Protista. Seaweeds, diatoms, and dinoflagellates are examples.

Angiosperm. Complex vascular plants with roots, leaves, flowers, and seeds. These common land plants are rare in the ocean. Sea grasses, mangrove trees, and salt marsh plants are examples.

Autotroph. An organism capable of manufacturing it own food. Most autotrophs accomplish this by photosynthesis, although some bacteria and Archaea use chemosynthesis.

Compensation depth. The depth where food produced by photosynthesis of autotrophs is equal to the food they utilize in their own respiration. This is their break-even depth.

Diatom. One of the most abundant groups of phytoplankton. Diatoms are unicellular, have highly efficient photosynthesis, and possess a glassy silica covering called a frustule.

Eutrophication. The addition of plant nutrients to aquatic environments. Excessive eutrophication can cause algae blooms that block light penetration and whose respiration and decomposition can create anoxic conditions.

Holoplankton. Organisms such as krill that live their entire life as plankton. Meroplankton, such as larval fish and larval oysters, are planktonic as larvae and later grow into nekton or benthos.

Light bottle/dark bottle experiment. A technique to measure rates of photosynthesis. Pairs of transparent and opaque bottles containing seawater and phytoplankton are suspended at various depths. Changes in oxygen production or use or carbon fixation reflect rates of net photosynthesis in the light bottle and respiration in the dark one.

Picoplankton. Among the smallest of plankton. From largest to smallest, plankton are sometimes grouped as macroplankton, net plankton, nanoplankton, picoplankton, and ultraplankton. The magnitude of importance of picoplankton in ocean ecology is a recent discovery.

Plankton bloom. A sudden increase in the abundance of plankton. When certain species of toxic dinoflagellates and cyanobacteria bloom they may kill fish, marine mammals, humans, and other animals. The toxic red tides are an example.

Zooplankton. Animal plankton. These include the major grazers of the ocean's phytoplankton such as foraminifera, a type of amoeba, and copepods and krill, which are Crustacea.

Test Your Learning

After working through these questions, check your answers against the key at the end of this book. If you answered any of the questions incorrectly, review the relevant sections of the text and video episode.

Multiple Choice

1. Toxic blooms, bioluminescence, locomotion, and primary production are associated with:
   a. dinoflagellates
   b. diatoms
   c. foraminifera
   d. radiolaria

2. The isotope most commonly used to measure rates of photosynthesis is:
   a. C-12
   b. C-14
   c. O-16
   d. O-18

3. Primary production in the _______ zone is often characterized by two blooms—one in the spring and another in the fall.
   a. tropical
   b. temperate
   c. north polar
   d. south polar

4. Excessive nutrient runoff from the land into coastal waters is called:
   a. carbonation
   b. nitrification
   c. nutrification
   d. eutrophication

5. The drifting larva of clam would be considered:
   a. meroplankton
   b. holoplankton
   c. macroplankton
   d. nanoplankton

6. The largest (in size) of the marine primary producers are in the _______ group.
   a. macroplankton
   b. gigaplankton
   c. brown algae
   d. green algae

7. Which plankton group has high photosynthetic efficiency, a silica frustule, and auxospores?
   a. diatoms
   b. foraminifera
   c. dinoflagellates
   d. coccolithophores

8. Which of these is not produced by photosynthesis?
   a. light energy
   b. chemical energy
   c. sugar
   d. oxygen

9. Which of these is not an autotroph?
   a. kelp
   b. mangrove
   c. coccolithophore
   d. krill

10. A light/dark bottle experiment measures:
   a. carbon dioxide
   b. standing crop
   c. productivity rates
   d. biomass changes

11. The main factor regulating phytoplankton productivity rates in tropical oceans is:
   a. light
   b. temperature
   c. nutrient availability
   d. zooplankton grazing

12. Which of these is false regarding the compensation depth?
   a. primary producer's photosynthesis and respiration rates are equal
   b. it corresponds to the bottom of the euphotic zone
   c. light has become too dim for photosynthesis to occur
   d. it is always deeper than the zone of maximum photosynthesis

Short Answer

1. A student driving to campus displays a bumper sticker that reads: "Everyone loves dolphins, but who cares about diatoms?" Later, she eats a tuna fish sandwich for lunch. Explain clearly what primary production is and relate it to her bumper sticker and her lunch.

2. A significant change in our understanding of marine primary production has emerged with the discovery of the role of picoplankton. Discuss the role of these organisms in ocean ecology.

3. Describe and explain the general pattern of primary production in temperate oceans.

Supplemental Activities

1. Study global ocean productivity at this Website: http://marine.rutgers.edu/opp/Production/Production1.html

   Use a world map to find the regions that follow. Record the approximate production rate for the spring season (April–June) vs. the winter season (January–March). Prepare a bar graph illustrating the data. The y-axis is production rate and the x-axis is the regions. Each bar should consist of a pair of adjacent bars, side by side, one for spring and the other for winter. Plot a double bar showing the seasonal difference for all the regions and explain these differences for at least five regions. Terms such as coastal runoff, upwelling, light, nutrients, thermocline, boreal spring, austral winter, island upwelling, and others should be included in your explanations.
   a. north Atlantic between Canada and Scandinavia
   b. southern ocean from Argentina to New Zealand
   c. Seas of Japan/Okhotsk
   d. coastal Namibia
   e. Persian Gulf/Arabian Sea
   f. Gulf of Alaska
   g. Gulf of Carpenteria
   h. coastal Peru
   i. coastal Argentina
   j. Mediterranean Sea

2. Collect plankton from a lake, bay, or the ocean. Instructions for constructing a plankton net and viewing your catch are found at http://

er.jsc.nasa.gov/SEH/Ocean_Planet/activities/ts3meac3.pdf. Be careful around water; wear a life vest.

3. Grow your own plankton. Zooplankton: Inexpensive kits with instructions for culturing the brine shrimp, *Artemia salina*, (sea monkeys) can be obtained at a toy store or ordered online. Test how different factors such as temperature, salinity, pH (a few drops of white vinegar lower the pH and a few drops of dilute bleach will raise it) affect their hatching success. Phytoplankton: Live phytoplankton can be cultured with kits from an aquarium store or purchased online. Product information: www.dtplankton.com. Test the effects of low and high light levels, and how adding a drop of diluted houseplant fertilizer or a drop of Geritol affects growth.

Lesson 21

Survivors

Overview

In the first lesson of this series, the theory that life on Earth originated in the ocean was explored. According to that theory, the first organisms were probably tiny creatures that formed spontaneously in the ocean water and got their energy from organic molecules. As these organisms increased in number, they may have been in danger of using up their food supply had it not been for the evolution of the process of photosynthesis, which enabled some of them to use energy directly from the sun to make their own food. Not only did they incorporate the sun's energy into their food, but in doing so they released free oxygen into the environment. Up to that time, oxygen was bound up with other molecules and not available for use by organisms. With the so-called oxygen revolution, which occurred between two billion and 400 million years ago, the percentage of oxygen in the atmosphere gradually changed from about one percent to its present level of over 20 percent. This increase in oxygen availability made possible aerobic respiration, in which food molecules are broken down and energy released rapidly and efficiently.

The energy-storing autotrophs provided food for the energy-consuming heterotrophs, which probably began as single-celled animal-like creatures that obtained energy from eating the autotrophs or from eating each other. These simple animal-like organisms began to develop, evolve, and increase in complexity, eventually forming into the true multicellular heterotrophs we now call *animals*.

It is presumed that toward the end of the oxygen revolution there was a very rapid increase in the numbers and kinds of these animals, but the fossil record is rather sparse for that time because the early animals did not possess skeletons or other hard parts that would have left a record in the rocks. Fortunately, such areas as the Burgess Shale in British Columbia, and the Ediacara Hills in Australia—by virtue of their location in the early seas—possessed the right combination of sediment types and water conditions to preserve the remains of some of those very fragile animals. As science attempts to piece together the lineage of modern-day animals, some shapes and body architectures are encountered that have long been extinct, leaving no obvious connection to the animals of today. In trying to assemble an evolutionary history, scientists look for similarities of body architecture and body complexity. The present-day classification system includes both living and fossil animals. It groups those with similar body forms and complexities into phyla, then proceeds to subdivide them into more specific categories.

More than 90 percent of all living and fossil animals are *invertebrates*, or organisms that lack a supporting internal skeleton. Many of the invertebrates are small in size and might seem inconspicuous in the marine environment, but all are of interest to the specialists who study them and attempt to make biological and evolutionary connections.

Invertebrates are not a scientific category, but it is a convenient grouping in initial classification schemes and includes about 33 animal phyla of great diversity. Invertebrates possess some of the greatest adaptations for surviving in challenging environments. Nowhere is their adaptability and survivability more evident than in the *intertidal zone*. Invertebrates that live in this environment have evolved a multitude of ways to survive the extreme conditions.

In this lesson, nine of the better-known and more conspicuous phyla will be considered to represent the diversity and increasing complexity of the marine invertebrates. These nine phyla are presented in order of increasing complexity to address their evolutionary history.

Many biologists agree that the most primitive of the true animals are the sponges of phylum *Porifera*. Most sponges are marine dwellers, and live attached to the sea floor or to other surfaces there. Their bodies are composed of a few functional cell types, and they appear in several basic shapes. All sponges feed on plankton and nutrient particles suspended in the water, pulling them in using specialized cells to create filtering currents. There are no actual organs for complex body functions such as respiration and circulation; they simply gather food, process it to release energy, and pass out the waste. Special cells create their characteristic supportive structure, or skeleton, as elaborate needles, or spicules made of silica of calcite, or as protein fibers.

Unlike sponges, which have no body tissues in which specialized cells perform specific functions, members of phylum *Cnidaria*, which includes sea anemones, *corals*, and jellyfish, have two distinct tissue layers—the outer skin or epidermis, and the inner digestive layer or gastrodermis. Most cnidarians are marine, and all are functional carnivores, characterized by various types of stinging cells called *cnidoblasts*, for capturing prey and for defense. The size and type of prey they capture depends on the size of the cnidarian, but all feed in essentially the same way; they encounter and sting or stun their prey, capture it, bring it to their mouth, digest it, and eliminate the waste. All cnidarians possess the same basic body form, called *radial symmetry*, and all have some kind of nerve network (not present in sponges), but with no area of sensory concentration like a brain. There are two fundamental body plans—the attached *polyp* such as anemones and corals, and the free-swimming *medusa*, such as jellyfish. The anemone-type polyps attach to the sea-floor substrates and are generally soft-bodied, while the coral-type polyps create cup-like calcite skeletons, forming either solitary corals or huge reefs. The medusoid cnidarians—jellyfish and their relatives—are also simple in structure, with no real organs or central nervous system. These are mostly free-swimming, able to move in the water and to capture their prey with long tentacles that are armed with stinging cells.

When animals began to become mobile, one of the most efficient body forms was probably the long and slender worm. There are three worm phyla to consider, all with an advanced *bilateral symmetry*, some kind of head-oriented nervous system, and structures for circulation, digestion, and elimination of waste. The most primitive of the worms are the flatworms of phylum *Platyhelminthes*, although they are not all flat. Most of the marine forms are small and free-living; a few are parasitic. They have one of the most primitive central nervous systems, including a rudimentary brain and some light sensors.

The next major structural advance is seen in phylum *Nematoda*, the so-called roundworms or threadworms. This phylum is represented in both marine and terrestrial environments, in both free-living and parasitic forms. They inhabit both environments in astounding numbers, but are probably best known to us as parasites of fish and other marine organisms.

By far the most advanced of the worms are in phylum *Annelida*; they are most familiar to us as garden earthworms, but are also very important in the marine environment. Their body structure is segmented, which is an apparently good evolutionary strategy for increasing body size by simply adding more identical units or segments. The body systems are becoming more specialized for specific tasks. Some species crawl over the seabed or burrow into it and others live in tubes attached to rocks or in the sediment. They employ a wide variety of feeding strategies. The class *Polychaeta* is the most important marine worm, occupying a variety of habitats.

Members of phylum *Mollusca*, such as snails, clams, and squid, are probably the most numerous group in the marine environment. They exhibit some remnants of segmentation, suggesting a common ancestry with the annelid worms. Like the annelids, they have well-developed body systems and occupy many different marine habitats. There are a number of classes in this phylum, three of which are most familiar to us. The *Gastropods*—snails and snail-like species—are characterized by having a conspicuous foot and some kind of shell. They use the foot to crawl, burrow, or attach to hard surfaces. This is probably the best-known and largest mollusk class, noted for their abundance, beauty, variety of shells, and the role of some species as food for humans.

The *Bivalvia*—clams, oysters, mussels—have two shells (valves) that open and close for feeding, and may be either free-living or attached to hard surfaces. The octopus, squid, and nautilus belong to the class *Cephalopoda*, so named because of the unique connection between the head and the foot. They all have tentacles of some kind, and either a conspicuous shell like the nautilus, a thin remnant of a shell like squid, or no shell at all like the octopus. All are free-living, moving between the sea floor and the surrounding water, and feed by capturing prey with their tentacles.

Phylum *Arthropoda* is the largest group of animals, in numbers of both species and individuals. They occupy a large variety of habitats—terrestrial, marine, and atmospheric. Like the annelids and mollusks, the basic arthropod body plan exhibits segmentation, and the nerve and muscle system is more advanced than the annelids, but not as specialized as in some of the mollusks. Notable arthropod features are the strong, flexible *exoskeleton*, the highly-specialized muscle groups that allow for rapid movement and flight, and the characteristic for which the group is named—jointed, articulated appendages (arthron = joint, pod = foot). The exoskeleton is made of a very tough, light material called *chitin*. In order to grow they must molt, or shed the exoskeleton briefly, then replace it to accommodate the new body. The largest class (Insecta) is poorly represented in the marine environment, while the next largest class (*Crustacea*) has over 30,000 primarily marine species such as shrimps, crabs, and lobsters. The variety of habitats occupied and the different feeding habits represented are exemplified by the diversity of sizes, body plans, and behaviors. It could be said that the crustaceans dominate the world of marine animals.

Phylum Echinodermata, meaning spiny-skinned, is exclusively marine and includes the familiar sea stars, sea urchins, brittle stars, and sea cucumbers. The body plan of the adults is radial symmetry, like the Cnidaria, and is in five parts (five arms, five body sections, etc.), with some variations. They lack eyes or any kind of brain—a seeming throwback from the well-equipped mollusks and crustaceans. Most echinoderms have some kind of locomotion capability, with a variety of suction-cup tube feet, operated by an internal water vascular pumping system. The sea stars are mainly predators, the sea urchins can be both scavengers and grazers, the brittle stars are mainly scavengers, and sea cucumbers feed by consuming bottom sediments and digesting out the nutrients. As a group, the echinoderms occupy many habitats and depths, but all are essentially bottom dwellers.

The most advanced in many ways is phylum Chordata—the true *vertebrates*—most familiar to us for the birds, mammals, and fishes, which will be covered in the next lesson. There are, however, two groups of little-known chordates that exhibit the basic chordate body plan at some time in their development, but as adults are so different that they are called invertebrate chordates. All of the chordates have several structures in common as juveniles, but some will lose them or modify them as adults. The common chordate structures are a notochord, a dorsal tubular nerve cord, and gill slits. The *notochord* is a stiffening internal structure that serves to give the body support and form. All chordates have a notochord as juveniles, but most lose it as adults when it forms into the skeletal vertebrae (hence vertebrates). The two groups of invertebrate chordates also have a larval notochord.

The *tunicates* are an unusual group of organisms that are unique in having a strong, flexible outer covering, or tunic, covering the living body. They may be planktonic or may be attached to the rocks of the sea floor, and as adults they have lost all but two of the chordate hallmarks. Tunicates are basically filter feeders, bringing in microscopic plankton and bits of organic material.

Another invertebrate chordate is the small, elongate, fish-like *Amphioxus*, which lives in shallow sand bottoms, can move a little, and feeds by gathering plankton. Amphioxus retains all of the chordate characteristics, including a well-developed dorsal tubular nerve cord, which is thought to have evolved into the spinal cord of the higher vertebrates.

Focus Your Learning

Learning Objectives

After completing this unit you should be able to do the following:

1. Describe Earth's earliest life forms, and how they got their energy.
2. Understand the evolution of photosynthesis, and how it led to the oxygen revolution and the subsequent proliferation of animal forms of life.
3. Define the terms autotroph and heterotroph as they relate to the oxygen revolution and the evolution of early animals.
4. Recognize the term phylum as it is used to classify animals.
5. Explain why the Burgess Shale is an important location for discovering the evolutionary history of animals by studying their fossils.
6. Understand the term invertebrate as it is used to describe groups of animals, both living and fossil.
7. Discuss the eight invertebrate phyla presented in this lesson, noting their basic taxonomy, ecology, and unique structural characteristics.
8. Describe the two invertebrate chordate groups, their characteristics, and how they seem to make the transition between the true invertebrates and the true vertebrates.
9. Discuss the concept of the backbone and how it may be used erroneously to separate more advanced from less advanced invertebrate animals.
10. Describe the intertidal zone of the marine environment, including both rocky and sandy/muddy shores—what physical conditions prevail there, and how the organisms that live there survive.

Assignments

This lesson is based on information in the following text and video assignments. The key terms, focus points, and practice test are intended to help ensure mastery of the material presented in this unit of study.

Text: Pages 402–418 in Chapter 15, "Marine Animals"

Video: Episode 21, "Survivors"

Video Focus Points

Review these points before watching the video assignment for this lesson to help you focus on key issues.

— The ocean provides Earth's largest living space, and supports the largest communities of living organisms, both in biomass and in numbers of species and individuals.

— The challenges of Earth's early environments caused the organisms living there to adapt to meet those challenges—adaptations that are still seen in organisms living today. Examples of some innovative adaptations in animals include the highly-toxic, shallow-water cubomedusae, or jellyfish of Australia, and the behavior of some midwater organisms that envelope an attacker with a distracting cloud of luminescent mucus.

— Such innovative adaptations for survival were thought to have resulted from what is called the oxygen revolution, which began some two billion years ago. Before that, early life was very simple and got its energy from the sea's organic molecules. Then the process of photosynthesis evolved in some organisms, resulting in the release of free oxygen. These organisms store and use energy from the sun, and are called autotrophs.

— This led to the evolution of organisms that could use the free oxygen to make more efficient use of the energy they got from eating autotrophs. These organisms, called heterotrophs, are best represented by what are now called animals.

— The first of these animals had simple body plans and no supporting internal structure, and are called invertebrates. Most of Earth's animals are included in this group, which is often subdivided into groups called phyla.

— Some animals are clearly different from others, and are easily grouped into phyla, but some—especially fossils—are not easily separated.

— One of the prime study sites for invertebrate fossils is the British Columbia Burgess Shale, thought to be over 700 million years old. Here conditions in the early seas were right to fossilize many of the developing animals, providing science with insight into their evolutionary history. There are a number of fossil types found there that appear to have no direct evolutionary link to present-day invertebrates. Scientists are interested in the appearance, evolution, and extinction (or not) of these early animals. Some phyla, however, appear to have survived the many environmental changes over geologic time, and are part of today's marine fauna. Sponges, jellyfish, and cephalopods are good examples.

— Considered to be the most advanced of the animals are those in the phylum Chordata, which includes the familiar sharks, fish, birds, and mammals, but also a group of primitive or invertebrate chordates, which are very different from the familiar vertebrate chordates but still have some of the same characteristics. The primitive chordates are the tunicates or sea squirts, which are strictly marine and occupy both mid-water and sea-floor habitats.

— There is some discussion about the separation of invertebrates from the true vertebrates, sometimes implying that the vertebrates are more advanced because they have a backbone. It is possible that the organisms that evolved in the ocean, being less affected by the force of gravity on their body structures, evolved other kinds of supporting structures but are still advanced.

— To demonstrate the variety of adaptations seen in animals, one need only to study the intertidal zone, especially the rocky shore. Here conditions change constantly, from being covered with water to drying out, from calm water to crashing waves, and from needing to breathe water to being uncovered and in the air for hours. These and other conditions found in the intertidal zone create special stresses for the resident organisms, which had to evolve a variety of ways to cope.

— Some of the more mobile creatures in the intertidal, like crabs and fishes, can move up and down with the water level to stay wet. Others, like snails, can hide in shaded crevices until the water returns, while still others, such as barnacles and mussels, simply close up and wait. This variety of survival techniques tends to result in concentrations of certain types of organisms into lateral bands or zones in the intertidal zone.

— These differences among species based on physical and physiological adaptations are of interest to scientists who study those adaptations, as well as behavioral factors such as feeding strategies, competition for resources, and reproduction. Also of interest are the types of organisms that have adapted to life in the sandy or muddy intertidal zones.

— While the intertidal zone has been studied for many years, it is only recently that technology such as remote and manned undersea vehicles has enabled scientists to actually visit and study in the deep sea.

Text Focus Points

Review these points before reading the text assignment for this lesson to help you focus on key issues.

— With the evolution of photosynthesis, some types of organisms (autotrophs) were able to capture energy from the sun and, in the process, release free oxygen into the atmosphere, allowing a much more efficient use of energy by living organisms. This oxygen revolution (about two billion to 400 million years ago) resulted in a virtual explosion of complexity and diversity of life forms.

— Unfortunately, most of these organisms had no skeletons or hard parts, and left only a sparse fossil record. At such places as the Burgess Shale in British Columbia and the Ediacara Hills in Australia, however, sediment types and water conditions in the early seas were such that some fossils of those early organisms were preserved.

— Due to this sudden increase in oxygen availability and food, animal-like organisms proliferated. These organisms, called heterotrophs, ate the energy-producing autotrophs and used the free oxygen to break down their energy molecules.

— In studying these animals, both fossil and modern, science attempt to assemble their common evolutionary histories as part of a classification scheme. One of the basic concepts of such a scheme is to categorize organisms according to similarities in body architecture and complexity. Those with similarities in those traits are grouped into phyla.

— More than 90 percent of all living and fossil animals have body plans that lack an internal supporting structure (skeleton), so they are called invertebrates. This is a descriptive, but not a scientific, term that currently includes about 33 phyla, comprising organisms of many sizes, body plans, and complexities. Nine of the better-known of those phyla are studied to represent invertebrate evolutionary history.

— Most scientists agree that the least complex animals are the sponges (phylum Porifera), which are mostly marine, bottom-dwelling (attached) animals with only a few types of specialized cells. To feed, they merely filter material out of the water, process it, and eliminate the waste.

— The next phylum (Cnidaria) shows more complexity, being formed of layers of specialized cells with definite functions. They have a simple nerve network and are characterized by stinging cells for protection and feeding. There are two basic body types—polyps (sea anemones, corals) and medusae (jellyfish). Their body shape, or symmetry, is radial.

— When animals began to move, the long, slender wormlike body shape was most efficient. Three worm phyla are considered here, from the more primitive flatworms (Platyhelminthes) to the more complex roundworms (Nematoda) to the most advanced segmented worms (Annelida). Changes are seen in their functional anatomy (tissues and organs) that made them progressively more efficient at adapting to their environments.

— Second in size only to the phylum Arthropoda, the Mollusca (snails, clams, squid) show great advances in specialization of organ systems, body structure, and the ability to occupy a diversity of habitats. Most species are marine, and they are found in both mid-water and bottom environments.

— Considered the most successful of Earth's animals (most diversity of habitat, body form, and feeding types, and largest in sheer numbers) is the phylum Arthropoda. This phylum contains both terrestrial (insects and spiders) and marine (barnacles, lobsters, shrimp) representatives. The arthropod body plan includes a hard outer covering, or exoskeleton, articulated appendages, and a highly efficient nerve/muscle network.

— Another invertebrate phylum worthy of note is the Echinodermata (sea urchins, sea stars). This odd phylum is advanced over other invertebrates in possessing an internal skeleton, but lacks the highly-developed sensory system seen in some of the more primitive invertebrates. All echinoderms are marine, and are associated with the sea-floor habitat.

— In piecing together the evolutionary history of animals, we find a possible transition from the invertebrates to what is considered to be the most advanced animal phylum—phylum Chordata. This group includes organisms with an internal supporting body structure and several other non-invertebrate characteristics. The transition mentioned is seen in two groups of primitive chordates (tunicates and Amphioxus), which possess the same basic chordate structures that distinguish the phylum, yet lack the more advanced body plans and supporting internal skeleton of the vertebrates.

Key Terms and Concepts

A thorough understanding of these terms and concepts will help you to master this lesson.

Amphioxus. A small, transparent invertebrate chordate, thought to be a link between true invertebrates and true vertebrates because it possesses a well-developed dorsal nerve cord similar to the spinal cord in vertebrates.

Animal. A multicellular organism unable to synthesize its own food and often capable of movement.

Annelida. The phylum of animals to which segmented worms belong.

Arthropoda. The phylum of animals that includes shrimp, lobsters, krill, barnacles, and insects. The phylum Arthropoda is the world's most successful.

Asteroidea. The class of the phylum Echinodermata to which sea stars belong.

Bilateral symmetry. Body structure having left and right sides that are approximate mirror images of each other.

Bivalvia. The class of phylum Mollusca that includes clams, oysters, and mussels.

Cephalopoda. The class of phylum Mollusca that includes squid, octopus, nautilus.

Chitin. A complex nitrogen-rich carbohydrate from which parts of the arthropod skeletons are constructed.

Cnidaria. The phylum of animals to which corals, jellyfish, and sea anemones belong.

Cnidoblast. Type of cell found in members of the phylum Cnidaria that contains a stinging capsule. The threads that evert from the capsules assist in capturing prey and repelling aggressors.

Corals. Any of over 6,000 species of small cnidarians, many of which are capable of generating hard calcareous (aragonite, calcium carbonate) skeletons.

Crustacea. The class of phylum Arthropoda to which lobsters, shrimp, crabs, barnacles, and copepods belong.

Echinoidea. The class of the phylum Echinodermata to which sea urchins and sand dollars belong.

Exoskeleton. A strong, lightweight, form-fitting external covering and support common to the animals of the phylum Arthropoda. The exoskeleton is made partly of chitin and may be strengthened by calcium carbonate.

Gas exchange. Simultaneous passage, through a semipermeable membrane, of oxygen into an animal and carbon dioxide out of it.

Gastropoda. The class of the phylum Mollusca that includes snails and sea slugs.

Hermatypic. Describing coral species possessing symbiotic zooxanthellae within their tissues and capable of secreting calcium carbonate at a rate suitable for reef production.

Holothuroidea. The class of the phylum Echinodermata that includes sea cucumbers.

Intertidal zone. The marine zone between the highest high tide point on a shoreline and lowest low tide point. Often subdivided into habitats, based on the types of organisms living there.

Invertebrate. Animal lacking a backbone.

Medusa. Free-swimming body form of many members of the phylum Cnidaria.

Metamerism. Segmentation; repeating body parts.

Mollusca. The phylum of animals that includes chitons, snails, clams, and octopuses.

Nematoda. The phylum of animals to which roundworms belong.

Notochord. Stiffening structure found at some time in the lifecycle of all members of the phylum Chordata.

Ophiuroidea. The class of the phylum Echinodermata to which brittle stars belong.

Phylum. One of the major groups of the animal kingdom whose members share a similar body plan, level of complexity, and evolutionary history (plural = phyla)

Platyhelminthes. The phylum of animals to which the flatworms belong.

Polychaeta. The largest and most diverse class of phylum Annelida. Nearly all polychaetes are marine.

Polyp. One of two body forms of Cnidaria. Polyps are cup-shaped and possess rings of tentacles. Coral animals are polyps.

Porifera. The phylum of animals to which sponges belong.

Protozoa. Multiphyletic animal-like group within the kingdom Protista. Protozoans include amoebas, Paramecium, foraminiferans, and radiolarians.

Radial symmetry. Body structure in which body parts radiate from a central axis like spokes from a wheel. Examples are sea anemones and sea stars.

Suspension feeder. An animal that feeds by straining and otherwise collecting plankton and tiny food particles from the surrounding water.

Tunicate. A type of suspension-feeding invertebrate chordate.

Vertebrate. A chordate with a segmented backbone.

Zooxanthellae. Unicellular dinoflagellates that are symbiotic with coral and that produce the relatively high pH and some of the enzymes essential for rapid calcium carbonate deposition in coral reefs.

Test Your Learning

After working through these questions, check your answers against the key at the end of this book. If you answered any of the questions incorrectly, review the relevant sections of the text and video episode.

Multiple Choice

1. Jellyfish have the strongest stinging cells and toxins because of:
   a. their voracious feeding habits
   b. their need to protect their delicate body construction from large prey
   c. the need to defend their young
   d. their need to defend themselves against other jellyfish

2. The so-called oxygen revolution was the result of:
   a. the evolution of photosynthesis
   b. collision of a comet with Earth.
   c. release of water into the atmosphere from volcanoes
   d. all of the above

3. The term phylum is used to classify:
   a. all animals
   b. all living organisms
   c. fossil animals only
   d. invertebrates only

4. The term invertebrate is used to describe organisms:
   a. lacking tissue layers
   b. lacking an internal skeletal structure
   c. with primitive organ systems
   d. all of the above

5. The most primitive "true animals" are the:
   a. jellyfish
   b. tunicates
   c. roundworms
   d. sponges

6. Cnidoblasts are:
   a. stinging cells found in sponges and corals
   b. cells that form the skeletons of sponges
   c. stinging cells found in cnidarians
   d. special cells found in roundworms

7. The body plan called radial symmetry is found in:
   a. sponges and roundworms
   b. sponges and sea anemones
   c. sea anemones and sea stars
   d. none of the above

8. Pelagic tunicates (salps) feed by:
   a. trapping and stinging their prey
   b. filtering particles out of the water using mucus nets
   c. engulfing small fish
   d. all of the above

9. The annelid worms modify their body structure by adding segments called:
   a. metameres
   b. cnidoblasts
   c. chitin
   d. exoskeletons

10. The Anisakis worm, which parasitizes fish, belongs to the phylum:
    a. Annelida
    b. Porifera
    c. Mollusca
    d. Nematoda

11. The most highly-evolved mollusks are in the class:
    a. Gastropoda
    b. Bivalvia
    c. Cephalopoda
    d. Polychaeta

12. The arthropod exoskeleton is composed mainly of:
    a. metameres
    b. cnidoblasts
    c. calcium carbonate
    d. chitin

Short Answer

1. For purposes of general description, the true animals have been divided into two categories, based on the presence or absence of an internal supporting structure (skeleton). The invertebrates lack such a structure, and comprise roughly 90 percent of all known living and fossil animals, representing 33 phyla. In this lesson we met eight truly invertebrate phyla, as

well as the invertebrate members of the phylum Chordata. Those eight invertebrate phyla were chosen to represent the invertebrates, and are presented to us in order of increasing complexity of body structure and function. Beginning with the most primitive, list each of those phyla in ascending order of complexity and briefly describe the distinguishing features of each, particularly those that define it as primitive or advanced.

2. The video presentation for this lesson states that "perhaps nowhere on Earth is the ability to survive better demonstrated than in the intertidal zone." List some of the characteristics of the intertidal zone, some of the types of organisms that live there and the mechanisms and strategies they use to survive.

3. The members of the phylum Chordata are considered the most advanced animals, yet that phylum also includes two interesting types of invertebrate chordates. Briefly describe and discuss those two types of primitive chordates, and the features that place them in the phylum Chordata.

Supplemental Activities

1. While presenting the nine representative animal phyla in this lesson, the text devoted an entire page (Box 15.1 on page 410) to a member of phylum Nematoda—the threadworms or roundworms. The subject is a worm in the genus Anisakis, which is parasitic in fishes and should be of interest to anyone who eats sushi or sashimi, which are raw fish. This worm has an interesting lifecycle in the marine environment, and sometimes infects humans as they eat raw or undercooked marine fish. Research this phenomenon, which some people consider an urban legend, and present the real story about it, confirming or dispelling the rumors. Begin your research with the presentation in the text, the Internet, or library. The sites listed below may be helpful. In your presentation, include the biology and lifecycle of the worm(s) (there may be more than one species), and how it infects and affects humans. Mention any recommended treatment, and how to avoid infection. If the opportunity presents itself, you may even speak to your local sushi/sashimi chef, to get opinions/reactions.

   http://www.fda.gov/food/ResourcesForYou/Consumers/ucm077331.htm

   http://www.dpd.cdc.gov/DPDx/HTML/Anisakiasis.htm

2. Your text (page 415) states that "large or small, obvious or retiring, crustaceans dominate the world of marine animals." It also lists some of the major evolutionary advances that made crustaceans so successful. As you have learned, the crustacean arthropods include lobsters, crabs, shrimps, barnacles, crayfish, and many other types. Take a look at a representative crustacean, noting the features common to all arthropods, as well as those peculiar to crustaceans. First, consult a text or lab manual for marine biology or zoology at your school or local library to learn about the functional anatomy of the crustaceans as a group. Next, choose a representative crustacean to study, preferably one from your area. If you live near a stream or lake, there may be crayfish. Bait stores often sell them. Put it in a small aquarium or a large jar, and watch how it moves and reacts. Coastal residents may have access to lobsters, crabs, or shrimp, often found in tanks in the supermarket or fish market still alive. Make some notes on their movements and behavior while they are alive, then obtain some terminated ones for a closer look. Try to get the entire animal, uncooked and unshelled. Then, referring to your lab guide, using your fingers and/or sturdy forceps borrowed from your school lab, or even large makeup tweezers or a set of small, needle-nose pliers, pry into the anatomy of the creature, removing appendages and mouthparts, again following you lab guide for names of parts and their functions. How is it put together, and how does it seem to work compared to your live observations? If convenient, and if you are following a good study guide, open up the carapace and see the muscle attachments and internal organs.

   Keep good notes on all of this, with accompanying sketches, and write a report about your adventure.

3. Early in this lesson you encountered one of the several theories about the origin of life on Earth, followed by the oxygen revolution and the ensuing evolution of Earth's animals. Learning about those events, and those organisms, relies heavily on information gathered from the fossil record. Although the best-known, and most useful fossil sites (Burgess Shale and Ediacara Hills) are not generally accessible, most universities and museums have some kind of fossil displays. The purpose of this exercise is to learn about the process of fossilization to better understand how we study them. The best resource for learning about the basic process will be a good geology or paleontology text, or an encyclopedia. The Internet sites below will also be helpful, although for an even broader perspective try doing a search on "fossilization process."

   Write a report on the basic sequence of events involved in fossilization, then describe the specimens you have looked at and how they were probably fossilized. Again, remember that you do not need to look at marine organisms to learn the basic process, although that would make it more pertinent to this course.

   www.zoomschool.com/subjects/dinosaurs

   http://www.isgs.illinois.edu/education/earth-sci/paleo.shtml

Lesson 22

Life Goes On

Overview

The animal kingdom is divided into about 34 major groups called phyla. Some are abundant, conspicuous, and widespread like Arthropoda and Chordata, while others are inconspicuous like phylum Cycliophora, which includes a single species of microscopic animal that lives on the hairs around the mouths of lobsters. This lesson deals with the phylum most familiar to us all, the Chordata.

Animals are often divided into the vertebrates and invertebrates. Vertebrates have a backbone and include the fish, amphibians, reptiles, birds, and mammals. Invertebrates do not have a backbone, or vertebral column. About 97 percent of animal species are invertebrates. The vertebrates are only one group within one phylum—phylum Chordata. And even this phylum has some members that are not vertebrates. All the rest of the Earth's animals are invertebrates. The vertebrates are a natural assemblage; the invertebrates are simply "everything else."

The chordate body has three salient anatomical features—a dorsal support rod called a *notochord*, a dorsal tubular nerve cord, and pharyngeal gill slits. The notochord is not a bone; it is more like a stiff, pencil-shaped sausage with a thick sheath. It provides support to the back and its presence allowed for the later evolution of a vertebral column with its attached muscles. The nerve cord or spinal cord distributes nerves to the muscles and contributes to efficient coordinated locomotion. This design led to efficient swimming and later walking and flying. The pharyngeal gill slits originated as feeding structures used to filter food from the water. As the chordates evolved into larger organisms, the respiratory function of gills became highly developed. Telltale remnants of ancestral gill slits typically appear during early embryological development in land vertebrates, but disappear well before birth. Gill pouches are present in a human embryo at about one month after conception.

An abundant group of invertebrate chordates is the sea squirts. These are globular benthic filter feeders that are unrecognizable as chordates in adulthood. Their microscopic swimming larva has all three chordate characteristics and somewhat resembles a miniscule tadpole. As the larva transitions to an adult, it loses its notochord and its nerve cord becomes reduced. The gills are part of a prominent pharyngeal gill basket used for filter-feeding.

Another invertebrate chordate is Amphioxus. This transitional form more closely resembles fish. All three chordate features are well developed in the Amphioxus, which also uses its gill slits to feed.

Ancestors of these invertebrate chordates gave rise to the vertebrate chordates. The vertebrates replace the notochord with an articulated chain of skeletal elements, the vertebrae, which form a flexible support for the animal and house and protect the spinal cord. In the most primitive fish, the class Agnatha, the notochord remains well developed throughout life while in higher vertebrates the notochord mostly disappears and is represented by the intervertebral discs. The first vertebrates were fish, which probably evolved more than 500 million years ago. They were jawless, heavily armored, and small and are members of class Agnatha. Most agnathans are extinct, but a few types, the eel-shaped hagfishes and the lampreys, remain. Many biologists do not include the jawless fishes among the vertebrates but put them in a separate group of their own.

Three classes of vertebrates are called fish. The jawless Agnatha are the rarest. The sharks, skates, and rays comprise the class Chondrichthyes, or cartilaginous fishes. There are fewer than 700 species in this class characterized by skeletons made of cartilage. The Chondrichthyes are mostly marine and include the largest of the fishes, the whale shark. This slow-swimming, plankton-feeding shark can reach lengths of 60 feet.

The class *Osteichthyes*, or bony fishes, is the largest vertebrate group. With more than 27,000 species, there are more bony fishes than all other vertebrates combined. They are widespread and abundant in both fresh and salt water. Two smaller primitive groups of bony fishes include sturgeons, gars, and a few others, but more than 90 percent of the bony fishes are in the most advanced order of Osteichthyes, the Teleostei. This order encompasses most of the familiar fishes including sardines, giant bluefin tuna, flying fish, flounders, sea horses, eels, and about 25,000 others.

It is difficult to make generalizations about a species greater than 27,000 in number. These fish are ectothermic, gill-breathing aquatic vertebrates with fins. Because they are aquatic the conditions that shaped their evolution are different from those that shaped their terrestrial relatives, the reptiles, birds and mammals. Water is denser, more viscous, and conducts heat better than air. It is more difficult to move through than air, but provides more support. It is difficult to maintain elevated body temperatures in an aquatic world and there is less oxygen than in the terrestrial world.

The fusiform shape of swimming animals permits efficient movement through water. Aquatic vertebrates have most of their muscle mass on their trunks and little on their limbs. This is generally reversed in land animals, where propulsion and support requires well developed and muscular limbs. In the dense watery environment, the support needed to counter gravity is achieved by regulating buoyancy rather than standing on legs. Bony fishes often have a gas-filled *swim bladder* that creates neutral buoyancy. Some deep sea fishes substitute fat for gas in their swim bladders. Fat is less dense than water but more dense than gas, and does not compress at high pressure. The swim bladder is absent in many types of fish such as flounders that live on the bottom, tunas that move rapidly between depths, and deep ocean species that live at very high pressures.

The Chondrichthyes lack a swim bladder. Sharks have very large oily livers that provide some buoyancy and their fins work like airplane wings to provide lift. The wing-like pectoral fins of a great white shark are clearly seen in Figure 15.24. Like an airplane, lift is only accomplished as they move forward so sharks will sink if they do not swim. A swim bladder makes bony fish more like blimps.

Collecting oxygen is accomplished by highly efficient gills. A very large surface area as well as a counter current flow of water over them and blood through them, increase the efficiency of the gills in extracting oxygen from the water.

Two more problems faced by all animals are finding and capturing prey while avoiding being found and becoming prey. In active mobile organisms such as fish, well-developed sensory structures and camouflage help solve both problems. Depending on habitat and life style, the senses of vision and smell are often important. Additionally, fishes have a lateral line system in their skin that senses low frequency vibrations such as those made by other organisms moving nearby. Sharks have a special electricity detecting system concentrated in their snouts that allows them to detect minute electrical discharges as weak as those made by the muscle contractions of a flounder buried out of view in the sand. Color patterns can serve to hide or advertise. Camouflage may be very precise, matching a single particular background, or more generalized such as that of a military uniform in which a variety of shapes and colors blend the body into the background. Most fish are countershaded with dark backs and light bellies. This is useful in the open sea where the sky above is light and the depths below are dark. Some fish are brightly and distinctly colored to attract mates or warn of their toxicity. Schooling behavior is quite common, with about 25 percent of fish species exhibiting this behavior during part or all of their life. There is safety in numbers with the darting and shifting of the school confusing some predators. Large tight schools may also appear as a single large organism and discourage the approach of some predators.

The amphibians are a transitional vertebrate class. Their name means "double life" and many spend their early life as gill-breathing aquatic larva and their adult life as lung-breathing terrestrial adults. Amphibians are almost totally absent from the marine environment because their skin is semi-permeable and they cannot tolerate the ocean's salt. Amphibians gave rise to the reptiles, the first group of vertebrates that were truly freed from the aquatic environment. Both the mammals and the birds evolved from reptile ancestors, and all three classes have some members that have re-entered the aquatic world.

All three major groups of reptiles—turtles, crocodiles, and snakes and lizards—are represented in the ocean. Marine turtles are large and streamlined with flippers rather than legs. This is a successful group with eight species that range worldwide in warm seas. Sea turtles return to land to lay their eggs and are noted for their remarkable powers of navigation, which allow them to travel at sea for years and return to the exact beach where they hatched to deposit the eggs of the next generation. Breeding occurs offshore; only the females come ashore to nest and lay.

Marine crocodiles are the largest extant reptiles, reaching lengths of 23 feet and weights exceeding a ton. They are noted for their aggressive behavior. These giants inhabit the waters of northern Australia and the Indonesian archipelago. The smallest marine reptiles are the sea snakes. They are generally about one meter long and inhabit the tropical Pacific and Indian oceans but are absent from the Atlantic. Related to the cobras, they produce a neurotoxin that is the most powerful snake venom known. They are not particularly aggressive and feed mostly on bottom fish. They cause more human fatalities than crocodiles because they are more widespread and sometimes bite fishermen attempting to remove them from their nets. In contrast to land snakes, the tail region of sea snakes is laterally flattened and paddle-like for swimming.

Many modern biologists include birds in the reptile class and it seems likely that birds are descendents of dinosaurs. For purposes of this course, we will classify them separately, as the Aves. Flight has been the dominant force shaping the evolution of birds. These warm-blooded, feathered creatures have several groups whose lives are intimately tied to the oceans. Their endothermic physiology permits them to range from pole to pole unlike their ectothermic reptilian relatives that are confined to warmer seas. The tubenoses include the albatrosses, petrels, and shearwaters and are probably the most oceanic of the birds. The tubular anatomy of their nostrils permits them to detect air speed, is the passageway for excreting excess salt from their *salt glands,* and is responsible for their acute sense of smell. Birds have superbly developed vision, but a generally poor sense of smell. The tubenoses seem able to smell schools of fish or squid from miles away. They are also masters of flight, spending extraordinarily long periods of time soaring over the oceans far from land. Their long pointed wings, perfect for gliding, are among their many adaptations for long distance soaring. The largest albatrosses have wingspans of 12 feet and feed by dipping rather than diving or landing on the sea. They return to land, often isolated islands, only to nest and reproduce.

Pelicans and their relatives—boobies, cormorants, and frigate birds—are tropical and subtropical species that have throat pouches. The pelicans and boobies are slower flying species that hunt by sudden crashing dives. The frigate bird is a magnificent soaring species that actually catches flying fish in flight, or steals food by attacking other birds in flight and forcing them to regurgitate their food. Cormorants are also fishers that dive for food. The gulls and terns are a familiar group of shore birds. Gulls are often seen inland near fresh water or scavenging at landfills. Terns are smaller and more delicate birds that dive for their prey. The Arctic tern has the longest migration of any bird species, flying 15,000 miles per year following the summer between north and south polar regions.

Penguins are among the most highly specialized members of the class Aves. If tubenoses are the masters of the air, penguins are masters of the sea. They are flightless, have oily peg-like feathers, short fin-like wings, and a layer of insulating blubber. They are restricted to the southern hemisphere, and the Emperor penguin lives and breeds in the frigid Antarctic winter. This species exhibits huddling behavior, crowding closely together to conserve heat. Clumsy on land, the penguins are graceful swimmers with extraordinary maneuverability. They feed on a variety of marine animals including krill and fish.

The mammals are also warm-blooded vertebrates. They generally have hair, live-birth from a uterus where the embryos receive nutrients from a placenta, and have mammary glands. The three groups of marine mammals arose from different land ancestors. The pinnapeds, which include seals, sea lions, and walruses as well as polar bears and sea otters, are members of the order *Carnivora,* which also includes dogs, cats, weasels, and skunks. The order Sirenia includes the manatees and dugongs; their closest land relatives seem to be the elephants. The whales seem most closely related to the ungulates or hoofed mammals such as cattle, pigs, and hippos.

The marine carnivores fall within two groups, the Fissipedia and the Pinnapedia. Fissiped means "divided foot" and this group includes the more familiar carnivores such as cats, dogs, and bears; they have toes. By contrast the seals, sea lions, and walruses are *pinnapeds* or wing-footed carnivores. Their feet are flippers. The two marine fissipeds are the sea otter and the polar bear. The sea otter is the smallest marine mammal and has the densest fur of any mammal, relying more on their air-trapping pelt than their thin layer of fat for insulation. Their dense fur was almost responsible for their extinction as they were heavily hunted until laws were enacted to protect them. They have made a good recovery and are fairly common along the Pacific coast of North America and Siberia. Sea otters are one of the few mammals that use tools, employing rocks as hammers to crack open mollusks and other shelled invertebrates. Sea otters stay at sea, but polar bears, generally included among marine mammals, do not. Excellent swimmers, they are confined to the north polar regions and are the largest of the bears. They wander the ice of the Arctic Sea feeding on seals and stranded whales.

The Pinnapedia are the seals, sea lions, and walruses. Among them, the seals have the most species, at 19. Seals also include the smallest species—the ringed seal weighing about 23 pounds—and the largest—the male southern elephant seals approaching four tons. The seals are streamlined, efficient swimmers. Their rear flippers cannot rotate forward and their front flippers are also of little use on land, so their locomotion on land is limited to a kind of clumsy undulating. They lack external ear flaps and have coarse hairs with no underfur. The sea lions are more dog-like in appearance and have coarse outer hairs underlain by a fine pelt. They have longer necks, external ear flaps, and can rotate their hind limbs forward and stand up on their front flippers for a somewhat less crude locomotion on land. The walruses are easily distinguished by their tusks. There is only one species of walrus; they live in the arctic, and may exceed 2,700 pounds. Walruses feed using their bristly whiskers to scour the ocean bottom for invertebrates. The tusks are used for defense and to help pull them up onto land or ice.

The Sirenia, the manatees and dugongs, are also called sea cows and are the only herbivorous marine mammals. They generally inhabit tropical regions, but the Steller's Sea Cow, a huge species reaching ten tons, once inhabited the Bering Sea. It became known to science in 1741 and was hunted to extinction by 1768. Sea cows are docile, ponderous animals. They have front flippers, but their rear is modified into a single rounded or forked tail fluke. All four of the remaining species are dangerously close to extinction.

The Cetacea, or whales, are the most highly adapted of the marine mammals. They have little or no hair, range far out to sea, and some species rarely come close to shore. Their nostrils are completely separated from their mouth and esophagus and have moved to the tops of their heads as the blowholes. They have no rear limbs although a remnant of their pelvis remains within their bodies where it serves primarily as a place for the attachment of the penis muscles. Their entire lives including copulation, birth, and nursing occur in the water. The order Cetacea is divided into two suborders, the Odontoceti and the Mysticeti. The *odontocetes*—the toothed whales—are hunters and include dolphins, porpoises, and orcas. They have well-developed echolocation systems. Like bats, the echolocation of odontocetes permits them to find prey and sense objects in their dim environments where vision is less useful. They are intelligent animals and use their sound production for communication as well. The sperm whale, romanticized in *Moby Dick*, is the largest toothed whale at up to 60 feet long.

The largest animal to have ever existed on Earth is the blue whale. It can exceed 100 feet in length and may weigh 200 tons. This giant belongs to the suborder Mysticeti, the baleen whales. Baleen whales have no teeth, but rather have rows of flat horn-like plates hanging from the roofs of their mouths. *Baleen* is composed of a protein similar to that in fingernails and it is used to filter food from the water. Most mysticetes feed on krill and other zooplankton. Because they strain rather than hunt, they do not have well-developed echolocation, and they do not make very deep dives because their food lives within the photic zone. They do vocalize for communication and this seems to be particularly sophisticated in humpback whales. Other baleen whales are the right whale, gray whale, and fin whale. Many baleen whales migrate between productive summer feeding ground in polar regions and warm breeding grounds in lower latitudes.

Focus Your Learning

Learning Objectives

After completing this unit you should be able to do the following:

1. Describe the three features of the chordate body plan and contrast the two groups of invertebrate chordates.
2. List the animals in this lesson in order according to their classification and understand both their common and scientific names.
3. Describe the three classes of fishes.
4. Describe the three major groups of marine reptiles.
5. Outline the general characteristics of the class Aves and compare the tubenose, gull, pelican, and penguin groups.
6. Discuss the general characteristics of the class Mammalia and discuss four features shared by marine mammals.
7. Describe the three orders of marine mammals and discuss the major groups within each order.

Assignments

This lesson is based on information in the following text and video assignments. The key terms, focus points, and practice test are intended to help ensure mastery of the material presented in this unit of study.

Text: Pages 417–433 in Chapter 15, "Marine Animals"

Video: Episode 22, "Life Goes On"

Video Focus Points

Review these points before watching the video assignment for this lesson to help you focus on key issues.

— Although they are a subgroup within one phylum among 34 phyla of animals, the vertebrates are one of the most conspicuous, successful, and advanced groups of living things on Earth.

— Chordates have a notochord, a hollow nerve cord, and gill slits in their pharynx. Some chordates lack a backbone, which makes them invertebrates, but the major groups—fish, amphibians, reptiles, birds, and mammals are the dominant chordates.

— The density, viscosity, thermal properties, salinity, and lower available oxygen create different challenges for marine animals than those faced by land animals.

— There are three distinct classes of vertebrates typically lumped under the category of fish. The first group is the jawless fishes such as the hagfish and lamprey, the second group is the cartilaginous fishes and includes sharks, skates, and rays, and the largest group is the bony fish with more than 25,000 species. Trout, tuna, sardines, eels, and most other fish are in this group.

— Some of the special adaptations seen among the fishes include streamlining and other adaptations for locomotion, camouflage, bioluminescence, counter-shading and other adaptations associated with hiding or advertising, highly developed sensory adaptations including the lateral line for sensing low frequency vibrations, swim bladders to adjust buoyancy, and gills highly adapted to extract oxygen from water are but a few.

— The reptiles, birds, and mammals are land creatures but all three classes have members that are adapted to ocean life. Among the reptiles are marine turtles, sea snakes, and marine crocodiles. Among the marine birds, the tubenose group includes the majestic albatrosses. These are highly adapted for extended periods of gliding and soaring, and some stay aloft for a month. By contrast, the penguins are flightless birds adapted for swimming in cold water. They have blubber, peg-like feathers, and short sickle-shaped wings.

— The marine mammals evolved from land ancestors within the last 50 million years. There are three major groups. The order Carnivora includes the sea otter and polar bear, and the seals, sea lions, and walruses. The order Cetacea is the whales and includes the Odontoceti or toothed whales such as the dolphins, orcas, and sperm whales and the Mysticeti or baleen whales such as the blue, gray, and humpback whales.

Text Focus Points

Review these points before reading the text assignment for this lesson to help you focus on key issues.

— Invertebrate chordates include the benthic sea squirts and the fish-like Amphioxus. Both use their pharyngeal gills for feeding.

— There are about 50,000 species of vertebrate chordates including the fish, amphibians, reptiles, birds, and mammals. About 27,000 of these are bony fishes.

— The earliest fishes were jawless, and little more than sucking digestive tubes. The class Chondrichthyes is the sharks, skates, and rays. Their skeletons are composed of cartilage. The largest class of vertebrates is bony fishes. They have skeletons composed of bone.

— Problems faced by fishes for survival in the ocean include those associated with locomotion through this viscous medium, maintaining neutral buoyancy, obtaining oxygen from water where it is scarcer than in air, dealing with the osmotic problems associated with salt water, finding prey, and avoiding predation.

— Marine reptiles include certain specialized turtles, crocodiles, and snakes. Sea turtles are large and have remarkable navigation abilities. Sea snakes are limited to the warmer Indo-Pacific region. They are the most venomous of all snakes. The marine crocodiles are the largest extant reptiles. They are aggressive and sometimes kill humans.

— Marine birds include the tubenoses, the pelican group, the gull group, and the penguins. The tubenoses, including the albatross, are superbly adapted for long distance soaring and gliding over the ocean, and have a well-developed sense of smell for locating prey. The gull and pelican groups are more closely tied to land while the penguins are highly modified for their flightless life. They have traded aerodynamics for hydrodynamics and are fast and highly maneuverable swimmers.

— Three major groups comprise the marine mammals. The Cetacea are the whales and include the toothed whales that are active hunters and the baleen whales that generally feed on krill and other small plankton using their baleen plates to filter their food from the water. The order Carnivora includes the dogs, cats, skunks, etc. Their close relatives, the polar bear and the sea otter, are marine species and their more distant relatives, the pinnapeds, include the seals, sea lions, and walrus. The Sirenia are the manatees and dugongs. These docile creatures are the only herbivorous marine mammals.

Key Terms

A thorough understanding of these terms and concepts will help you to master this lesson.

Aspect ratio. A geometric relationship between surface area and length of a wing or fin. Long thin wings have high aspect ratios and are the most efficient for long distance soaring and gliding flight. The wings of albatrosses and frigate birds have very high aspect ratios.

Baleen. The hard, flat, fringed, elongated plates that hang from the roof of the mouth in baleen whales. Food such as krill is collected as water is filtered through the rows of baleen.

Carnivora. An order of the class Mammalia that have specialized teeth for tearing flesh. Many meat-eating mammals are not Carnivora but the pinnapeds, sea otter, and polar bear are marine carnivores.

Lateral-line system. A special sensory system composed of a series of pores, canals, and sensory cells running along the sides of fish. It detects low frequency vibrations in the surrounding water.

Notochord. An elongated, cellular rod encased in a tough fibrous sheath that characterizes the phylum Chordata. It lies within the dorsal or back region and provides support for the body.

Odontoceti. The toothed whales such as the dolphins, porpoises, orcas, and sperm whales.

Osteichthyes. The bony fishes. This is the largest class of vertebrates.

Pinnapedia. A distinct suborder of the mammalian order Carnivora characterized by flap-like appendages. It includes seals, sea lions, and walrus.

Salt glands. Specialized glands for the extraction and removal of excess salt. They are found in many marine vertebrates. Sea turtles expel salt as tears from the eyes and in marine birds and iguanas the salt is expelled through the nostrils.

Swim bladder. A gas-filled sac within the bodies of many bony fishes. By adding or removing gas at different depths and pressures, the swim bladder allows the fish to maintain neutral buoyancy.

Test Your Learning

After working through these questions, check your answers against the key at the end of this book. If you answered any of the questions incorrectly, review the relevant sections of the text and video episode.

Multiple Choice

1. Which of these is not a characteristic of all members of phylum Chordata?
   a. notochord
   b. pharyngeal gill slits
   c. vertebral column
   d. dorsal nerve cord

2. Which of these is not a filter-feeder?
   a. sperm whale
   b. humpback whale
   c. blue whale
   d. whale shark

3. The sensory system of fishes that detects low frequency vibrations in the surrounding water is called the:
   a. counter-current system
   b. lateral line system
   c. Ampullae of Lorenzini
   d. semi-circular canals

4. Which of these is in the order Sirenia?
   a. sea lion
   b. sea otter
   c. manta ray
   d. manatee

5. Which of the following have specialized salt glands to concentrate and excrete excess salts from body fluids?
   a. whales
   b. sharks
   c. sea turtles
   d. manatees

6. Tuna, bass, eels, and swordfish are in the most advanced group of bony fishes, the:
   a. Holostei
   b. Teleostei
   c. Perciformes
   d. Salmoniformes

7. Which of these has the highest aspect ratio?
   a. tail fluke of a manatee
   b. wing of a pelican
   c. dorsal fin of a sail fish
   d. wing of an albatross

8. The largest living fish is the:
   a. great white shark
   b. beluga sturgeon
   c. whale shark
   d. goliath jewfish

9. Which of these is noted for its remarkable navigation abilities?
   a. basking shark
   b. banded sea snake
   c. green sea turtle
   d. blue-footed boobie

10. Which of these does not rely primarily on blubber for insulation?
    a. sea otter
    b. sea lion
    c. penguin
    d. orca

11. Echolocation is most well developed in:
    a. elephant seals
    b. sea snakes
    c. toothed whales
    d. baleen whales

12. Which of these probably has a swim bladder?
    a. tuna
    b. manta ray
    c. shark
    d. sea bass

Short Answer

1. List and discuss four characteristics shared by all marine mammals.
2. List four problems faced by marine fishes that are different than those faced by land vertebrates. Discuss each problem and its solution.
3. Compare and contrast the tubenoses and the penguins.

Supplemental Activities

1. Study the structure of a bony fish. Purchase a whole, uncleaned, fish from a seafood or grocery store and examine its external and internal structure. Dissect it, with caution, using a sharp knife.

   This Website provides a good study of the external anatomy:

   http://www.flmnh.ufl.edu/FISH/Education/Diagrams.htm

   Use it to write a report, including your own drawings, describing the features and measurements of your particular specimen. These Websites will be helpful for locating and identifying the internal and external structures.

   http://www.biologycorner.com/worksheets/fishcolor.htm

   http://www.fishanatomy.net/index.htm

2. Marine mammals exhibit a variety of adaptations for breath-hold diving. This response to diving is called the Mammalian Diving Reflex and is exaggerated in marine mammals but present to some extent in all mammals. Research the MDR on the Internet and write a short report outlining as many features of this reflex as you can find. Find information linking the MDR to 1) deep free-diving records in professional divers, 2) the Ama women of Korea, 3) the ability of humans, especially youngsters, to sometimes survive submerged in cold water for dozens of minutes, and 4) the recommendation of some physicians to patients with occasional tachycardia (speeded heart rate) to stick their faces in cold water. Demonstrate the MDR on yourself by measuring your heart rate (pulse) seated at rest at a table and then while dipping your face into a pan of ice water, also while seated.

3. It is sometimes said that the reptiles were the first vertebrates to be freed from the water and live entirely on land. In a sense this is untrue because the embryos of these animals develop in an ocean-like environment within an amniotic egg or uterus. Go to the University of California Museum of Paleontology Website http://www.ucmp.berkeley.edu/ and find your way to information about the Aves, amniote egg, placenta, and vertebrate flight. Write a short report about the structure and function of the amniote egg and compare it to a placenta and uterus. Also study the subject of vertebrate flight and write a short report about this subject.

   Explore the marine reptiles and birds. Go to the following Websites and write a short paragraph about three species of marine turtles, three species of penguins, three species of tubenoses, and three species of whales. At the marine turtle Website take the online quiz and print out your results. At the Antarctic Website be sure to listen to the sounds of the species chosen.

   Turtles: http://www.cccturtle.org/

   Penguins, seals, whales, and tubenoses: http://www.antarcticconnection.com/

   Tubenoses: http://www.coastalstudies.org/what-we-do/stellwagen-bank/fulmar.htm

Lesson 23

Living Together

Overview

The animals studied in Lesson 22 do not live in isolation, they interact with members of their own and other species to form ecological communities. Ecology is the branch of biology that studies the interactions of populations in communities; i.e., how populations interact and the consequences of the interactions. Ecology can also be defined as the study of the distribution and abundance of organisms. In other words, ecology also studies where organisms live, how many of them live there, and why. Understanding abundance and distribution involves understanding interrelationships such as competition, predation, parasitism, and disease and physical limiting factors such as light, temperature, and salinity.

Both competition and predation can occur among individuals of the same species or between individuals of different species. Competition among grasshoppers for grass is called intraspecific competition. When they compete with cows for that same resource, the competition is interspecific. Intraspecific predation is cannibalism. Less obvious than predation and competition is symbiosis. *Symbiosis* means "living together," and refers to interactions in which organisms live in very close association with each other; often with one species living on or in another. Symbiotic relationships include mutualism, commensalism, and parasitism. Ecologists are beginning to recognize that symbiotic associations can be as important as competition and predation in regulating the abundance and distribution of organisms. The malaria parasite, for example, affects the abundance and distribution of humans. Mutualism involves an association in which both species benefit. The relationship between zooxanthellae and coral polyps is an example of mutualism. In a commensal relationship the commensal is benefited while the host is indifferent. Pilot fish swim along with predatory sharks and feed on the scraps of food that result from the sharks' somewhat messy feeding habits; the shark is unaffected. Parasites have a deleterious effect on their hosts. Roundworms, tapeworms, pathogenic bacteria and fungi, and ticks and lice are examples. These organisms feed upon their hosts.

Parasite evolution is tricky. It is generally considered that the most advanced parasites do the least harm to their hosts because killing the host would be suicidal. This is a useful generalization, but there are many exceptions. Many parasites have complex lifecycles involving several different hosts. A particular parasitic worm may live in the intestine of a stork. The worm lays eggs, which enter the sea with the bird's feces and hatch into larvae. The larvae may bore onto the skin of a small fish and form black-colored cysts. The lifecycle is completed when another stork eats the small fish. If the cysts are black polka-dots on an otherwise well-camouflaged fish, the probability that a stork will see, capture, and eat the fish is increased. The parasite benefits if it disables the fish, but does not benefit if it significantly disables the bird.

Cleaning symbiosis is an interesting and widespread phenomenon involving parasitism, mutualism, and predation simultaneously. Some animals, often small fish or shrimp, feed by removing parasites from the skin and gills of larger species. The cleaners often establish cleaning stations where fish come to be cleaned. At the moment a cleaner fish consumes a parasite from the skin of a larger fish, predation, mutualism, and parasitism intersect.

Along with biological interactions, physical and chemical factors such as temperature, salinity, and light also affect abundance and distribution. For a review of these factors, see Lesson 19. Each species has a range of tolerance for any factor; they have an optimal range within which they flourish, suboptimal ranges within which they are under stress, and upper and lower limits of tolerance beyond which the excess or deficiency of the factor is lethal. Text Figure 16.2 illustrates an environmental tolerance curve for temperature. Ecologists often use the prefixes steno- and eury- to describe narrow and wide ranges of tolerance for particular factors. *Stenothermal* organisms have a narrow range of tolerance for temperature while eurythermal organisms have a wide range of tolerance. Similarly the terms stenohaline and euryhaline refer to salinity tolerances while stenobaric and eurybaric refer to ranges of tolerance to pressure.

The abundance of any species in its environment is largely determined by birth and death rates. If we disregard immigration and emigration, populations grow if births exceed deaths and shrink if deaths exceed births. The general pattern of population growth is illustrated by J-shaped and S-shaped population growth curves. If a new species is introduced into a habitat where it has infinite resources and can flourish, it will exhibit J-shaped population growth. The small population will take advantage of the environment, growing slowly at first because there are few individuals reproducing. As the population grows, the number of individuals producing offspring grows and the population numbers skyrocket. Of course there is no such thing as infinite resources so, in reality, as the population grows it will encounter environmental resistance. Factors such as competition for food, mates, and space increase as do disease, parasites, and predation. These factors work together to slow the rate of growth of the population by increasing the death rate. The population levels off at some maximum size that the environment can sustain. At this level, called the *carrying capacity*, births and deaths are approximately in balance.

Regardless of population density, individual organisms can be distributed in their environment in three general patterns. Random and uniform distributions are rare. A clumped pattern resulting from attraction to food, shelter, mates, or from social interactions such as schooling or huddling is most common. Phytoplankton blooms and intertidal zonation are also clumped distributions. The term "patchy" is often used to describe plankton distribution. These spatial distributions cannot be divorced from temporal ones. A new volcanic island, sunken ship, or hurricane-ravaged reef will be invaded by colonizing species that are slowly but surely replaced by others until a climax community is reached. This parade of changes in a community is called ecological succession and it marches toward a climax community. Unless the environment experiences a shift in conditions, climax communities are more stable and permanent than the successional stages that preceded them.

Several marine communities are surveyed in this lesson; some of them have been discussed individually earlier in the course. The communities are now viewed comparatively. Noteworthy and sometimes unexpected similarities among them as well as salient differences between them will become apparent. Each community is shaped by local environmental conditions such as temperature, salinity, and light. Each has some source of primary production and some level of biodiversity. The populations that make up the community may have high or low densities and be dominant or of lesser overall consequence.

Intertidal communities lie at the fringe of the ocean between the high and low tide levels. The rising and falling of the tides, the power of the waves and the slope of the substrate from almost horizontal beaches to vertical cliffs are some of the variables that shape intertidal communities. Additionally, the substrate may be rocky or composed of cobble, gravel, sand, silt, or clay. The upper intertidal zone is subject to long periods of exposure to air. Desiccation and temperature extremes are more important and biological factors such as competition and predation may be of less importance in this environment. In the lower regions of the intertidal the reverse is true. The rocky intertidal can be exposed to a constant pounding of waves but is nonetheless a place of abundant life. Organisms attach to the substrate or are able to hold on tightly. Nutrient availability is generally high in the intertidal zone because the shallow water is well mixed and nutrients also wash in from the land. Light is also readily available so productivity is high. The varying conditions from the higher to the lower intertidal create a great variety of niches, and well-defined zonation is almost always present in the rocky intertidal. Algae, snails, mussels, barnacles, and sea stars are often important members of intertidal communities. Birds and mammals forage when the tide is out and fish and other marine predators seek food in the intertidal when the tide returns.

Whereas the rocky intertidal provides a solid and reliable substrate, cobble, gravel, and sandy shores are among the harshest of marine environments. Their strong waves wash away the smaller silt and clay particles leaving an environment of pounding cobbles or scouring sands and gravels. Although there is an entire community of microscopic interstitial organisms living between the grains of sand, this fascinating community is invisible to the naked eye. Compared to the rocky shore where life abounds, the most obvious feature of these intertidal communities is the absence of life. The familiar sand crab, which filters food from the receding waves, and the colorful *Donax* clams are among the few species that inhabit this hostile environment. Ghost crabs live in burrows above the high tide and scavenge the beach for food that washes ashore. Gulls, sand pipers, and other shore birds also find food in this zone. Tropical black sand beaches may be the harshest of marine environments. The black sand heats intensely while the glassy volcanic grit blasts and scours any organisms that might live there. Virtually nothing can survive under these conditions.

By contrast, muddy beaches are much more hospitable. The lack of high-energy waves allows the accumulation of silt and clay on protected beaches. Estuaries are protected and are among the most productive marine environments. Their shores are stabilized by salt marshes or mangrove forests. Nutrients are delivered continuously from land drainage and are washed up and mixed from the shallow bottom. Phytoplankton productivity is high and the fringe of grasses or trees provides a rich soup of organic particles that forms the base of many estuarine food webs. The marshes of the temperate zones die back in the autumn and are mowed down by shifting ice in the winter while the tropical mangrove trees shed leaves year round. This harvest of organic particles and its decomposing bacteria is called *detritus* and is a major food source in estuaries. The protected characteristic and high food availability makes estuaries a prime nursery and feeding ground for many fish. Most shrimp, oysters, and crabs that are harvested commercially are estuarine species. Estuaries such as Chesapeake and San Francisco Bays and Long Island, Puget, and Pamlico Sounds are heavily populated and impacted by humans. In estuaries, the interests of commercial fishermen, sport fishermen, swimmers, boaters, and nature lovers often come into conflict with municipal, industrial, shipping, and military operations.

Like estuaries, coral reefs are shallow near-shore communities that are highly productive and highly impacted by human activity. But coral reefs do not exist where rivers mix with the ocean and they do not include species that tolerate low salinities; their water is clear and nutrient levels are very low. Detritus does not play a significant role in reef food chains, and whereas estuaries usually have low biodiversity, coral reefs are noted for their high biodiversity. The coral reef region may have sea grasses nearby, and calcareous and coralline algae can be important primary producers too, but the key to coral reef productivity lies in the relationship between the coral polyps and their symbiotic zooxanthellae. The zooxanthellae are photosynthetic and provide food and oxygen to the corals. They also increase the rate at which coral reefs produce their calcareous skeletons. In return the coral animal provides the zooxanthellae with carbon dioxide, nitrogen wastes, and protection.

Coral reefs and estuaries are very different but they share high productivity and high levels of degradation caused by humans. While not exempt, the open ocean is farther away from the harmful effects of humanity. Most of the ocean floor, the region of the abyssal plains, is sparsely populated with life. It is an extensive, constant, cold, aphotic, hypersaline, hyperbaric environment that depends on a drizzle of particulate organic food from the photic zone above. There are no plants and the animals are mostly deposit-feeders and scavengers and their predators. Metabolism is slow and the organisms are sometimes unusually large and long-lived. The hydrothermal vent community is also an abyssal benthic community but is, by contrast, highly productive and rich in life.

Whereas the coral reef is a highly productive community in well-lit but nutrient-poor oceans, the hydrothermal vent community is a highly productive community in a nutrient-rich but lightless ocean zone. This is a community whose primary production is accomplished in the absence of sunlight by chemosynthetic bacteria and Archaea using energy from sulfide and methane molecules to manufacture carbohydrate food. The vents are heated by magma close to the surface at spreading centers of mid-ocean ridges. Hot, mineral-rich water emerges from seeps and cracks at temperatures as high as 350°C. The organisms live in nearby water at temperatures sometimes exceeding 100°C and frequently as warm as the water around coral reefs. Not far from the vent community the water quickly cools to the 1–4°C characteristic of most deep ocean. Like the coral reef, the hydrothermal vent community is dependent on mutualistic symbiosis. The abundant tube worms, *Riftia*, have no mouth, intestine, or anus. An internal sac is filled with chemosynthetic bacteria that produce food for the worms. The red plume of *Riftia* is a gill. It is filled with hemoglobin-rich blood that picks up CO_2, H_2S, and O_2, the raw materials for chemosynthesis, and circulates them to the symbiotic bacteria. In this peculiar, hot, lightless world, the water contains high concentrations of these chemosynthetic bacteria playing the role taken by the photosynthetic phytoplankton in the sunlit world above.

Most of the ocean world lies in the open water pelagic realm. About 83 percent of the oceans' life is found in the upper 200 meters associated with the photic zone. Less than one percent is found deeper than 3,000 meters. The phytoplankton-based community of the photic zone was discussed earlier (Lessons 9 and 20). Lurking just below the photic zone is the *deep scattering layer*. This is a relatively dense collection of fish, squid, and other animals that hide in the safety of the darkness during the day and migrate upward to feed in the photic zone at night. The largest community of the ocean is the vast *bathypelagic zone*. It is often hypoxic and there is no light, no photosynthesis, no chemosynthesis, and almost no life. The creatures of this region include some of Earth's strangest. Gulper eels can swallow prey larger than themselves and may only find food once or twice in a year. Flashes of bioluminescence are omnipresent, serving to attract mates or food or to startle predators. The bathypelagic is vast but low in abundance, productivity, and diversity. As you approach the end of the course and are able to integrate the ocean's chemistry, physics, geology, and biology you should appreciate the various marine communities and the organisms that populate them.

Focus Your Learning

Learning Objectives

After completing this unit you should be able to do the following:

1. Define ecology and ecological community and relate the terms niche, habitat, and biodiversity to the study of communities.
2. Draw and explain an environmental tolerance curve and give examples of how the prefixes eury- and steno- relate to the abundance and distribution of organisms in a community.
3. Compare intraspecific and interspecific competition and illustrate how these regulate the distribution of *Chthamalus* and *Balanus* barnacles in the rocky intertidal community. Distinguish between exploitative and interference competition and give examples.
4. Compare and contrast J-shaped and S-shaped population growth curves and describe how environmental resistance converts J-growth into S-growth and establishes a carrying capacity.
5. Discuss symbiosis by comparing mutualism, commensalisms, and parasitism. Give examples.
6. Describe the major features of each of the following communities and be prepared to compare and contrast any one of these with the others.
   a. rocky intertidal
   b. sandy intertidal
   c. salt marsh/estuary
   d. coral reef
   e. pelagic photic zone
   f. deep scattering layer (DSL)
   g. bathypelagic zone
   h. abyssal plains
   i. hydrothermal vent/cold seep
   j. whale fall

Assignments

This lesson is based on information in the following text and video assignments. The key terms, focus points, and practice test are intended to help ensure mastery of the material presented in this unit of study.

Text: Chapter 16, "Marine Communities," pages 434–459

Video: Episode 23, "Living Together"

Video Focus Points

Review these points before watching the video assignment for this lesson to help you focus on key issues.

— Ecosystems are dynamic assemblages of both living organisms and their surrounding non-living environment. The interacting organisms of an ecosystem comprise a community.

— Tropical reefs are constructed on a foundation of calcium carbonate secreted by corals, coralline red algae, and calcareous green algae. They have high biodiversity and are restricted to warm, clear, shallow oceans. The relationship between the coral animals and their symbiotic zooxanthellae is a major factor explaining reefs' high productivity in well-lit but nutrient-poor seas.

— The corals and zooxanthellae have a mutualistic relationship but as in most communities, all three types of symbiosis—mutualism, commensalisms, and parasitism—along with predation and competition, regulate and shape the nature of the community.

— Because of their proximity to land and human populations, coral reefs and estuaries are particularly vulnerable to human activities. Both are highly productive, shallow-water ecosystems. But estuarine water is nutrient-rich and turbid and the estuarine ecosystem relies primarily on phytoplankton-based and detritus-based food webs. Compared with coral reefs, estuaries have lower biodiversity.

— The rocky intertidal community is also a coastal community with high productivity. Nutrient run-off from land and the strong mixing of the adjacent shallow ocean by waves and tides keeps nutrients abundant and available. The periodic rise and fall of the tides, continuous wave stress, and intense competition for space on the rocks are important in shaping this community.

— The discovery of the hydrothermal vent community surprised oceanographers in 1977. This community was first observed near a mid-ocean spreading center where magma superheats water that seeps into cracks and crevices. The hot water leaches minerals from the bedrock and delivers sulfides and other energy-rich molecules to the water around the vents as it emerges. Like the coral reef, this highly productive community seemed enigmatic at first. The deep oceans are generally cold, hyperbaric, lightless, and not productive. Photosynthesis is not possible.

— Primary production by chemosynthetic bacteria provides the base of the vent food web. These bacteria utilize the leached energy-rich sulfides to provide energy to produce glucose.

— As with the coral reef, symbiosis is a key feature of the hydrothermal vent community. The impressive tube worms, large clams, and certain other animals of the hydrothermal vents contain mutualistic chemosynthetic bacteria that produce food and receive a safe living place in exchange. Other animals filter, scrape, and otherwise feed directly on the rich soup of bacteria that flourishes within this community.

— As one moves from the hot water of the vents to the frigid deep ocean water nearby the community changes. Vent communities also seem quite ephemeral, lasting only perhaps a few years as tectonic activity alters the subsurface plumbing providing the hot water. All biological communities are dynamic and change continuously in space and time.

Text Focus Points

Review these points before reading the text assignment for this lesson to help you focus on key issues.

— A community is the interacting organisms within an ecosystem. The individual species each have a habitat and niche influenced by such factors as competition, predation, and symbiosis and physical factors such as temperature, salinity, light, and nutrients. All communities rely on some type of primary production.

— Population growth over time often exhibits an S-shaped curve where environmental resistance slows the growth rate until the population stabilizes at the environment's carrying capacity for that species.

— Organisms may be found in uniform, random, or clumped distribution patterns. Clumping might be caused by food, light, or nutrient availability and by schooling, attraction to mates, and other social interactions. There are many reasons why populations exhibit clumped distributions and few that result in random or uniform distributions.

— The following marine communities are examined and compared: 1) rocky intertidal, 2) seaweed, 3) sand and cobble beaches, 4) salt marsh and estuary, 5) coral reef, 6) pelagic photic zone, 7) deep scattering layer (DSL), 8) bathypelagic zone, 9) deep-ocean floor, 10) hydrothermal vent/cold seep, and 11) whale fall.

— The structure of each community is influenced by light and nutrient availability and other local conditions as well as by the interactions of the organisms within the community.

Key Terms and Concepts

A thorough understanding of these terms and concepts will help you to master this lesson.

Alien species. Also called introduced species or exotics, these are non-native species introduced into an area where they did not exist before. They may travel in the ballast water of oceanic ships, by aquarists, or other modes. They can upset existing ecosystems and displace local species.

Bathypelagic zone. The vast deep ocean zone between the deep scattering layer and the seafloor. It is the largest living space in the ocean but it is constantly dark, cold, hyperbaric, and generally contains little life because food is scarce.

Black smoker. A plume of sulfide-rich water emerging from a hydrothermal vent. The dissolved sulfides precipitate from the hot water and appear black.

Carrying capacity. The maximum size of a particular population that a particular environment can support, or "carry," for a sustained period of time. Births by reproduction are balanced by deaths caused by competition, predation, disease, and other factors causing the population to stabilize.

Deep Scattering Layer. A relatively dense collection of small fish, squids, and other animals that aggregate below the photic zone in open water. As sonar sound waves from depth finders hits them, it scatters. They rise each night to feed in the photic zone and return to the cover of darkness in the day.

Detritus. Bits and pieces of decomposing organic matter that serve as the base of several important marine food webs. Decomposing salt marsh plants, mangrove leaves, and sea grasses are important in estuaries and the rain of detritus from the photic zone is important to the deep ocean benthic community.

Interference competition. Competition in which one organism secretes a chemical that kills nearby competitors or drives them away. This is in contrast to exploitative competition, in which one competitor exploits a resource more efficiently and thus precludes others.

Riftia. The giant tube worm of the hydrothermal vent community. They seem to be related to members of the phylum Pogonophora, or beard worms, and are characterized by the absence of a mouth, intestine, or anus. *Riftia* house symbiotic bacteria that produce its food.

Stenothermal. A species with a narrow range of temperature tolerance. A eurythermal species has a wide range of tolerance to temperature.

Superheated water. When water is heated under pressure such as occurs in hydrothermal vents, its temperature can exceed the sea-level boiling point of 100°C. Hydrothermal vent water often exceeds 350°C.

Symbiosis. A close interaction of two species in which one species often lives on or in another's body. Parasitism, commensalism, and mutualism are the three major types of symbiosis.

Test Your Learning

After working through these questions, check your answers against the key at the end of this book. If you answered any of the questions incorrectly, review the relevant sections of the text and video episode.

Multiple Choice

1. The most common spatial distribution of organisms is:
   a. random
   b. scattered
   c. clumped
   d. uniform

2. The resource that is typically in the greatest demand in the rocky intertidal is:
   a. nutrients
   b. space
   c. light
   d. moisture

3. A J-shaped population growth curve is converted to an S-shaped population growth curve by:
   a. environmental resistance
   b. increased birth rates
   c. light attenuation
   d. increased resource availability

4. An organism that can move freely between surface waters and great depths would have to be:
   a. stenothermal
   b. euryhaline
   c. stenotrophic
   d. eurybaric

5. Which of these is not a significant member of the hydrothermal vent community?
   a. shrimp and crabs
   b. calcareous algae
   c. sulfur bacteria
   d. octopi

6. Which of these would probably have the lowest biodiversity?
   a. a black sand beach
   b. an estuary
   c. a hydrothermal vent community
   d. a kelp forest

7. An interrelationship between two species in which one is benefited and the other neither benefited nor harmed is called:
   a. inquilinism
   b. commensalism
   c. indirect symbiosis
   d. exploitative competition

8. Which of these communities would have the least variation throughout all ocean latitudes?
   a. coral reef
   b. estuarine
   c. deep ocean
   d. rocky intertidal

9. In which of these regions is life least abundant?
   a. bathypelagic zone
   b. sandy beach
   c. disphotic zone
   d. benthic abyssal plains

10. The deep scattering layer community is found:
    a. in the anoxic regions below fine sediments world wide
    b. migrating daily between the disphotic and euphotic zones
    c. confined to the abyssopelagic realm
    d. between the thermocline and the oxygen minimum layer

11. Which of the following are sometimes referred to as "marine nurseries"?
    a. coral reefs
    b. open ocean photic zones
    c. hydrothermal vent communities
    d. estuaries

12. Periwinkle snails and blue mussels have specialized respiratory systems adapted for life in:
    a. the upper intertidal zone
    b. the oxygen minimum zone
    c. anoxic sediments
    d. thermal plumes

Short Answer

1. Compare and contrast the hydrothermal vent community and the abyssal plain community and explain how the whale fall community might be related to both.

2. Compare the coral reef community with the estuarine community.

3. A useful approximation of the pattern of growth of populations is the S-shaped growth curve. Make a drawing illustrating the features of an S-shaped growth curve and explain it.

Supplemental Activities

1. Construct an imaginary hydrothermal vent community. Design your community so it illustrates the main features of the vent community and present it in a way an elementary school child would understand. The following Websites will help you.

   http://library.thinkquest.org/18828/

   http://www.ocean.washington.edu/courses/oc411/intro.htm

   http://seawifs.gsfc.nasa.gov/OCEAN_PLANET/HTML/oceanography_recently_revealed1.html

2. At the University of Puerto Rico Website:

   http://geology.uprm.edu/Morelock/coral-reef.htm

   go to Reef Controls and explore how macro-scale, meso-scale, and micro-scale factors affect coral reef ecology and then write a short report predicting how you think similar factors would affect an estuarine ecosystem.

3. Some parasites are passed from aquatic organisms to humans. Produce an information pamphlet for distribution at a doctor's office describing the lifecycle, symptoms, dangers, prevention, and treatment of the following parasites: Diphyllobothrium; Trichobilharzia; Chlonorchis.

Lesson 24

Treasure Trove

Overview

The ocean has been viewed as an unlimited treasure of natural resources and wealth. As you will see, this optimistic outlook is on a collision course with human population growth, greed, and mismanagement. The ocean has several possible futures, but many of the resources of the ocean are headed in a direction that can only be described as bleak.

This lesson surveys the four categories of marine resources—physical, biological, energy, and nonextractive. Physical resources are nonliving matter such as oil, salt, and metal ores, while biological resources are living matter such as fish and seaweed. Biological resources include fish, crustaceans, and seaweed. Energy in the form of heat or water motion can be extracted from the ocean to supply energy needs on land. Nonextractive resources do not involve removing matter or energy from the ocean, but refer to use of the ocean for endeavors like transportation—such as the shipping industry—and recreation—such as pleasure boating, sport fishing, beach going, and whale watching. Adding matter or energy to the ocean—as in using it as a waste receptacle for matter or heat—also falls in the nonextractive category.

The ocean's resources can also be categorized as renewable or nonrenewable. Renewable resources can replace themselves through processes such as reproduction. Most biological resources are renewable. Non-renewable resources can't be replaced except by processes that occur over geologic time scales. Most mineral resources are nonrenewable. This lesson will examine marine resources carefully for such details as its economic worth, means of extraction, distribution, abundance, current status, and future prospects.

Oil and natural gas are the ocean's most valuable mineral resources. These hydrocarbons are primarily used as fuels to drive the industrialized world, although the list of products that are manufactured from petroleum—such as plastics—is enormous. Oil from the continental shelves accounts for about one-third of all oil used and oceanic natural gas represents about one-fourth of the world's total production.

The action of waves and currents is a natural sorter of sand and gravel, and beach mining and offshore dredging of sand and gravel produces profits second only to oil and gas. Extraction is often controversial because it may damage beaches, increase the water's turbidity, and increase shoreline erosion, thus damaging coastal ecosystems.

Some of the salts dissolved in seawater are collected profitably by simple evaporation or other extractive processes. The most abundant salt in seawater and the one humans have been collecting for the longest time is table salt, or sodium chloride. Evaporation of seawater collected in shallow ponds leaves a mixture of salts composed of about 78 percent sodium chloride. The evaporating process can be manipulated to yield gypsum ($CaSO_4$), which is used in wallboard, and a variety of other salts. Magnesium is one of the most abundant elements in seawater and is extracted profitably along with magnesium compounds. Magnesium is a lightweight metal used in aircraft and other construction applications. Magnesium chloride ($MgCl_2$) is used to melt ice; magnesium sulfate ($MgSO_4$) is Epsom salt, which has a variety of medical uses. These two salts are also used in the production of tofu from bean curd. Magnesium hydroxide, $Mg(OH)_2$ is used as an antacid and a laxative and also in heat-resistant linings for furnaces. During the evaporation of seawater, water vapor may be collected and condensed to yield fresh water. As the human population grows, potable water is becoming a scarce commodity. Desalination of seawater by evaporation, freezing, and reverse osmosis will become an ever-increasing necessity as existing ground water, streams, rivers, and lakes become depleted or polluted.

While oil, gas, sand, and gravel are the most profitable, and table salt is the oldest, methane hydrate and manganese nodules are perhaps the most interesting of the ocean's mineral resources. The deep sedimentary deposits of *methane hydrate* are the largest reservoir of hydrocarbons on Earth. The deposits are composed of methane and ice and are found buried under about 200 to 500 meters of sediments on the continental slopes. Brought to the surface, methane hydrate sediments fizz as they release methane, and ignited this green, muddy sediment burns vigorously. But a safe and profitable means of extracting this methane has not yet been devised. The extent of these deposits makes them a potential source of clean-burning natural gas, but because methane is a greenhouse gas, its release into the atmosphere could aggravate global warming.

While methane hydrate lies in sedimentary deposits, manganese nodules precipitate from seawater. These lumpy accretions of manganese, copper, nickel, and other metals litter the abyssal seabed. Their location at great depth, international debate over rights to this open sea resource, and the availability of manganese on land presently make their retrieval unprofitable.

Phosphorite is a sedimentary mineral deposit rich in phosphorus and composed of chemicals similar to those that harden your teeth and bones. It is used to make fertilizer, and deposits are found at lesser depths than manganese nodules. Like manganese, the availability of land-based deposits of phosphorite makes its extraction from the ocean unprofitable. This is also generally true for metallic sulfides and muds, which collect around hydrothermal vents. Superheated water percolating through the crust at spreading centers emerges from the vents rich in metal sulfides. Sulfides of zinc, iron, silver, cadmium, and other metals precipitate from hydrothermal waters as they mix with the surrounding cold water. In other areas the metal-rich compounds emerge dissolved in hypersaline water and collect in muddy pools around the rifts. Hydrothermal precipitates and muds are mined in the Red Sea, Gulf of California, and along the Juan de Fuca Ridge off Oregon.

Along with extractable matter, the ocean contains a great deal of energy in its waves and currents that can be harnessed to generate electricity. The physics of using moving water to generate electricity is fairly simple, but the practice of operating large-scale generating facilities in the ocean is not. The future of profitable, environment-friendly generation of electricity from waves and currents remains uncertain. The thermal gradient between cold deep water and warm surface water is also a potential source of energy generation. The *Ocean Thermal Energy Conversion Project* (*OTEC*) in Hawaii demonstrated that electricity can be generated this way. This experimental facility was disassembled in 1997 because it, too, was less profitable than conventional methods of electricity generation.

The biological resources of the sea include fish, crustaceans, mollusks, whales, seaweeds, and others. These are renewable resources but the demand for them by the burgeoning human population has resulted in overfishing and the collapse of several of the world's fisheries. Present methods of food production and distribution from land and sea combined have not been able to prevent widespread global malnutrition and famine. Modern fishing techniques are sophisticated and highly efficient, but it is unlikely that yields of food from fishing can significantly reduce world hunger. The oceans are probably already being harvested near or above their sustainable limits.

Historically, fisheries are a common property resource. In principle, this means that every fish in the sea belongs to everyone collectively until it is caught; it then belongs to the individual who catches it. Common property resources like fisheries are difficult to manage and conserve because although some *fishers* are conservation-minded and responsible in limiting their own catches, others will take as much as possible. When the resource is exhausted, they take their profits and switch to a different source of income. The goal of fisheries management is to control over-fishing so fish stocks can produce at high levels far into the future. There is a maximum level of fishing that any stock of fish can endure and still reproduce sufficiently to replace what is caught. Called the *maximum sustainable yield*, this is the cornerstone of fisheries management. Fisheries management has two major goals: to determine how many fish of each stock may be harvested annually on a sustained basis and to prevent more than that from being taken.

Fisheries management primarily involves managing the people, not the fish. There are two main ways to do that. Fishing can be regulated as to where, when, and how fish may be caught, thereby reducing efficiency. Or limited entry techniques can be employed to limit who is allowed to catch fish. This is achieved by licensing or leasing designated regions of the ocean to a limited number of people. In essence, either anyone who wishes to catch fish is allowed do so but in such a manner that they cannot destroy the stock, or the number of people allowed to catch fish is so limited that they cannot destroy the stock.

Of course, rules are useless unless they can be enforced and that is problematic on the open oceans. Similarly, determining how many fish can be harvested without depleting the stock must be based on scientific data; data that the fishers themselves must often supply in order to regulate their own livelihood. One hopeful strategy designed to reverse the ravages of overfishing is the establishment of marine sanctuaries. The United States has established 13 of these refuges in its territorial waters and the International Whaling Commission has banned whaling in eight million square miles of the Southern Ocean to protect the great whales that migrate there to feed.

The maximum sustainable yield of ocean fisheries is probably between 100 and 135 million metric tons and harvests are already near that level. Approximately half of the fish stocks whose yields are known are already overfished. The doubling time of the human population seems to be about 40 years. Even if the harvest of the oceans can be sustained at 135 million metric tons, the growth of human population will shrink the *per capita* amount of fish available.

The orange roughy became a desired food species in the early 1980s and was fished to *commercial extinction* in 13 years. This is a dramatic, but not isolated case. The amount of fish caught is called the yield or catch and is usually expressed in dollars of profit or tons of fish. By contrast, the cost of catching them is called the effort and might be expressed in dollars of cost, man-hours, or vessel-days. In the early days of exploitation of a stock of fish such as the orange roughy, the yield per effort is high. The yield per effort generally maximizes when the level of fishing approximates the maximum sustainable yield. If fishing pressure continues to increase, overfishing occurs and the catch per unit effort declines. Profits fall and the species may be driven to economic extinction and occasionally to biological extinction. In 1995, the fishing industry expended $124 billion of effort to produce $70 billion of yield. Government subsidies, price supports, and other artificial means encouraged over-fishing resulting in destruction of the stocks, damage to marine ecosystems, and yields that were anything but sustainable. This irrational gluttony has been aptly dubbed "madhouse economics."

In addition to the destruction of desired stocks of fish, fishing poses another difficult problem. It is an imprecise activity and efforts to catch one species often result in collateral damage to others. These unwanted species are called *bycatch* or bykill. Worldwide bykill of unwanted and discarded marine life is about one-third of the total catch. Marine turtles killed in shrimp trawls, dolphins killed in tuna nets and innumerable fish, squid, and marine mammals trapped in drift gill nets are a few examples. Drift gill nets can be 50 miles long. This and other highly efficient contemporary fishing techniques are relentless. Overfishing, bykill, and ecosystem destruction are frequently the result.

Although the destruction of any type of animal life is troublesome to some people, the destruction of marine mammals, and the great whales in particular, is troublesome to most people. These animals seem to have a majesty and, perhaps, phylogenetic closeness to us that makes their destruction seem particularly wanton. The blue whale is the largest animal that has ever lived on Earth. It has been hunted to near extinction even though there is nothing obtained from whales that cannot be obtained elsewhere. After most of the large whales were slaughtered to near extinction, the International Whaling Commission ordered a moratorium on their harvest in 1986. Japan never abided by the spirit of this moratorium and Norway resumed whaling in 1993. These whalers concentrate on the minke whale, the smallest of the baleen whales. Pirating and poaching continue and the toothed whales, including dolphins, are still harvested by Japan and other countries. In the United States, the Marine Mammal Protection Act of 1972 and additional policies and acts of congress have protected these small whales. The California gray whale, an odontocete that spends most of its life near shore within the protected waters of Canada, the United States, and Mexico has made a very strong recovery under this protection.

Other marine biological resources include fur-bearing mammals, seaweeds, algae, and other marine organisms. In the United States a combination of societal pressures and strong marine mammal protection laws have greatly diminished or halted the hunting of seals, sea lions, and sea otters for fur, and the formerly declining species have made good recoveries. Anyone who has visited the waterfront in San Francisco can attest to the abundance of sea lions. Japan leads the world in the utilization of seaweeds for food. Additionally, various chemicals extracted from marine algae have pharmaceutical uses and other chemicals such as algin have a wide variety of applications. Algin, a large polysaccharide molecule, is used as a thickening agent, a stabilizer, a thawing agent, and is found in or used in the manufacture of salad dressings, ice cream, beer, cosmetics, and a seemingly endless parade of other products. It is a sizing used to improve the printing of the college logo on your tee-shirt and the pimentos in green olives are made from a pimento-algin gel. Along with these food and industrial applications, algae and other marine organisms yield an array of molecules of medical interest.

The diversity of organisms in marine ecosystems produce a great diversity of chemicals, many having clinically useful properties. Acyclovir, isolated from a sponge, is an antiviral compound used in treating herpes. A variety of useful and promising antiviral, antitumor, and anti-inflammatory agents are being studied, tested, or used. The combination of marine pharmacognosy, the search for medically useful chemicals from sea life, and biotechnology, the manipulation of the genes that cause the production of these chemicals, holds great promise for fighting disease.

Fish and other marine organisms are the only wild organisms still hunted in significant quantities for human consumption. Agriculture long ago replaced hunting and gathering as the major mode of food production on land and a similar farming revolution is occurring in the sea. On a global scale, food production by aquaculture may soon overtake cattle ranching. China has practiced low-technology freshwater aquaculture for centuries and Asia leads the world in fish-farming. The aquaculture of marine organisms for food and other uses is called *mariculture*. It is a rapidly growing industry led by the heavily-populated island nation of Japan. Shrimp, oysters, mussels, abalone, milkfish, salmon, yellowtail, and algae are among the growing list of species being successfully cultured. Salmon are also farmed. Because they return to their natal stream to spawn, salmon can be raised to fingerling size in the protected environment of a hatchery and then released to migrate out to sea to feed, grow, and mature. In a few years the survivors of this ocean odyssey return to freshwater and are harvested on their spawning runs.

Damage to marine ecosystems is probably at or close to crisis levels in most parts of the world ocean. A rational, enforceable international law of the sea is needed. The 1609 *Mare Liberium* forms the basis of modern international conventions on ocean law. The 12-mile territorial limit and 200-mile exclusive economic zone are useful in fostering regulation of about 40 percent of the ocean. The high seas and the open ocean seabed remain common property and their regulation is presently problematic. Both national and international marine sanctuaries offer some hope but only if enforcement can be successfully implemented.

Focus Your Learning

Learning Objectives

After completing this unit you should be able to do the following:

1. Distinguish among marine physical, biological, energy, and nonextractive resources and give several examples of each. Describe the characteristics and extent of utilization of each resource.
2. Describe how the ocean's thermal gradient is used to generate electricity.
3. Discuss fisheries and their management including the concepts of common property, maximum sustainable yield, overfishing, the relationship between effort and yield, bykill, madhouse economics, and marine sanctuaries.
4. Briefly discuss the history, regulation, and future prospects of the whaling industry.

Include your own thoughtful opinions on this subject.

5. Describe, discuss, and compare aquaculture, mariculture, and marine ranching.

6. Outline and discuss the history and current status of the United Nations International Law of the Sea Convention.

Assignments

This lesson is based on information in the following text and video assignments. The key terms, focus points, and practice test are intended to help ensure mastery of the material presented in this unit of study.

Text: Chapter 17, "Marine Resources," pages 460–487

Video: Episode 24, "Treasure Trove"

Video Focus Points

Review these points before watching the video assignment for this lesson to help you focus on key issues.

— As the human population continues its unbridled growth, humanity is turning to an increasing exploitation of marine resources.

— Seaweeds, including the beds of giant kelp along the California coast, yield a variety of products including foodstuffs and chemicals with pharmaceutical and industrial uses.

— Ocean sediments yield a variety of mineral resources such as sand and gravel, tin and titanium, phosphates, manganese nodules, and even diamonds.

— The most important and profitable mineral resources from the seabed are oil and natural gas. Modern extraction procedures such as horizontal drilling make extraction of submarine oil more efficient.

— As on land, biodiversity itself can be considered a resource. The array of biochemicals produced by marine organisms has provided many useful pharmaceutical compounds already and many more remain to be discovered.

— In contrast to the marine pharmaceutical industry, which is underutilized and in its infancy, the marine fisheries industry is well developed and probably over utilized.

— Overfishing, depletion of stocks, and bycatch plague the world's fisheries.

— Alleviation of these problems requires effective fisheries management but the sampling, modeling, and other activities required for management of wild fish at sea pose a daunting challenge. Establishing marine sanctuaries seems to be a hopeful advancement in fisheries conservation and management.

Text Focus Points

Review these points before reading the text assignment for this lesson to help you focus on key issues.

— Marine resources include physical resources, biological resources, energy resources, and nonextractive resources. Several examples of each of these are defined, discussed, and evaluated.

— In terms of worth, coal and oil and sand and gravel are the most profitable physical resources. Humans have been extracting salt longer than any other ocean mineral and methane hydrate and manganese nodules are perhaps the most interesting mineral resources as yet not widely utilized.

— With continued uncontrolled human population growth, potable water produced by a variety of desalination methods may become the most important ocean resource.

— Moving water in waves and currents and the ocean's thermal gradient are potential sources of energy for electricity generation. Their present utilization is minimal.

— Most fishery resources are probably already being harvested at levels near their maximum sustainable yields. Overfishing, mismanagement, and bykill plague the fishery industry.

— The history of the whaling industry is a classic example of driving species to economic and perhaps even biological extinction. Most of the large baleen whales hunted throughout the history of whaling reached their maximum harvests between 1945 and 1975 and then crashed to economic extinction. Whaling remains controversial and problematic.

— Aquaculture has been successfully practiced for millennia, especially in fresh water. Modern techniques of mariculture or ocean farming provide some hope for supplementing fisheries products obtained by hunting the wild stocks.

— Managing the resources of the open ocean is difficult. Most of the ocean is considered high seas and owned collectively by all nations. The United Nations International Law of the Sea Convention forms the basis of management of open waters. Recognition of a 12-mile territorial zone and a 200-mile exclusive economic zone for coastal nations permits more efficient regulation by placing approximately 40 percent of the oceans under the regulation and responsibility of individual nations.

Key Terms and Concepts

A thorough understanding of these terms and concepts will help you to master this lesson.

Bycatch. When one type of organism such as tuna is harvested, another type of organism such as dolphins may be unintentionally killed. The unintended kill is called bykill or bycatch.

Commercial extinction. Depletion of a fishery species to such low levels that it is not profitable to harvest. Overfishing is the major cause although this can be exacerbated by habitat destruction and pollution.

Exclusive economic zone. An ocean region within which a coastal nation retains sovereignty over the resources, environmental protection, and economic activity. It generally encompasses a band extending 200 miles from the nation's coast.

Fisher. The gender-neutral term used to describe people who hunt and gather fish and other aquatic animals. They may use primitive or modern techniques and were formerly called fishermen. People who farm the sea are called mariculturists.

Hydrocarbon. The class of organic molecules made of hydrogen and carbon. We mostly use hydrocarbons as fuels. Methane, propane, kerosene, and oil are examples.

Mariculture. Mariculture is ocean aquaculture or marine farming. Oysters, milkfish, shrimp, and salmon are among the most important mariculture species.

Maximum sustainable yield. The maximum number or weight of a particular fisheries species that can be harvested annually without depleting the stock.

Methane hydrate. A widespread, deep, sedimentary deposit of methane mixed with ice. It is the largest known reserve of methane but its extraction is currently not feasible.

Natural gas. Gaseous hydrocarbons. Natural gas is composed of methane, CH_4, the simplest hydrocarbon molecule. It is often found associated with oil and is mostly used as fuel in homes, industry, and occasionally in vehicles.

Ocean Thermal Gradient Conversion Project (OTEC). The Ocean Thermal Gradient Conversion Project is an experimental facility on Hawaii that demonstrated the feasibility of utilizing the difference in temperature between the deep and surface ocean to generate electricity.

Stock. A group of fish managed as a unit. Stock is a management term and may or may not be the same as a biological population. Occasionally a stock may include two similar species managed as a unit.

Test Your Learning

After working through these questions, check your answers against the key at the end of this book. If you answered any of the questions incorrectly, review the relevant sections of the text and video episode.

Multiple Choice

1. Which of these marine products can be found in ice cream, cosmetics, paints, shaving cream, and salad dressing?
   a. algin
   b. phosphorite
   c. iron sulfide
   d. gypsum

2. Second to oil and gas the most profitable marine mineral resource is:
   a. phosphorite
   b. iron and other ferrous metals
   c. coal
   d. sand and gravel

3. The International Law of the Sea Convention established the size of exclusive economic zones at ____ nautical miles.
   a. 3
   b. 12
   c. 200
   d. 360

4. Present levels of growth and exploitation of marine resources are
   a. sustainable
   b. unsustainable
   c. endorsed by the U.S. EEZ project
   d. decreasing annually

5. The country where fresh water aquaculture is most widespread and successful in producing food for human consumption is:
   a. Israel
   b. China
   c. Japan
   d. Holland

6. The fishing industry today faces many challenges including all but one of the following:
   a. it is approaching the maximum sustainable yield
   b. overfishing
   c. excessive mariculture
   d. destructive fishing methods

7. Which of these is a nonextractive marine resource?
   a. electricity generated by thermal gradient technology
   b. manganese nodules
   c. fresh water produced by desalination
   d. recycling ocean water to cool a coastal power plant

8. About ___ percent of the world's oil production comes from the seabed
   a. 5
   b. 17
   c. 25
   d. 34

9. The two major whaling nations are Japan and:
   a. China
   b. Russia
   c. Norway
   d. Iceland

10. The major species of fish harvested from the sea is:
    a. pollock
    b. anchoveta
    c. tuna
    d. whiting

11. A common Bryozoan (a small encrusting invertebrate) has been found to produce a chemical useful
    a. in controlling the growth of cancer cells
    b. as a dental cement
    c. as an antibiotic
    d. to reduce blood pressure

12. Which of these is a hydrogenous sedimentary deposit that precipitates from seawater?
    a. manganese nodules
    b. phosphorite
    c. gypsum
    d. diamonds

Short Answer

1. Outline and discuss several aspects of the fisheries management problem.
2. Explain how the ocean's thermal gradient can be used to generate electricity.
3. Fishing is aquatic hunting while aquaculture is aquatic farming. Discuss ocean aquaculture.

Supplemental Activities

1. It should be clear to you that we stand at a crossroads with regard to the oceans and that they are in peril unless several harmful trends are reversed. Download the executive summary of the Pew Oceans Commission from their Website at http://www.pewoceans.org/. Explore this Website and read the parts that interest you. Read the executive summary and choose five points from it that were not discussed in video Episode 24 or Chapter 17 and

choose five points that were discussed but are expanded in the summary. Briefly discuss each of the ten points.

2. Go to the Smithsonian/NASA Ocean Planet Website. Enter the exhibit and go to the Sea Store at http://seawifs.gsfc.nasa.gov/ocean_planet.html. Explore the store and find 20 interesting products, services, or uses of sea resources that were not mentioned in the text or video. Write a few sentences about each resource.

3. Go to the National Marine Fisheries Service Website at: http://www.nmfs.noaa.gov/. Click on each of the categories in the "Features" Section and find one fact in under each topic that relates to the material in Lesson 24. Write each fact and briefly describe its relationship to this Lesson.

Lesson 25

Dirty Water

Overview

Marine pollution involves materials or energy that affect the ocean's physical, biological, or chemical environment in a negative way. There are, and always have been, naturally occurring agents of pollution such as volcanic gases, animal waste, and natural terrestrial runoff that enter the world ocean. What makes even these substances dangerous is volume; being present in excess of normal levels. In addition to naturally occurring substances, there are materials generated by humans which, when introduced into a natural environment in excess of the assimilative capacity of that environment, can have negative effects. It is estimated that about 75 percent of ocean pollution comes from human activities on land, and that there are now few, if any, marine ecosystems that are free from *pollutants*.

Other criteria for evaluating the effects of pollutants include the toxicity of the agent and the sensitivity of the affected organisms or ecosystems to those toxins. Length of exposure to the pollutant is an important factor, and related to its *biodegradability* or the time needed for it to decompose and lose toxicity.

Because of its tendency to cover large areas of the ocean surface and the nearshore, pollution by oil seems to receive the most attention, and the public image of ocean pollution most usually recalls massive oil spills, ships aground, and well blowouts. Even though there are numerous natural oil seeps, they account for only about 10 percent of the oil released into the ocean; most of the rest being from urban runoff and vessel spills. There are substantial differences in the effects of the different types of oil—crude, refined, and used refined (motor) oil—the last being the most toxic due to the additional chemicals it picks up as it goes through an engine. The effects of oil on marine organisms and habitats depend on the location and size of the area impacted, and the type and amount of oil released. Depending on the circumstances, organisms can be poisoned, smothered, or have certain biological functions temporarily or permanently impaired.

Mitigation and cleanup procedures related to oil spills also vary with the location and the type of oil. Oil spilled at sea will usually begin to dissipate within a few days, with the lighter components evaporating into the air or forming a surface slick, and the heavier ones sinking to the bottom. If the oil reaches the intertidal zone, cleanup can range from steam cleaning the rocks with hot water to the use of chemicals. As a result of the *Exxon Valdez* disaster in 1989, it has become clear that in many cases the cleanup procedures are more harmful than leaving the oil to the natural processes of bacterial and chemical degradation.

Although oil spills receive the most media attention due to their magnitude and obvious effects on wildlife, there are other categories of marine pollution that are not so spectacular, but can be equally insidious. Heavy metal contaminants, including mercury, lead, copper, and tin, can enter the marine ecosystems through industrial and urban runoff, landfills, gasoline leakage, marine antifouling paints, and a number of other sources. Even small quantities of these elements can be toxic, interfering with an organism's cellular metabolism. Humans can also be affected by consuming contaminated seafood, as exemplified by the effects of mercury on the residents of Minimata, Japan in the mid-1900s. Synthetic compounds can be especially toxic even in small amounts, because the materials of which they are composed resemble nothing present in nature and therefore tend to resist biodegradation. Synthetic hydrocarbons are the best known of these, and are found in many common household, agricultural, and industrial chemicals. The pesticide DDT and its relatives were used heavily in the mid-1900s, and found to have such wide-ranging negative effects that its use was banned in the United States. DDT did most of its damage through the process of *biological amplification*, which occurs when a toxic chemical enters a food web and works its way up the food chain through consumption and concentration from the lowest to the highest trophic levels.

While many pollutants can eventually biodegrade and dissipate, there is a category of solid waste that does not break down for many years. The most insidious of these materials are the plastics, which usually make their way to the ocean from landside trash dumps and careless discarding. These pieces of plastic can smother, strangle, or choke marine creatures, continuing to be a threat for many years due to their lack of biodegradability.

As noted earlier, one of the main descriptive criteria for a pollutant is that it is present in excess of the ability of the environment or organism to assimilate it. This can even occur when excessive nutrients are present, as is the case during *eutrophication*. The addition of excessive amounts of nutrients into an ecosystem stimulates the rapid growth and expansion of some organisms to the detriment of others, thus upsetting the natural balance in the system. This selective growth of some organisms can affect others by using up their oxygen, releasing toxins, and by physically smothering or choking some of the resident species.

Some sources of these extra nutrients in the marine environment are urban runoff, wastewater treatment plant discharge, agricultural fertilizers, and excessive soil erosion due to heavy rain. They can enter the ocean from storm drains, streams, and river deltas, and are especially threatening in enclosed bays and estuaries because the poor circulation of these configurations tends to concentrate the materials and retain them for long periods. The recent trend toward development of natural bays and estuaries into recreational boat marinas exacerbates this problem because of the release of oil and other pollutants from the numerous vessels.

The disposal of animal wastes—human, pet, and agricultural—is directly related to the overpopulation of coastal areas. Although this waste, collectively called *sewage*, is mostly water, it is the solid material, or *sludge*, remaining after treatment that is the source of concern. Millions of tons of sewage are processed annually in the United States, and much of the effluent, the liquid component, is discharged into the ocean through pipelines, after various levels of treatment. Most treatment attempts to kill the disease-causing microbes and to reduce the nutrient levels, but even with that the discharged effluent can affect nearshore ecosystems and cause eutrophication. Sludge disposal is much more complicated: the solid waste is subjected to digestion (breakdown) processes, then it is dried, burned, or dumped into landfills, depending on the local, state, and federal regulations.

Another source of pollution is the discharge of waste heat energy from coastal power-generating plants that use seawater as a coolant. Some power plants are now using alternating heating-cooling procedures to help mitigate the warming effects on the surrounding waters.

Excessive amounts of natural sediments (sand, silt, mud) from agricultural and urban runoff, mining and forestry activities, and erosion from heavy rains, can affect coastal ecosystems by burying them or otherwise altering their natural configurations.

A fairly recent concern is the introduction of new, *exotic species* into an ecosystem through the flushing of ballast water from foreign-based transport ships. These organisms, usually in larval form, are taken into the ships ballast tanks as they are filled prior to a voyage. When the ship reaches its destination port and prepares to take on cargo it pumps its ballast water into the destination port harbor, releasing the transient larvae. Often these organisms will mature and adapt rapidly to the foreign waters, overcoming resident species.

The world's coral reefs have been under attack by incidents of coral bleaching, apparently caused by water temperatures in excess of normal summer highs for more than a few weeks. Added to that are activities like introducing chemicals to anesthetize fish for aquarium collection, dynamiting the reefs to kill fish for food, and destroying the reef structure for construction materials.

Global environmental issues related to human activities include ozone layer depletion and global warming. Both of these conditions are linked to the intimate relationship between the ocean and the atmosphere, causing atmospheric pollutants to be introduced into the oceans. *Ozone* is a gas composed of three atoms of oxygen, which occurs naturally as part of a multi-gas layer that surrounds Earth at an altitude of about 12 to 25 miles. This *ozone layer* is important to Earth's ecosystems because it interrupts some of the high-energy ultraviolet radiation coming from the sun before it reaches Earth. Over time the organisms exposed to the radiation that naturally reaches Earth have evolved protective mechanisms, but certain types of atmospheric pollution have recently reached the ozone layer and caused it to weaken and decline. This allows abnormal amounts of ultraviolet radiation to strike Earth's biota, and to damage living things by breaking their nuclear DNA strands and causing abnormal cellular activity. All organisms exposed to sunlight are potential victims, with the amount of injury being related to their location on the planet and the amount of sunlight received. Holes in the ozone layer have been detected over the United States, Australia/New Zealand, and Antarctica since the 1960s, with the largest being over Antarctica. Humans are affected indirectly through injury to both terrestrial and marine ecosystems, and directly through an increased threat of skin cancers. The chemicals that affect the ozone layer are found in many industrial and household chemicals, and are converted by sunlight into atmospheric gases, which can attack and deplete the ozone. The actual amount of depletion depends on the location of the pollutants and the amount of sunlight an area receives. A number of nations, including the United States, have signed treaties agreeing to reduce or ban the production of these pollutants.

Historic records indicate that after each ice age, Earth's temperature has risen slowly. Recent records, however, indicate an unnaturally rapid acceleration in temperature rise since the last ice age, and it has been related to a phenomenon called the greenhouse effect. A combination of atmospheric gases, collectively called *greenhouse gases*, has interfered with the natural radiation/absorption/reradiation cycle of the sun and Earth, causing abnormal amounts of the sun's energy to be trapped in Earth's atmosphere instead of being reradiated back to space. This entrapment is called the *greenhouse effect* because of its similarity to the way the glass in a greenhouse traps the sun's energy to warm its interior. Some greenhouse effect is necessary to maintain Earth's heat balance, but the amount of heat energy now being retained is upsetting that balance. The main source of the excess greenhouse gases is industrialization, particularly the use of fossil fuels. Power plants and other human-related industries and activities have contributed to the increased levels of these gases, which occur naturally in the atmosphere. Carbon dioxide is the most troublesome. Under normal conditions, the ocean is able to absorb and balance the carbon dioxide in the atmosphere, but it is now being produced in amounts too great for the ocean to absorb.

Concerns about the overall effects of global warming include the global rise in sea and land temperatures and the rapid rise in sea level as the ice sheets melt. These temperature rises affect both terrestrial and marine ecosystems, and the rapid rise in sea level could cause destruction in the heavily populated coastal areas. Since much of the problem is due to the release of greenhouse gases during the burning of fossil fuels, alternative power sources are being sought. Nuclear power is a logical alternative, but poses the problem of how to dispose of the spent fuel. Remedies for these and other global concerns eventually focus on over-population, and the inability of Earth to be a dump for unlimited amounts of human waste. Earth's environment is a delicate balance that must be maintained if it is to survive and flourish.

Focus Your Learning

Learning Objectives

After completing this unit you should be able to do the following:

1. Recognize and understand the terminology used to define and discuss the human activities that impact natural processes and ecosystems.
2. Compare and contrast natural pollutants with human-generated pollutants.
3. Discuss oil pollution—types of polluting agents, their sources, and the habitats and organisms affected.

4. Define and explain eutrophication, including the substances involved, their sources, and how organisms and habitats are affected.
5. Discuss pollution by heavy metals—the types of metals involved and their sources, organisms and habitats affected, short- and long-term residual effects.
6. Recognize the various types of synthetic products that can pose environmental threats, particularly the organic chemicals and plastics; their short- and long-term effects on ecosystems, and the concept of biological amplification.
7. Understand the potential environmental impact of sediments, from both natural and human-related sources.
8. Discuss sewage pollution—types, sources, potentially harmful components, effects on organisms and ecosystems.
9. Recognize and understand the environmental threats posed to marine ecosystems by waste heat energy and by introduced species.
10. Explain the concepts of environmental alteration at the global level, especially ozone layer depletion, global warming/greenhouse effect, nuclear energy/ionizing radiation, sea level rise.

Assignments

This lesson is based on information in the following text and video assignments. The key terms, focus points, and practice test are intended to help ensure mastery of the material presented in this unit of study.

Text: Chapter 18, "Environmental Concerns," pages 488–523

Video: Episode 25, "Dirty Water"

Video Focus Points

Review these points before watching the video assignment for this lesson to help you focus on key issues.

— Oil pollution in the ocean comes from urban runoff, tanker spills, and oil released as a result of recreational boating.

— In offshore oil spills, the oil may disperse at sea and have minimal impacts; close to shore the oil may affect sea birds and mammals, mid-water and benthic fishes, and invertebrates, plankton, and algae.

— The oil and/or its various chemical components can narcotize and poison living organisms, upset their short- and long-term metabolic activities, and lower reproductive potential.

— Oil spills that come ashore can also affect the biology of the intertidal organisms, in both the short and long term, but often the cleanup procedures are more destructive than the actual oil contact. Sometimes it may be better to let natural processes decompose and disperse the oil.

— Because they are partially enclosed by definition, the highly-productive estuary ecosystems may be particularly affected by oil pollution and by the substances brought into their waters by freshwater urban runoff.

— Eutrophication is caused by abnormally high levels of nutrients, which may cause the overpopulation of an area by undesirable species.

— Heavy metals and other human-generated compounds may also cause damage to marine environments.

— The phenomenon of global warming has also been related to human activity, particularly the greenhouse effect caused by an excess of carbon dioxide in the atmosphere, which traps the sun's energy close to Earth and causes Earth to heat up.

— The threats of global warming include the effect of increased water temperatures on marine ecosystems and the global rise in sea levels due to melting ice at the poles.

— Global warming and the resulting temperature changes will also affect the global climate and weather patterns.

Text Focus Points

Review these points before reading the text assignment for this lesson to help you focus on key issues.

— As exemplified by the history of civilization on Easter Island, we must realize that humans have always exploited our resources, and now that exploitation and its after-effects are having a global impact. Of particular concern is the introduction of excessive amounts of natural and synthetic compounds to the oceans, many of which are toxic to marine ecosystems at some level.

— Marine pollution is the "introduction into the area, by humans, of substances or energy, that change the quality of the water . . ." A pollutant is the particular substance that causes the detrimental effects.

— Pollution by oil is currently causing the most concern, since in its various forms it is the most widespread and can damage the largest variety of ecosystems.

— Crude oil, refined oil, and used oil all have different compositions and therefore have different potential negative effects on marine organisms.

— Depending on the type of oil, its method of release into the ocean, and the habitats it encounters, oil will have various effects and will require specific cleanup procedures. Often, however, the cleanup procedure can be more destructive than allowing natural processes to decompose and disperse the oil.

— Heavy metals from industrial processes, antifouling paint for ships, urban refuse disposal, and air-based pollutants can also enter the sea and damage marine organisms and habitats.

— Synthetic organic chemicals and plastics are often introduced into the sea through spills, leaks, and illegal disposal, but may not decompose and disperse by natural processes because they are composed of nothing resembling nature. These can get into marine food webs and cause damage at all trophic levels, particularly through the process of biological amplification.

— Eutrophication is the abnormal population growth of certain species, to the detriment of the natural system balance, due to excessive amounts of certain nutrients entering the water. These nutrients are usually the product of terrestrial runoff (both natural and human-related) into the sea, and can be especially devastating in semi-enclosed bays and estuaries.

— The category of solid-waste pollution refers mainly to plastics, which may take hundreds of years to decompose and which, depending on form and composition, can strangle or choke organisms that encounter them.

— Abnormal amounts of sediment (silt and mud) can smother marine habitats and their occupants when introduced into the marine environment from agricultural runoff, mining and forestry activities, and urban runoff. Even excessive sediment resulting from natural processes like heavy rains can have such effects.

— Disposal of animal waste (sewage) has always been a problem, now exacerbated by over-population of humans and, often, their pets and farm animals. Treatment and disposal of these excessive amounts of waste continues to impact the world oceans.

— Of somewhat less magnitude, but still of concern, is the effect of waste heat energy (from coastal power plants), and of the numerous exotic/introduced species that enter the oceans from the ballast water of ships and often overwhelm local species.

— Increasing chemical pollution may be affecting the health of coral reefs, although some of their demise has resulted from other types of human activity such as fish collection and extraction of construction materials.

— Pollution of the oceans and the atmosphere have been linked to the depletion of the ozone layer and to the global warming phenomenon. Of particular concern is the increase in exposure to damaging ultraviolet radiation (due to depletion of the ozone layer) and the warming of Earth's lands and waters due to the greenhouse effect. This rise in temperature can affect the global climate and weather patterns, and cause significant rises in sea levels due to melting ice at the poles.

— Reference is made to a paper by Garrett Hardin, outlining the limits of Earth's capacity to support life. Solutions to these problems are suggested—cutting back on resource use, addressing over-population as the basic root cause, and acknowledging the reality of the population/resource conflict.

Key Terms and Concepts

A thorough understanding of these terms and concepts will help you to master this lesson.

Biodegradable. Able to be broken down by natural processes into simpler compounds.

Biological amplification. Increase in the concentration of certain fat-soluble chemicals such as DDT or heavy-metal compounds in successively higher trophic levels within a food web.

Chemical dispersants. Chemicals designed to break down the molecular structure of a material (oil, in this case), causing it to disperse and to speed up its natural decomposition.

Chlorinated hydrocarbon. The most abundant and dangerous class of halogenated hydrocarbons; synthetic organic chemicals hazardous to the marine environment.

Chlorofluorocarbons (CFCs). A class of halogenated hydrocarbons thought to be depleting Earth's atmospheric ozone. CFCs are used as cleaning agents, refrigerants, fire extinguishing fluids, spray-can propellants, and insulating foams.

Eutrophication. A set of physical, chemical, and biological changes brought about when excessive nutrients are released into water.

Greenhouse effect. Trapping of heat in the atmosphere, incoming short-wavelength solar radiation penetrates the atmosphere, but the longer-wavelength outgoing radiation is absorbed by greenhouse gases and re-radiated to Earth, causing a rise in surface temperature.

Greenhouse gases. Gases in Earth's atmosphere that cause the greenhouse effect, including carbon dioxide, methane, and CFCs.

Introduced (or exotic) species. In the present context, organisms, usually in larval form, that are transported from the seas on one country to another via ships ballast, and released into foreign port waters.

Ionizing radiation. Fast-moving particles or high-energy electromagnetic radiation emitted as unstable atomic nuclei disintegrate. The radiation has enough energy to dislodge one or more electrons from atoms it hits to form charged ions, which can react with and damage living tissue.

Marine pollution. The introduction by humans of substances or energy into the ocean that change the quality of the water or affect the physical and biological environment.

Narcotization. The deactivation of an organism's nervous system and senses by chemical means; designed to desensitize but not to kill.

Nuclear energy. Energy released when atomic nuclei undergo a nuclear reaction such as the spontaneous emission of radioactivity, nuclear fission, or nuclear fusion. About 17 percent of the electrical power generated in the United States is provided by the nuclear fission of uranium in civilian power reactors.

Ozone. The tri-atomic form of oxygen. Ozone in the upper atmosphere protects living things from some of the harmful effects of the sun's ultraviolet radiation.

Ozone layer. A diffuse layer of ozone mixed with other gases surrounding the world at a height of about 12 to 25 miles.

Pollutant. A substance that causes damage by interfering directly or indirectly with an organism's biochemical processes.

Polychlorinated biphenyls (PCBs). Chlorinated hydrocarbons once widely used to cool and insulate electrical devices and to strengthen wood or concrete. PCBs may be responsible for the changes and declining fertility of some marine mammals.

Sewage sludge. Semisolid mixture of organic matter, microorganisms, toxic metals, and synthetic organic chemicals removed from wastewater at a sewage treatment plant.

Test Your Learning

After working through these questions, check your answers against the key at the end of this book. If you answered any of the questions incorrectly, review the relevant sections of the text and video episode.

Multiple Choice

1. One of the negative effects of coastal power generating plants is:
   a. eutrophication
   b. elevated water temperatures due to cooling process
   c. decreased water temperatures due to cooling process
   d. attraction for invasive species

2. The process of progressive concentration of materials in the marine food web is called:
   a. eutrophication
   b. disruptive metabolism
   c. biological amplification
   d. selective biodegradation

3. In the greenhouse effect, excess amounts of certain atmospheric gases trap what kind of radiation?
   a. nuclear (from uranium)
   b. ultraviolet
   c. infrared
   d. all of the above

4. In the 1950s and 1960s the residents of Minimata, Japan were poisoned by ________ released into the ocean.
   a. tin
   b. copper
   c. lead
   d. mercury

5. During the early stages of an offshore oil spill, many organisms can become ____________ by high concentrations of benzene, xylene, etc.
   a. narcotized
   b. paralyzed
   c. euthanized
   d. impotent

6. Which of these is not a major effect of plastic waste?
   a. disruption of reproductive activities
   b. a Texas-sized mass of debris in the Pacific ocean dubbed the "Eastern Garbage Patch"
   c. thousands of albatross chicks die each year when their parents feed them bits of plastic mistaken for food
   d. fish and sea birds are strangled by six-pack rings

7. A substance is considered to be biodegradable if:
   a. it can be broken down by natural processes
   b. it spends no more than one year in the environment
   c. it can be used as a nutrient by the local organisms
   d. none of the above

8. Often it will take ____________ for a plastic six-pack holder to decompose in the marine environment.
   a. 10 years
   b. centuries
   c. 150 years
   d. none of the above

9. About _______ percent of pollutants reaching the ocean come from land-based sources.
   a. 62
   b. 91
   c. 77
   d. 47

10. Which of these is the ideal "method of best choice" for cleaning up the shoreline after an oil spill?
    a. steam cleaning
    b. natural degradation
    c. compatible chemicals
    d. none of the above

11. Eutrophication may occur when high levels of ___________ are introduced into an environment.
    a. toxic chemicals
    b. refined oil
    c. nutrients
    d. metabolic waste products

12. Concerning oil spills, the greatest habitat destruction has occurred:
    a. in highly productive intertidal regions
    b. at the air-water interface
    c. when the heavy oil contacts the sea floor
    d. in bays and estuaries

Short Answer

1. Using the information from the text chapter, compare the toxicities and potential for environmental damage of crude oil, refined oil, and used refined (motor) oil.

2. Discuss the process of eutrophication: its definition, potential causes, and types of organisms and habitats affected.

3. At the beginning of the text chapter for this episode, the tragic history of Easter Island sets the stage for our discussion of human-related environmental disasters. List, describe, and discuss briefly the sequence of events involved in that history.

Supplemental Activities

1. Your text (p. 530) discusses a pivotal paper by Garrett Hardin (*Science*, 1968), entitled "The Tragedy of the Commons" dealing with the carrying capacity of Earth, and its limitations. All through the present lesson the causes of the various types of pollution inevitably point to one thing: overpopulation. Over 200 years ago (1798) Thomas Malthus recognized this same problem and wrote a lengthy essay entitled "An Essay on the Principle of Population," which has been called the most significant essay on population ever written. Even today, Malthus' paper is considered to be a landmark, and serves as a logical basis upon which Hardin and others could expand. The Malthus essay is very long and detailed, but for most purposes the important concepts are found in the last part of Chapter 1 and in Chapter 2. Chapter 2 can be retrieved from http://www.marxists.org/reference/subject/economics/malthus/index.htm. This chapter will segue nicely into Hardin's 1968 article, obtainable at http://www.constitution.org/cmt/tragcomm.htm.

   The above Website also includes Hardin's extensive paper in its entirety, followed by a number of reviews and comments from various authors, which can serve as summaries for his original paper. However, for our present purposes, it is probably sufficient to refer to the comments on Hardin's paper in our text (p. 59–520). Read both of these papers (Malthus Ch. 2 and the 1968 Hardin paper and/or the text summary), then write a summary report tying them together and relating their concepts/theories/philosophies to the present lesson, and selecting one or two of the main pollution issues as examples (oil, greenhouse gases). This exercise will also be an introduction to some of the major historical information sources that can serve you well in future discussions about everything from birth control to solar-powered vehicles.

2. The descriptive term "biodegradable" is used a number of times in this lesson, and has become one of the watchwords for environmental awareness. It has also been said that of environmental buzzwords, "biodegradable" has perhaps been the most misused and is perhaps the most difficult to understand. The text presented one definition for the term, although most of us probably already know what it means and which things we use in our daily routines are, or are not, biodegradable. There may be some surprises in store, however, and it is definitely worthwhile to revisit the concept and get an update. First, visit the Websites below.

   http://ecomall.com/greenshopping/organic-sol.htm

   http://environment.about.com/od/earthtalk-columns/a/biodegradable.htm

   Then, visit your local supermarket and make some notes on which products (cat litter, laundry and dish soaps) are labeled as being biodegradable. Most of those labels will have a home page for further information. Find out what you can about them. Then, inventory your home supplies of these materials, and write a short report about your general findings on the mechanisms of biodegradability, the general percentage of supermarket products that profess to be biodegradable (either totally or partially), and then how your household inventory fits into all this. In addition to above Website, just searching on the term "biodegradable" may lead to some more interesting facts.

3. So, you got your car washed and waxed, but after only a few days, when you went to clean your windshield, you came away with an absolutely filthy paper towel. What is all of this dirt, and where did it come from? The last answer is easy—it is atmospheric pollution, and travels in the air. But have you ever really looked to see what it is made up of? Of course, this varies depending on where you live, but most of it is probably just plain dirt (dust and fine sand carried on the breezes), and there are most likely

bacteria and mold spores (which are everywhere but which you can't see—yet!). To get an idea of what comprises your resident atmospheric debris (except for the gases), try this simple experiment. First, obtain an ordinary disposable plate—paper is good, but the surface of a foam plate tends to hold fine material a little better; plastic does not work well. Place the plate on a flat outdoor surface, where it will be the least likely to be moved or blown away, and weigh it down with a small rock. After a few days, remove the weight, cover the plate carefully with foil or plastic wrap (so your debris is not disturbed) and take it indoors to a level surface. If you have access to your school's microscopes, look at the debris on the plate (a good hand lens, such as the one you used to make your sediment display in an earlier lesson, will also work) and see what you can identify (sand, dust, metal particles, soot, etc.). If you wish to quantify your effort, mark the surface of the plate into a grid before you place it outdoors, then you can note the numbers of each different kind of particle per grid unit. Since this is an experiment, write your report along the lines of a scientific journal (see your text pages 5–8, especially the format in Figure 1.4, although you won't be able to formulate a law just yet).

As mentioned earlier, there are probably some bacteria and mold spores that you cannot see, but which are always there. To see those, visit your school's biology or microbiology lab and ask for a couple of Petri dishes filled with any kind of nutrient agar. With these, we will capture the air debris and try to grow these living microbes so we can see them. To do this, choose a level outdoor space, preferably away from direct sunlight, which will dry up the agar and inhibit the growth of the microbes; use your paper plate site if it fits these requirements. Place the Petri dishes on the surface and remove the lids. Let the Petri dishes sit for about 24 hours, then replace the lids and let the dishes incubate (indoors preferably) for another 48 hours. The molds and bacteria from the air will grow on this nutrient media, and should begin to show colonies within that time. The bacteria will show up as small, usually shiny, dots (which will continue to spread over time), and the molds as fuzzy spots. Identification beyond the bacteria/mold stage requires special culturing methods and is not necessary for this experiment). If you do this experiment, document it along with your paper-plate work. So, this is what you breathe every day, and what goes into your oceans and water supplies, and the reason that all of this does not kill you is because your body has accommodated to most of it. However, when the pollutants are found in excess of what your systems can handle, they may overcome our resistance and cause discomfort or illness (allergies, smoggy eyes).

Lesson 26

Hands On

Overview

This guide begins by stating that oceanography is the scientific study of the oceans. The course has surveyed oceanography and has not been spare in the number of facts, ideas, and major theories explored. Science is defined as a body of publicly-verifiable facts, ideas, and theories, but scientists emphasize the inquiry by which these facts are discovered. In other words, oceanography is not just the facts published in your textbook and elsewhere; it is also the processes used to discover them. Speaking about his students who spend a full semester at the marine lab on Catalina Island in California, Anthony Michaels notes:

> "...it lets them...get fully immersed in the projects, fully immersed in the subject and to really understand science as a creative process. They don't just do the normal lessons... follow the directions through a book. They do experiments on things that have never been done before. They look at the physiology of an organism that isn't known. And by doing that, they see science as exploration, science as creating new knowledge."

This course is concluded by reexamining the question, "What is oceanography?" Your answer to this question now should be significantly expanded from what it was at the outset. The questions of who does oceanography, why they do it, and why they seem to be having so much fun are also addressed. Furthermore, you are invited to remain aboard for the endless voyage as an interested citizen or, perhaps, a professional oceanographer.

Science is characterized by empiricism and curiosity. It is also marked by rigor, fervor, and joy. Like all scientists, oceanographers are people who find joy in seeking answers to questions that interest them. And contrary to the caricature of an isolated individual in a lab coat, science is a community process. It is fostered by cooperation yet driven by competition. And it is fun. J. William Schopf states:

> "Science is really good stuff ...it's fun to discover new things and it's fun to do the work. I'm just one of these fortunate people that...they actually pay me money to do what I think is just really fun."

The facts of science must be at once empirically verifiable and potentially falsifiable. That is, they must be able to be proved correct if they are correct and proved wrong if they are wrong. If a scientist succeeds at the former and fails at the latter, then he or she may proclaim with pride and certainty to be tentatively sure they are correct. Science does not have absolute truth. Even the most well-established theories are always open to re-interpretation if new, conflicting information comes to light. Science recognizes those who add supporting evidence to strengthen existing theories, but it stands in ovation for those who turn science on its head and make it abandon its most cherished paradigms. Wadati, Benioff, Hess, and Wilson are famous for their contributions to plate tectonic theory but Wegener, the revolutionary, will always be the most famous, unless, of course, another scientist shows him to be wrong.

Humans are innately curious. We want to construct knowledge, understand the universe, and find truth. Science is one way to do those things. Faith and the arts and humanities are other ways of knowing. But science differs from faith and art in its empiricism. The facts of science cannot be accepted on faith. It may seem that science is limited by the boundaries of empiricism. It is. It may seem like a box within which is found only the knowledge that is empirically verifiable and potentially falsifiable. But it is not. The boundaries of science's box are fuzzy and we are commended for thinking outside them. For example, it seems unlikely that any of the several hypotheses about the origin of life will ever be verified empirically, and most are not presently falsifiable. Does this mean that study of the origin of life by science is impossible or a waste of time? On the contrary, because it lies at the boundaries of science where verification and falsification are difficult, it captures our attention and curiosity much more than questions that are more mundane and answerable.

The boundaries of science are not only fuzzy, they are ever-expanding. Some oceanographer somewhere will add a new fact to science today. Certainly many of the scientists in *The Endless Voyage* videos and their students as well, have made discoveries since the time the videos were recorded. One may isolate a marine biochemical that retards the growth of tumors and appear in newspapers worldwide. Another may discover a new species of worm that will be of interest to fewer than a dozen other scientists on Earth. One may be famous and the other not. But both will have made significant contributions and be pleased that they can begin tomorrow on the next problem.

There may be a tendency to diminish the role of the affective realm in the empirical world of science but it is there. The single consistent message in Episode 26 is just that. Oceanographers love their work, they are in awe of that which they study, and for many there is a mystical dimension to what they do. Inspiration and creativity are as essential for science as they are for art.

The scientists in *The Endless Voyage* study a wide variety of questions and reemphasize the point that oceanography is multidisciplinary. But the fact that they came to the profession of oceanography for many reasons is demonstrated as well. In addition to a spiritual or romantic dimension, oceanographers are certainly motivated by curiosity, whether it arose during a memorable experience at the seashore as a child, or on the job as a young adult. They are motivated by inspirational mentors. They even have practical and pragmatic motivations. Whatever the motive, the quest to understand the oceans is universal. Any oceanographer would agree that what we don't know is always more interesting than what we do. As Kenneth Coale states:

> "As much as we think we know about the oceans, there's still a lot more we don't know about the ocean. It's up to us to teach our students to have an imagination so that they can push back the frontiers of marine science everywhere."

These frontiers lie in the deep oceans, the polar oceans, and all oceans. They lie in sea, land, air, and space. And they lie in the quest to provide food and water to the burgeoning human population with its increased consumption. The necessity of understanding the global environment has never been more urgent. The world population was three billion in 1960 and six billion in 1999. If it doubles again in 39 years we will have 12 billion humans on the planet before 2040. Preserving both humans and their humanity is a daunting task and oceanographers are in the middle of any contributions science can offer to this task. Improved technologies are vital to their efforts.

Today, entire oceans can be studied from space. A vast array of remote sensing technologies are utilized, including satellites, ROVs, and drifting and attached buoys that monitor ocean changes on a global scale. Potentially, instruments can upload their data from any place in the ocean to any scientist in a lab on land via satellite telemetry. The H.M.S. *Challenger* set sail a mere century and a half ago. The *Challenger* expedition sampled depths, sediments, and organisms at 362 stations in three and a half years. That is an average of fewer than three stations every ten days. The Jason-1 satellite, launched in 2001, can measure sea surface changes to within one inch over 95 percent of the ice-free ocean every ten days. The development of the Global Positioning System, GPS, has revolutionized navigation. The Global Ocean Ecosystems GLOBEC is taking inventory of global ocean biodiversity, JGOFS is studying the global carbon cycle, and the Ocean Drilling Project begun in 1968 with the R/V Glomar Challenger continues vigorously today aboard the R/V JOIDES Resolution. SeaWiFS, CLIVAR, TOGA, WOCE, and RIDGE are other global oceanographic projects and tools that have moved oceanography from the twentieth to the twenty-first century. And there are more, some just beginning, others on the drawing board, others in the imaginations of contemporary oceanographers, and others as yet unimagined. The voyage is, indeed, endless.

Focus Your Learning

Learning Objectives

After completing this unit you should be able to do the following:

1. Understand the general process of science, and illustrate this process at work in oceanography using at least three examples from the course.
2. Discuss some of the motivations that drive oceanographers to choose their profession. Be aware of the personal characteristics and training needed to pursue a career in oceanography.
3. Describe the purpose of each of the following global oceanographic initiatives and understand the importance of each to oceanography: SeaWiFS, TOPEX/Poseidon and Jason-1, GPS, CLIVAR, TOGA, WOCE, GLOBEC, JGOFS, ODP, and RIDGE.
4. Describe at least four areas of ongoing oceanic research and understand their importance.
5. Describe at least four scientific discoveries and describe why they are important.

Assignments

This lesson encompasses information found in all chapters of the text, but particular attention should be paid in reviewing the following pages.

Text: pages 5–8, 53–57, and 525; Appendix X, pages 543–546

Video: Episode 26, "Hands On"

Video Focus Points

Review these points before watching the video assignment for this lesson to help you focus on key issues.

— As scientists, oceanographers are driven by curiosity. But they are also influenced by altruism, selfishness, a sense of adventure, childhood experiences, cooperation, competition, the inspiration of gifted mentors, and the need to solve practical problems facing humanity. Furthermore many confess a spiritual/romantic/mystical dimension of their work.

— Like their predecessors, many contemporary oceanographers concentrate on particular, localized problems. Technologies refined in the latter half of the twentieth century allow other oceanographers to study entire oceans using remote sensing. These global initiatives hold the promise of expanding understanding of the dynamics of the world ocean and its effect on climate, food production, environmental preservation, and other issues of planetary scope.

— Advances in technology have fostered this global approach to oceanography.

Text Focus Points

Review these points before reading the text assignment for this lesson to help you focus on key issues.

— Oceanography is a multidisciplinary science including major contributions from the worlds of physics, chemistry, geology, biology, engineering, and others.

— Ocean science includes established facts, tentative facts, and untested hypotheses. It is a body of knowledge and the processes used to create that knowledge. Like all science, oceanography attempts to exploit curiosity based on observations of the natural world to construct hypotheses and test them with experiments and, where possible, propose theories that attempt to explain nature.

— Along with laboratory experiments and investigations and field work from shore and at sea, contemporary oceanography exploits a variety of remote sensing and other technologies to investigate the oceans on a global scale.

— Satellite technology, fixed and drifting buoys, and manned and unmanned submarines contribute to these global initiatives.

— SeaWiFS, TOPEX/Poseidon and Jason-1, GPS, CLIVAR, TOGA, WOCE, GLOBEC, JGOFS, ODP, and RIDGE are examples of modern oceanographic satellites, tools, and initiatives.

— Change has been a recurrent theme in this course. Earth's land, sea, and air are intimately interrelated and dynamic. Science, including oceanography, is also in a state of flux as our understanding of the ocean is continuously refined, revised, and expanded.

— Understanding the ocean fosters a growing appreciation of it. It inspires art, literature, and song and it provides food, leisure, sport, relaxation, and romance.

— The road to a career in oceanography is not an easy one. A rigorous pursuit of achievement in math and science is mandatory. A spirit of adventure, the ability to engage in hard work at sea, and the acquisition of a variety of skills useful in the laboratory, on board a research vessel, and in the office dealing with institutions, agencies, and colleagues are beneficial.

Key Terms and Concepts

A thorough understanding of these terms and concepts will help you to master this lesson.

CLIVAR. The Climate Variability and Predictability Program is part of the wider World Climate Research Programme (WCRP). Its goals are to study global climate variability and the effects of human activity on climate in the 10–100 year range.

Fact. An established piece of information. Scientific facts must be empirically verifiable and potentially falsifiable.

Global Oceanographic Initiative. Any research program that investigates oceanography on a global scale. Remote sensing by satellite is among the technologies that facilitate these studies.

GLOBEC. The Global Ocean Ecosystem Dynamics Program studies how global change will affect the abundance, diversity, and productivity of marine populations. GLOBEC researchers focus on the herbivorous zooplankton and their carnivorous predators. These second and third trophic levels constitute the major food of larval fish.

GPS. The Global Positioning System has revolutionized navigation. A series of 24 satellites in precise orbits form a "constellation" of artificial stars, which serve as reference points. A GPS receiver detects radio signals from three GPS satellites and precisely calculates position using triangulation.

Ichthyologist. Ichthyology is the study of fish. An ichthyologist is a scientist who studies fish.

Jason-1. This satellite succeeded TOPEX/Poseidon and precisely measures ocean surface topography. It studies global ocean currents and sea surface winds, and improves understanding of ocean circulation and its effect of global climate. It has been used to forecast El Niño events.

JGOFS. The Joint Global Ocean Flux Study studies physical chemical and biological processes at work in the ocean that affect the cycling of carbon. The role of the ocean in the uptake and release of carbon dioxide are critical to understanding climate and global warming.

Sea surface topography. The shape of the surface of the ocean. Thermal expansion and contraction, upwellings and downwellings, surface currents, winds, and seafloor bathymetric features can affect sea surface topography.

TAO-TRITON. An array of ocean sensing buoys extending across the equatorial Pacific Ocean used primarily to forecast El Niño and La Niña climate events.

Test Your Learning

Being an epilogue for the series, this lesson encompasses information found in all chapters of the text.

Multiple Choice

1. Which of these programs concentrates on studying global climate change?
   a. CLIVAR
   b. RIDGE
   c. SeaWiFS
   d. JGOFS

2. Which of these programs concentrates on studying the global carbon cycle?
   a. GLOBEC
   b. CLIVAR
   c. SeaWiFS
   d. JGOFS

3. In the video, George Matsumoto states that "...we're facing a huge shortage of people skilled in the art of taxonomy." He is referring to those who studied:
   a. the laws and principles of the classification of organisms
   b. the biology of fish
   c. the ways in which the ocean affects the Earth's climate
   d. the relatively new specialty of hydrothermal vents

4. According to the video, nearly all oceanographers:
   a. do the majority of their research in a lab
   b. have an interest in the environment and a desire to spend large amounts of time working outdoors
   c. spend more then 50 percent of their working time aboard ship
   d. are involved in studying the biological aspects of the ocean

5. The study of fish is called:
   a. herpetology
   b. osteology
   c. ichthyology
   d. limnology

6. The ______ system has revolutionized navigation by using 24 satellites to triangulate latitude and longitude.
   a. TOPEX
   b. GPS
   c. SeaWiFS
   d. GLOBEC

7. The ______ project studies ocean biodiversity and is centered on zooplankton.
   a. GLOBEC
   b. CLIVAR
   c. BioRAD
   d. JGOFS

8. Which of these does not have the study of global climate change as a primary component of its mission?
   a. RIDGE
   b. CLIVAR
   c. WCRP
   d. GLOBEC

9. Employing satellites, buoys, and ROVs has expanded our ability to conduct ocean research. Such tools are categorized under:
   a. robotics
   b. electronic detection
   c. remote sensing
   d. environmental modeling

10. The oceans cover 70 percent of Earth's surface but represent about ____ percent of the living space.
   a. 50
   b. 75
   c. 90
   d. 99

Short Answer

1. Remote sensing and sophisticated research submarines have become an integral part of modern ocean science. Review the course and find one example of the use of a manned or unmanned submersible (ROV), a satellite, and ocean buoys to study an oceanographic problem. Describe each.

2. Oceanography is a science and at its core driven by the innate curiosity of the human species. But several other motives also drive oceanographers in their work. Discuss several of these and give examples.

3. The practicing oceanographers in the video paint a romantic, adventurous, and generally inviting picture of their vocation. The road to a career in oceanography is actually a long one requiring hard work and dedication. Discuss marine science as a profession including training requirements, job opportunities, and factors that affect employability of prospective oceanographers.

Supplemental Activities

1. One of the most compelling illustrations in the textbook is figure 18.33. Paul Jokiel of the Hawaii Institute of Marine Biology claims there are three major problems assaulting Earth's environment: uncontrolled human population growth, increased *per capita* consumption, and ignorance and/or denial of the two previous points.

   Choose five different environmental problems affecting the oceans or the Earth that are most disturbing to you. For each, decide if human population growth and increased consumption exacerbate that problem in any way. For each item you choose write one or two sentences describing how the problem is related to population growth or consumption.

   Regardless of social, religious, political, or economic issues that might be attached to any discussion of human population growth, the underlying and undeniable arithmetic of this growth is easily, although not widely, understood. Three values of particular interest in populations are current size, rate of growth, and doubling time. You can calculate the future size of a population with the exact same equation you would use to calculate the future size of your bank account given the principle and interest rate. The doubling time can be approximated with a little trick. Divide 69 by the growth rate or interest rate. If you have $1,000 invested at 3.5%, it will double to $2,000 in 69/3.5 = 19.7 years. If you have 6 billion humans reproducing with a 2% growth rate, the population will double in 35 years. Some underdeveloped countries have growth rates approaching 6%. What is their doubling time?

   Visit this Website and explore links that flow from it: http://www.census.gov/main/www/popclock.html.

   Choose ONE of the following points of view and write a one page essay supporting that option.

   a. Jokiel is what former NY Congressman Jack Kemp would call a "Malthusian Alarmist." Despite current trends the population problem is not nearly as bleak as it seems because... or
   b. This population problem does, indeed, paint a bleak picture for the oceans. If I were in charge of Earth I would adopt the following policies because... or
   c. The situation is even bleaker than Jokiel projects because they are only talking about resources, infrastructure, pollution, etc. Many other social problems could result from overpopulation and make the situation worse because...

2. Select one major oceanographic topic that interests you such as global climate change, plate tectonics, marine productivity, or ocean resources. Choose information from any three different *Endless Voyage* lessons, any three major oceanographic institutions including one from outside the U.S.A. if you wish, and any three federally sponsored global ocean science initiatives and write a summary paper defining the topic, integrating the information from the various sources to demonstrate interrelationships.

   Some Institutions:

   Scripps Institution: sio.ucsd.edu/
   Lamont Doherty: www.ldeo.columbia.edu/
   Woods Hole: www.whoi.edu/
   List of Ocean Research Labs:
   www.openseas.com/oceanweb.htm

   Some Sponsors:

   The National Science Foundation:
   www.nsf.gov/
   NOAA: www.noaa.gov/
   NASA: http://www.jpl.nasa.gov/earth
   Naval Research Laboratory:
   www.nrl.navy.mil/

3. Observe a satellite. Visit these Websites to find out how to observe one: www.heavens-above.com/ and www.satobs.org/satintro.html. Register yourself at the heavens above Website and log on. Find a satellite that can be observed from your location. A simple magnetic compass and pair of binoculars will be helpful. Print out the star chart with the satellite's pass trajectory and observe it. Find the following satellites by doing an Internet search for each. Write a short description of the mission of each and how it relates to oceanography: NOAA-17, Jason-1, NAVSTAR, Oceansat, Okean-o, Nimbus-7, Argos.

Answer Key

Lesson 1: The Water Planet

1. c page 9; Video
2. b page 10; Video
3. c Video
4. a page 13; Video
5. b page 17, Video
6. a page 13; Video
7. a page 17
8. a page 14
9. b Video
10. c page 15, Video
11. d Video
12. c page 9, Video

Short Answer

1. Consider the following terms: curiosity, experiments, hypothesis, law, observations and measurements, and theory. Rearrange these terms in order from the one with the least amount of supporting evidence to the one with the most, and explain each term. Briefly describe one specific example of each of these terms that was discussed in the reading or video for Lesson 1.
   - Curiosity: Innate human curiosity causes a question to arise as to when, where, how, or why a phenomenon or process occurs or an object exists. (Humans curiosity led them to study how the universe began.)
   - Observations and measurements: A scientific investigation germinates. Investigators use their senses (sight, touch, smell, etc.) to observe the phenomenon and collect data. Numerical measurements are made where possible. (Hubble observed stars and measured their movements, discovering that the universe is expanding.)
   - Hypothesis: A suggested answer to the question is formulated. It is based on the observations and measurements. A good hypothesis is testable and should lead the investigator closer to an answer through its acceptance or its rejection.
   - Experimentation: The hypothesis is tested with an experiment(s) conducted under controlled conditions in nature or in the laboratory. The experiments must be repeatable and the results consistently reproducible.
   - Theory: Inductions or generalizations (patterns) may emerge from the previous steps. If these patterns are consistent and permit predictions that are generally correct, the theory may become widely accepted as a best explanation and be elevated to a theory. (The mass of empirical evidence including expansion of the universe, background microwave radiation, etc., verifies the big bang theory while current attempts to falsify it largely fail.)
   - Law: With further scrutiny and testing, theories may be elevated to laws. Laws are larger scientific constructs that explain natural phenomena. Laws predict the behavior of the natural world with unvarying results under identical conditions. (The law of gravity explains the accretion of atoms and particulate matter to construct planets. It also explains their orbits.)

2. Although it is a non-scientific statement, one might say "Stars have died so we (humans) may live." Outline the major phases in the life of a typical star and explain how the death of ancient stars contributed to your life.
   - Stars form from clouds of hydrogen and dust called nebulae, which contract by gravitational attraction.
   - As they contract they heat. When hot enough to emit heat and light, they are called protostars.

- Further contraction and heating to temperatures of about 10 million degrees Celsius triggers thermonuclear fusion as hydrogen atoms fuse to make helium. The protostar stops contracting, releases large amounts of energy, and is now a star.
- As fusion proceeds, the star stabilizes and releases energy at a steady rate. Our sun is presently in this stage of the stellar lifecycle.
- With continued fusion a large amount of the hydrogen is consumed and converted to elements as large as carbon and oxygen. The star's subsequent consumption of these heavier atoms increases its energy output.
- In a medium sized star like the sun this late period in the star's life is characterized by expansion into a red giant. This will happen to the sun in about five billion years and Earth will be consumed.
- Some of the heavier elements will be thrown into space as the star contracts, pulsates, and shrinks. Most of the elements of the dead star remain in the cooled, contracted mass.
- Much larger stars burn faster and hotter and produce elements as heavy as iron in their fusion reactions.
- These stars die as their cores compress. The massive explosion of this highly compressed and heated core produces a supernova and it is in this sudden explosion that nuclear fusion creates the elements heavier than iron. The explosion is sudden and cataclysmic and spreads the various elements into interstellar space. All the elements heavier than hydrogen were produced in stars.
- All the atoms in the universe except hydrogen, the simplest one, were produced in stars. The carbon, and other elements like oxygen, nitrogen, sulfur, and phosphorus, from which living thing are constructed were produced during some phase of the nuclear reactions that occur in stars.

3. Explain why a biosynthesis similar to that which produced the simple organic building blocks which enabled the early steps in the evolution of living organisms cannot occur today.
   - The process requires an absence of oxygen, which is now plentifully provided in the atmosphere by green plants
   - This oxygen in the atmosphere blocks most of the dangerous wavelengths of light from reaching the surface of the ocean.
   - Large organic molecules would be consumed by the many tiny organisms present in today's ocean.

Lesson 2: First Steps

1. a page 47, Video
2. c page 42
3. a page 49, Video
4. b page 40
5. d pages 45–46
6. d page 51
7. d page 31
8. d page 53, Video
9. b pages 49–50, Video
10. d pages 50–51
11. a page 47
12. c page 49

Short Answer

1. Discuss the role of Claudius Ptolemy (90–168 A.D.) in the discipline of cartography, including the famous "Ptolemy's Error" and its later consequences.
   - Ptolemy improved on Eratosthenes imaginary latitude/longitude chart lines by orienting those lines in standard north/south and east/west configurations.
   - Ptolemy also further refined on Hipparchus' 360-degree division of the planet's surface by dividing each degree into minutes and seconds of arc.
   - Unfortunately, in attempting to improve on Eratosthenes surprisingly accurate estimate of Earth's circumference, he used flawed calculations of atmospheric refraction and estimated Earth's size to be about 70 percent of its true value.
   - Ptolemy also overestimated the size of Asia, reducing the calculated width of the part of the world between the Orient and Europe (unexplored and uncharted at the time).
   - More than 1,500 years later Columbus used Ptolemy's erroneous calculations of the Orient-Europe distance to convince people that they could reach Asia by sailing west!

2. Captain James Cook made three major voyages of discovery and science, all of which were highly successful and productive. Although he had well-trained crews and excellent scientific personnel, much of the credit for the success of these cruises is attributed to the man himself, with his many skills and talents. List and

briefly discuss those skills and talents as they applied to these voyages.

- Cook was an intelligent and patient leader. He ran what would now be called a clean ship, knowing that a healthy and contented crew would do a good job. He was particularly concerned and knowledgeable about living conditions, and insisted on cleanliness and ventilation in the crew's quarters.
- Cook was also knowledgeable about disease, and watched the diets of his personnel when possible to prevent such things as scurvy.
- He was also an accomplished diplomat, and established good relations with the people of the islands he visited.
- Although he was not a trained scientist, Cook was knowledgeable in a great many scientific fields, and made extremely accurate observations on the natural history of what he saw at sea and on land.
- Cook was also a skilled artist, and undoubtedly used that talent to illustrate many of the specimens of plants and animals he collected.
- His skills as a navigator and cartographer were legendary, and his charts of the Pacific Ocean were accurate enough to be used to plan beach invasions in World War II.
- All of this added up to an extremely versatile and talented man, earning him a place in history as the first marine scientist.

3. Although the voyages and accomplishments of Cook, Wilkes, Maury, and Darwin are highly significant in the history of oceanography and marine science, the four-year *Challenger* expedition still stands alone as being the most important. List and discuss briefly some of the reasons for that recognition.
   - It was the first sailing expedition devoted completely to marine science. Even though the ship and crew was Royal Navy, the mission of the expedition was to voyage across the oceans of the world, gathering numerous types of oceanographic data and natural history samples.
   - One of its goals was to investigate the claim of Professor Edward Forbes that life could not exist in the deep sea (below 549 meters), due to the high pressure and lack of light. A mechanized winch and ingeniously made nets allowed them to bring up deep-sea samples from depths as great as 8,125 meters, disproving the Forbes theory.
   - Using advanced-design dredges and grabs, they occupied 362 deep stations, yielding nearly 5,000 species new to science.
   - At each sampling station they took oceanographic data, including salinity, water temperature, and density. This added immeasurably to our knowledge of the physical structure of the deep ocean.
   - They recovered some of the first deep-sea manganese nodules ever seen, sparking an interest in deep-sea mining.
   - All told, the expedition was an unqualified success and set numerous records for depth of soundings and samples, and new species collected. It also remains history's longest continuous scientific oceanographic expedition.
   - However, it is actually the magnificent 50-volume *Challenger* Report that provided the foundation for the new science of oceanography, even more than the actual voyage.

Lesson 3: Making the Pieces Fit

1. a pages 75–76, Video
2. d page 64, Video
3. c Video
4. d page 75
5. a page 72
6. c page 67
7. b pages 69–70
8. b page 90, Video
9. c Video
10. a pages 75–76, Video
11. d page 64, Video
12. a pages 74–75, Video

Short Answer

1. Prepare two labeled drawings of the Earth sliced in half. On the first illustrate and describe the layers of the Earth based on their composition and on the second illustrate the layers of the Earth based on their physical properties.

   The drawing based on composition should show:
   - continental crust (composed of granite)
   - oceanic crust (composed of basalt)
   - mantle (complex composition, mostly peridotite-like minerals)
   - outer core (liquid and composed of iron and perhaps sulfur)
   - inner solid core (solid and composed of iron and some nickel)

The drawing based on physical properties should show:

- lithosphere (solid outermost layer includes crust and solid uppermost mantle)
- asthenosphere (hot partially melted fluid upper mantle below lithosphere)
- lower mantle (although hotter than upper mantle, the lower mantle is solid because of the greater pressure in deeper mantle)
- outer liquid core (a viscous liquid layer with temperatures exceeding 5000°C
- inner solid core (hotter because of depth but solid because of greater pressure)

2. Wegener's idea of continental drift was not widely accepted when he proposed it. Briefly outline the arguments, presented by Naomi Oreskes in the video, between American and European scientists over how the scientific method should be conducted. Comment briefly on your views regarding this argument.
   - Oreskes argues that American views of isostasy were inconsistent with Wegener's so they had a more difficult time accepting his theory. She states, "So, in a sense, it really was that the Americans had to give up more than the Europeans did."
   - There was more to the argument than a disagreement over isostasy. There was a fundamental difference in the approaches of Americans and Europeans to the scientific method.
   - Americans argued that extensive empirical investigation and intense observation must precede the formulation of any theory.
   - Europeans, the Germans in particular, viewed the American approach as a random endeavor and suggested that having an organizing idea in advance of and during the accumulation of empirical evidence is a more fruitful pathway to the creation of new knowledge.
   - Students should find at least some merit in both approaches and recognize, as Oreskes states, that Wegener's proposal of continental drift followed the German model to a greater extent than the American.
3. The debate between the ideas of catastrophism and uniformitarianism in the eighteenth century influenced the twentieth century debates about continental drift. Discuss these two contrasting ideas, explain their effect on the continental drift proposal of Wegener and discuss your views as to whether continental drift is a uniformitarianist or catastrophist idea, or both or neither.
   - Hutton's idea of uniformitarianism or gradualism suggested that Earth was very old and the features of present geomorphology were a result of a long period of processes identical to those at work today (1788). Wind water and other erosive factors, sedimentary accumulation, and volcanism would be among these.
   - Catastrophists believed that the Earth's geomorphology resulted from a series of global catastrophes and that each successive catastrophe destroyed all evidence of the preceding one. The great flood of Noah was the most recent catastrophe and it was that flood that carved, built, and even scrambled the Earth's exterior to produce the world as we see it.
   - Gradualism demands a very old Earth, suggests that slow gradual change is normal and implies, although does not state, that interpretations of the Judeo-Christian bible other than absolute literal ones are possible.
   - Catastrophism permits a very young Earth. Ussher's calculation of the biblical date of creation at 4004 B.C. is often given when discussing catastrophist ideas. It suggests that Earth is a rather static place punctuated by short periods of great change and it draws upon an absolute literal interpretation of biblical accounts for support.
   - Supernatural intervention aside, plate tectonic theory has both catastrophic and gradualistic features. Students should develop ideas that the entire process is a gradual one, but account for such catastrophic events as earthquake belts, explosive volcanism, mountain building. And, whereas neither of these two ideas worked well to explain the collected data and observations of Earth's structure, plate tectonics does.

Lesson 4: World in Motion

1. b pages 79–85, Video
2. b pages 92–93, Fig. 3.33
3. a page 85, Video
4. c page 79, Video
5. d page 78, Video
6. d page 85, Video
7. b page 94, Video
8. d Video
9. d pages 93–94

10. b pages 75–76, Video
11. c pages 85–89, Table 3.1 on page 89, Video
12. a pages 79–83, Video

Short Answer

1. One key to understanding plate tectonics is symmetry. Certain features of the seafloor are symmetrical on opposite sides of spreading centers at mid-ocean ridges. Describe at least three of these symmetrical features.
   - Paleomagnetic striping is symmetrical in polarity and age.
   - The age of the seafloor is symmetrical around the mid-ocean ridges. The newest basalts are forming at the central rift valleys of the ridges and the crust gets older moving away from the ridge. The oldest seafloor is 200 to 250 million years old. Older sea floor has subducted and melted.
   - The thickness of seafloor sediments is symmetrical. As predicted, the thinnest (youngest) deposits are close to the ridges and the thickest (oldest) are farthest away.
2. Discuss the relationships among the Hawaiian Islands-Emperor Seamount chain, the Aleutian Trench, and plate tectonics.
   - The Hawaiian Island-Emperor Seamount chain has formed over a mantle plume or hot spot. This is Earth's longest continuously active hot spot and demonstrates the existence, rate, and direction of movement of the massive Pacific Plate.
   - The Loihi Seamount and the active volcanoes of Hawaii (Kilauea and Mauna Loa) are presently over the hot spot.
   - The other features (islands, seamounts, and guyots) of the Hawaiian-Emperor chain formed over the past 70 million years as the Pacific Plate moved over the hot spot.
   - Loihi, a seamount southeast of Hawaii, is the newest peak and Meiji Seamount at the northern tip of the Emperor Chain is the oldest remaining one. It has taken Meiji and the Pacific Plate about 70 million years to move from the hotspot to its present location poised to subduct into the Aleutian Trench where subduction melts the plate and its seamounts, forms the Aleutian Trench and generates the Aleutian island arc.
   - Darwin Point, or the Hawaiian-Emperor Bend, is approximately 40 million years older than Loihi and marks the time and place where the movement of the Pacific Plate over the hot spot changed from generally north to its present northwest direction.
3. Outline the stages of the Wilson cycle. List the motion, features, and an example of each stage.
   - The (J. Tuzo) Wilson Cycle synthesizes the long term cycling of crustal plates from rifting open of a continental land mass, through formation of an ocean to the oceans closing and disappearance as continental masses collide. The seven stages are:
     a Conception Stage. Crustal upwarp over rising magma with the beginning of rifting and volcanism. There are presently no good examples although parts of West Africa seem to be showing signs of conception.
     b Embryonic Stage. Uplift, rifting, and continued volcanism produce a system of rift valleys and lakes on a continent. (East African Rift Zone)
     c Juvenile Stage. Divergence produces a narrow sea with a central ridge and matching opposing coasts. (The Red Sea)
     d Mature Stage. Continued divergence and widening of the ocean at the mid-ocean ridge; separate continental masses now exist. (Atlantic Ocean)
     e Declining Stage. Convergence and subduction begins. Island arcs and trenches form around perimeter of ocean; trenches form at the subduction zone. Ocean narrows. (Pacific Ocean)
     f Terminal Stage. Convergence continues, continental margins begin to collide, ocean narrowing continues producing irregular seas and young mountains. (Mediterranean Sea)
     g Suturing Stage. Convergence and uplift continue as continental masses collide, compress and override. Intervening sea is obliterated. (India-Eurasia collision, Himalaya Mountains)

Lesson 5: Over the Edge

1. b page 104
2. d page 105, Video
3. c page 109
4. a Video
5. b pages 111–112, Video
6. c page 122, Video
7. b page 115
8. d page 116

9. b page 118
10. c page 124
11. a page 114
12. d page 116

Short Answer

1. Since their discovery in 1977, deep-sea hydrothermal vents have attracted the attention of many types of scientists. Now, with the study and description of the large seafloor structures called black smoker chimneys, there may be another player in this drama—commercial interest in those structures as a resource for highly purified minerals. Think of all the different factions with potential interest in the hydrothermal vents' amazing features, and comment briefly on each and what their particular focus might be. Consider the federal government (both military and civilian), academic and private research (with their wide range of specialties), and the more commercial interests just mentioned.
   - The actual military interests, focused mainly through the Office of Naval Research, might be rather minimal. They would want to know sizes and locations to add to their databases. If a controversy about ownership or mining rights ever arose, the military might have to step in. Also, their submarines and submersibles have been and would be useful in doing research and surveys.
   - Civilian government agencies (NOAA, National Science Foundation) should be most interested in studying both the physical and biological aspects of the vents as part of their ongoing basic research programs and those involving funding for universities and research institutions.
   - Both public and private universities and research laboratories encompass a wide variety of scientific interests in both the physical and biological fields, and would probably be most interested in the structures for pure research on how they were formed and the vast array of new species and associations found there.
   - If there are newly interested commercial factions, they would most likely focus on the more physical aspects of the structures—how they are formed, the types and purity of the chemicals and minerals involved, and the possibility of mining them profitably. Of course, any activities like mining would involve concerns about how they would affect the more scientific studies on structure formation and the biological communities.

2. Turbidity currents are blamed for a lot of things, mainly for transporting significant amounts of sedimentary material on and around the undersea features. Do a basic description of a turbidity current, then list some of the seabed features and the areas where turbidity currents are most active and involved.
   - A turbidity current is best described as an undersea avalanche of sediments.
   - They are thought to be the main sculpting agent for submarine canyons, carrying large amounts of abrasive sediments down continental slopes, cutting and eroding as they move.
   - Submarine canyons are thought to be the main funnel for bringing loose material off the continental slopes and onto the abyssal plain, with the main force being turbidity currents.
   - Turbidity currents are often started by earthquakes, mixing sediments with the surrounding water above a sloping surface.
   - Turbidity currents are often called density currents, and are known to travel downslope at up to 50–60 miles per hour.
   - Depending on the type of material available, and the steepness of the slope, turbidity currents can carry not only sand but also such coarse-grained material as cobbles, gravel, and pebbles.
   - Most of the sediments that form the continental rise are transported to the area by turbidity currents.

3. A general description of the deep ocean floor is a mixture of ocean ridge systems and adjacent sediment-covered plains. The ridge systems draw much interest because of their active rift zones, spreading centers, faults, and vents, making the abyssal plain seem dull and featureless by comparison. But since the abyssal plains and the abyssal hills cover about one-fourth of Earth's surface, they are worthy of some attention. Briefly describe the abyssal plains and the abyssal hills.
   - Abyssal plains are basically flat areas of sediment-covered seafloor, which are found in all oceans. They are most common in the Atlantic Ocean, with fewer in the Indian and Pacific.

- Abyssal plains lie between the continental margins and the oceanic ridge systems.
- The largest plain (Canary Island Plain) covers about 350,000 square miles.
- Most abyssal plains are flat; the plain in the Atlantic varies only a few meters in depth over its entire area.
- The flatness is not due to the underlying basalt seabed, but to the very thick layer of sediment covering it and smoothing out the smaller relief features.
- Most of the plains sediments appear to be of terrestrial or shallow-water origin, transported by surface winds or turbidity currents.
- In areas where sediment is not thick enough to cover all the basalt floor features, one may encounter abyssal hills—small extinct volcanoes partially covered with sediment and usually less than 200 meters high.
- Together, abyssal plains and abyssal hills account for nearly all of the seafloor not incorporated in the oceanic ridge systems.

Lesson 6: The Ocean's Memory

1. a pages 148–150, Video
2. b page 137, Video
3. b page 134
4. d page 134
5. c page 150
6. d page 142
7. d page 150
8. b page 150, Video
9. c page 142
10. b pages 148–150, Video
11. d page 146
12. d page 146, Video

Short Answer

1. Discuss the growing commercial focus on marine sediments, noting particularly the different types of resource materials involved, the methods, and economics of their retrieval.
   - There are certain kinds of deposits found in ocean sediments that are, or could be useful and valuable. Presently, the most important seem to be involved with energy—the oil and natural gas that are formed in, and trapped in, marine sediments.
   - Organic material becomes buried in that sediment, and with time becomes buried deeper and deeper. The deeper it is, the warmer it becomes, causing the organic molecules to break down and eventually become crude oil. If the heat and pressure continue to increase, that oil can be further broken down into natural gas.
   - Other resources associated with marine sediments are sand and gravel, tin and titanium, manganese and phosphorite (nodules), and even coal and diamonds. All have economic potential, depending on the technology available to mine and retrieve them.
   - Manganese deposits continue to be of major interest, but presently the cost of mining and refining exceeds that of terrestrial deposits.
   - The process of economically extracting resources from marine sediments continues to be a major challenge. Current oil and gas retrieval still depends on drilling into sedimentary rock and freeing the oil and gas, which under present technology may still leave up to 60 percent of it in the field. There has been some success using injection techniques—hot water, steam, or chemicals are injected into the drill holes, to melt out the oil and force it into a producing well.

2. Interest in the composition of the sea floor probably began with the early voyagers as a navigational aid, then became more focused with the first marine science expeditions, and is currently "big science" due to major advances in sampling and study methods. Document some of the milestones in this sequence of events, culminating with the state-of-the-art technology being used today.
   - In an earlier lesson you learned that James Cook collected samples of the seabed, using wax-coated weights lowered on lines (mid-to-late 1700s).
   - Sir John Ross and Sir James Ross (early 1800s) used a clamshell-type sampler to obtain bottom samples from a depth of nearly 2000 meters.
   - The now infamous bathybius involved (1857, H.M.S. *Cyclops*) samples of bottom sediments that were preserved in alcohol and stored in bottles. Sir Thomas Huxley later examined them and discovered a grayish ooze coating the mud sample. He thought it was a primitive life form, but further examination showed it to be a slime resulting from a chemical reaction between the preservative and the mud and seawater.
   - On the *Challenger* expedition (1872) samples were taken with wax-tipped poles, mechanical grabs, and dredges, using a

steam-powered winch for cable retrieval. Samples were retrieved from over 8000 meters depth, and in some of those samples were found the first recorded manganese nodules.

- Current techniques involve a variety of bottom-sampling devices—clamshell grabs, piston corers, and rotary drilling and coring tools, operated from drillships originally designed for oil and gas recovery. These ships have drilled thousands of sampling holes and have retrieved massive cores (some over 1000 meters in length!). These operations essentially began in 1968, with the Deep Sea Drilling Project using the *Glomar Challenger*, later replaced (1978) by the Ocean Drilling Project and the drillship *Resolution*.
- Supporting and supplementing the sampling and information gathered by the drillships, current technology for locating and mapping marine sediment deposits includes satellite imaging and continuous-recording seismic profilers.

3. The first recorded retrieval and identification of what we now call hydrogenous sediments was on the *Challenger* expedition. Although not considered rare, they are thought to cover less than one percent of the total ocean floor, and their chemistry and the mechanics of their formation is still not entirely clear. Summarize what is known about the composition and formation of these sediments.
   - Hydrogenous materials seldom form particles by themselves, but are usually associated with other seafloor components (terrigenous or biogenous sediments).
   - While the best-known hydrogenous sediments are the manganese nodules, even those are not pure, but will also contain iron oxides and small amounts of other metals.
   - One of the reasons for the incomplete understanding of their formation is possibly their rate of formation, thought be only 1–10 millimeters every million years; one of the slowest chemical reactions known in nature.
   - Manganese nodules vary in size, depending on their stage of growth; some are barely visible, most are about fist-sized, and some may exceed a meter in diameter.
   - Manganese nodules usually form around, and through chemical reactions with particles of the dominant sediment type of the area, but they are also found enclosing shark teeth, microscopic algae, and even bits of bone. Bacterial activity may also be involved, but this possibility is still being researched.
   - Nodules composed mainly of phosphates have also been found, and their formation is assumed to be similar to that of the manganese.
   - Other types of chemically-formed hydrogenous sediments can build up around hydrothermal vents where hot, metal-rich brines are ejected from the sea floor and rapidly cooled, causing the metal sulfides to precipitate out.
   - Evaporites are a unique type of hydrogenous material, formed (as the name implies) not in the deep sea but by precipitation, as shallow oceanic bays, landlocked seas, and inland lakes evaporate, leaving large deposits of salt-rich crystals.
   - Another special case is the formation of ooliths (oolitic sands) when seawater becomes slightly alkaline, causing molecules of calcium carbonate to precipitate out around tiny shell fragments or other particles. These are especially abundant in some of the warm, shallow tropical waters, where there is high biological productivity.

Lesson 7: It's in the Water

1. a page 156, Video
2. b page 158, Video
3. b Video
4. a page 156, Video
5. c page 193
6. c page 192
7. d page 186
8. d page 195
9. c page 186
10. a page 192
11. b pages 191–192
12. a Video

Short Answer

1. Explain the reasoning that makes some scientists think that adding iron to certain regions of the ocean can reduce global warming.
   - Carbon dioxide is a greenhouse gas and as the level of carbon dioxide increases in the atmosphere, the atmosphere traps more heat and global temperature rises.
   - Carbon dioxide is very soluble in water and the ocean holds a great quantity of this gas. Uptake and release of CO_2 by the ocean is an

important factor controlling levels of CO_2 in the atmosphere.
- Large portions of the tropical and Antarctic Pacific Ocean are unusual in that iron is the major factor limiting phytoplankton production. Over most of the oceans nitrogen seems to be the major limiting factor.
- Scientists have demonstrated both in the lab and *in situ* that adding iron will stimulate phytoplankton growth in these iron-poor regions.
- The reasoning those who support enriching the seas with iron is that the stimulation of phytoplankton blooms will increase photosynthetic utilization of CO_2 in the sea, increase the flux of CO_2 out of the atmosphere and into the ocean and reduce or reverse global warming.

2. Figure 7.8 illustrates the general distribution of carbon dioxide and oxygen in seawater. Explain why these two curves are roughly opposite. Also explain why the concentration of dissolved oxygen is least at around 500 meters and increases in deeper water.
   - Photosynthesis uses up CO_2 and produces O_2 in the photic zone. Oxygen also enters seawater by diffusion from the atmosphere above. Thus the highest levels of oxygen are in the upper layers.
   - Respiration produces CO_2 and uses up O_2. This respiratory production of CO_2 by decomposition and bacteria, other microorganisms, and animals cause this gas to accumulate in deeper zones. The higher pressures and lower temperatures in deeper waters also allow these waters to hold more dissolved gas.
   - In most of the ocean a well-defined oxygen minimum zone exists at about 500 meters. The utilization of oxygen by decomposition and respiration produces this hypoxic zone. The deepest oceans would be anoxic but the thermohaline circulation carries dissolved oxygen as cold, oxygen-rich water sinks to the bottom from the Polar Regions.

3. Discuss how hydrogen bonding contributes to the thermostatic effects of water.
   - The thermostatic effects of water are the moderating effects of water on global climate.
   - Several properties of water allow it to absorb and release large quantities of heat without large changes in temperature. These include its high heat capacity, high latent heat of fusion, high latent heat of vaporization, and the fact that ice floats.
   - To raise the temperature of water the molecules must be caused to move faster. Much of the heat added to water is used to break hydrogen bonds rather than speed up molecules and elevate temperature.
   - As ice thaws, it absorbs 80 calories of heat per gram. The molecules of ice are all interconnected by hydrogen bonds in the crystal lattice. Many of these bonds must be broken to convert the solid to liquid and this absorbs heat. The reverse is true as water freezes.
   - To evaporate, all the hydrogen bonds in liquid water must be broken as the liquid turns to gas. The latent heat of vaporization of water is 540 calories per gram. As ice melts and thaws it absorbs and releases this large amount of heat without changing temperature.
   - The fact that ice is less dense than water causes ice to float. Thus ice is always at the surface when it is in the oceans. And of course it lies on the surface of land as glaciers and ice caps. This widespread exposure to the atmosphere of both land ice and sea ice insures that it is available to absorb and release latent heat and thus moderates atmospheric temperature change.

Lesson 8: Beneath the Surface

1. c Video
2. d pages 175–176, Video
3. b Video
4. d page 172, Video
5. c pages 174–175, Video
6. a page 170
7. b page 170, Video
8. c page 174, Video
9. b page 179
10. b page 172
11. b page 176, Fig. 6.25
12. d page 171

Short Answer

1. Describe where the sofar layer and the shadow zones are located and explain how refraction of sound waves creates these two peculiar ocean regions.
   - The speed of sound in seawater is directly proportional to temperature and pressure. Because sound waves refract toward regions

of lower velocity two interesting acoustic layers are produced in the ocean.
- The shadow exists near the seasonal thermocline at about 80 meters depth. Here, the region above has lower pressure and the region below has lower temperature so the sound velocity both above and below the shadow zone is lower and sound refracts away from this layer. It is difficult for sound to enter the shadow zone so clever submarine commanders can hover at this depth and reduce the chance of being detected by enemy sonar.
- The sofar layer or deep sound channel exists near the bottom of the main thermocline at depths of 600–1,200 meters. This is the depth of minimum sound velocity. Waters above have higher temperatures and waters below have higher pressures, both of which have higher velocities. Sound generated in the sofar layer refracts back into this layer of minimum velocity and tends to get conserved. Therefore the sound energy is trapped by its refraction and can travel great distances with minimum attenuation.

2. Describe and define the following ocean zones and prepare a cross-section diagram of the ocean illustrating them: epipelagic, mesopelagic; bathypelagic; abyssopelagic, and hadopelagic.
   - The epipelagic zone corresponds to the euphotic zone and extends down to about 200 meters. This is the zone where most of the food and oxygen of the ocean are produced. Photosynthesis generates most of the ocean's organic carbon and a combination of atmospheric mixing and photosynthesis provide most of the ocean's oxygen.
   - The mesopelagic zone corresponds to the disphotic zone and extends down to about 1,000 meters. The bottom of the mesopelagic zone also corresponds roughly with the lower reaches of the main thermocline. This is the twilight zone, an important link for carbon cycling between the photic zone and the deep realm.
   - The bathypelagic, abyssopelagic, and hadopelagic zones are the deep zones. They are perpetually dark, cold, and under high pressure. The bathypelagic extends roughly between 1,000 and 4,000 meters. The abyssopelagic zone lies between 4,000 and 6,000 meters and corresponds approximately to the upper boundary of Arctic and Antarctic bottom waters. Ocean regions deeper than 6,000 meters form the hadopelagic zone and are found in the deep-sea trenches at subduction zones.

3. Compare and contrast the salinity, temperature, and precipitation balance in the surface waters of tropical, temperate, and polar oceans.
   - All surface regions of the oceans are subject to the vagaries of local conditions that vary on long and short time scales. In general the following are true.
   - Tropical oceans have higher rates of evaporation than precipitation and generally have high salinities in the range of 34–36 ppt. Particularly high salinities (> 36 ppt) are found within the gyres of the central north and south Atlantic where evaporation is high and the water is entrained within the gyre. Similarly high rates of evaporation exist in tropical north Indian Ocean between the Arabian Peninsula and the Indian sub-continent. In general, evaporation exceeds precipitation in tropical oceans and precipitation exceeds evaporation in the temperate and polar oceans.
   - The highest ocean temperatures are found on the western sides of the Atlantic and Pacific in the regions of South East Asia, the Gulf of Mexico, and the Caribbean. Here surface currents (North Equatorial Currents) have traveled across the tropical Atlantic and Pacific and have been warmed.
   - Temperate oceans have wider annual fluctuations of temperature than either polar or tropical seas. In general, temperate oceans vary seasonally between 10 and 25°C while polar oceans are rarely higher than 5°C although they may reach 10°C where remnants of the Gulf Stream brings warm water to the region of Greenland. Cooler water is carried toward the tropics by eastern boundary currents and warmer water is carried toward the poles by western boundary currents. This creates a general east to west warming in the Atlantic and Pacific Oceans.

Lesson 9: Going to Extremes

1. d page 381, Video
2. c page 250
3. d page 509
4. b page 408, Video
5. c page 408, Video

6. d page 392
7. b Video
8. a pages 504–505, Video
9. b page 392, Video
10. b synthesis, Video
11. d page 390
12. a Video

Short Answer

1. Compare and explain the general patterns and causes for the differences in primary production in polar vs. tropical oceans.
   - Primary productivity in polar oceans is regulated primarily by light availability while in tropical oceans it is regulated primarily by nutrient availability.
   - In both polar oceans high levels of primary production are limited to the summer season when light is available. Polar regions lack a thermocline so there is no physical barrier to nutrient upwelling, but the magnitude of upwelling is generally much greater in the Southern Ocean than in the Arctic Ocean so rates of primary production in the south are greater than in the north.
   - Most of the tropical oceans are strongly stratified and nutrients become trapped below the thermocline making open tropical oceans biological deserts. Notable exceptions exist in several places. In eastern tropical oceans, wind-driven surface currents and Ekman transport create coastal upwelling systems such as the one off the coast of Peru. The upwelling of nutrient-rich deep water feeds a flourishing phytoplankton community with its resultant food chain.
   - Diverging surface currents form a thin band of equatorial upwelling across the tropical open oceans. Nutrients from deeper water create high rates of productivity in these well-lit waters.
   - On the western sides of oceans north and south equatorial currents that have traveled across the ocean above the thermocline are largely depleted of nutrients as they approach continents. In these regions, such as the Caribbean, northeast Australia, Philippines, and Brazil, highly productive coral reefs flourish in nutrient-poor seas. The key to the high productivity here is the tight and efficient nutrient cycling between hermatypic corals and their mutualistic zooxanthellae.

2. Discuss man-induced problems contributing to the global decline in coral reef ecosystems.
   - Coral reefs flourish in shallow, clear, warm, tropical seas. They exhibit a dramatic biodiversity but are at the same time delicate and susceptible to an almost endless number of perturbations. More than 25 percent of the world's reefs disappeared in the past century and scientists estimate that, perhaps, 70 percent of the remaining reefs near population centers could disappear in the next 50 years.
   - With their symbiotic zooxanthellae, hermatypic corals require sunlight. Excess sedimentation, beach mining for sand and gravel, algal blooms stimulated by cultural eutrophication and any other forces that increase turbidity and reduce light penetration stress and kill these corals. Overfishing that causes the removal of herbivorous fish can result in overgrowth of attached algae and similarly reduce available light.
   - Direct physical damage to reefs is caused by the practice of dynamiting reefs to kill fish for consumption and to stun fish for the aquarium trade. Channels are cut and dredged in reefs to permit boat passage and reef limestone is mined for construction material. Snorklers and SCUBA divers visit reefs in ever-increasing numbers with associated accidental breakage and physical damage to corals by anchors and continuous abrasion by curious humans.
   - Living in regions of high insolation, corals have evolved pigments to protect them from the high levels of tropical UV. Continued depletion of stratospheric ozone may be subjecting shallow water species to damaging levels of UV. Global warming seems to be affecting corals and is implicated in coral bleaching.
   - The use of cyanide, rotenone, and other toxic chemicals to collect fish for the aquarium trade devastates fish and invertebrate populations and alters food webs. A wide variety of chemical pollution, including oil and pesticides stresses or kills reef organisms.

3. Discuss the causes, extent, and consequences of ozone depletion.
   - Ozone, O_3, is an unstable form of oxygen gas (O_2) that is produced in the upper atmosphere as ultraviolet radiation splits O_2 and the free single O molecules react with O_2 to make O_3.

- Manufactured chlorofluorocarbons (CFCs) such as Freon escape into the atmosphere and find their way to the ozone layer where they are also split apart by ultraviolet radiation. The resulting free chlorine and fluorine atoms catalyze the removal of O from O_3 (producing O_2) and rapidly destroy ozone and reduce the layers ability to block UV radiation from reaching the Earth's surface.
- In lower and mid-latitudes ozone depletion has been about two to four percent but in the high latitudes over the poles, the destruction has been much greater (50 percent) creating the so-called ozone holes.
- Ultraviolet radiation damages the molecular structure of DNA and proteins and thus can seriously damage living things. One would predict that terrestrial plants including food crops and phytoplankton in the upper two meters of ocean would be damaged and their productivity reduced. Shallow water corals, and other marine species near the surface where ultraviolet radiation is strongest would be damaged and human skin cancers, cataracts, and other conditions caused and exacerbated by excess UV would increase. Instances of all these things have been documented.
- By international treaty, many of the nations of the world are working to reduce the production and release of CFCs into the atmosphere.

Lesson 10: Something in the Air

1. c page 202
2. d page 204
3. c page 205
4. b page 206
5. d page 206, Video
6. a page 209, Video
7. b page 215
8. b page 213, Video
9. b page 215, Video
10. a page 211
11. d page 202, Video
12. c page 202, Video

Short Answer

1. Much of the discussion and explanation of the global climate system is based on an understanding of a basic air mass. Define and discuss the air mass—its composition, characteristics, behavior, and interaction with Earth's land and water.
   - Air is composed of nitrogen and oxygen, with a small percentage of other gases and varying amounts of water (either as vapor or a liquid).
   - Air has mass, and therefore density. The closer it is to Earth's surface, the closer packed are its molecules and the denser it is. As one moves farther away from Earth's surface (sea level is used as an index point and standard), air is subjected to less of its own density/pressure and will expand as the molecules are allowed to move farther apart.
   - As a volume of air warms, heat energy pushes its molecules farther apart, causing it to occupy more space but become less dense and, therefore, to tend to rise. Warm air can also hold more water vapor than can cooler air.
   - In a volume of cooling air the molecules are closer together, occupying less space and losing the ability to hold water as vapor. This may result in a coalescing or condensation of the water vapor into its solid form—droplets, dew or snow—which are then heavy enough to fall to Earth's surface.
   - An air mass tends to form as a large body of moving air, having an essentially uniform temperature and humidity throughout. As these masses move they are either heated and humidified or cooled and dried, based on the characteristics of the land or sea over which they are passing.
   - As air masses with different densities and humidities approach each other, they may not mix, causing a front to form at the boundary, characterized by excessive turbulence as the air masses interact. Internal disturbances within a single air mass can also result in turbulence. Both types of turbulences may result in one of the most common types of storms—the cyclone.

2. Meteorologists agree that there is still a great deal to be learned about weather, climate, and atmospheric circulation, with an eye toward more accurate forecasting. In the video, one scientist discusses several of the ongoing research projects—tornado detection, tracking, and warning; computer modeling using weather satellite data; hurricane detection and tracking. Briefly summarize the discussions of those research areas.

- For tornado detection and tracking, Doppler radar systems are becoming more common, used both by federal weather services nationwide, and by many local media stations.
- Time frames between detection of a tornado and the alerts/warnings have improved from a negative 8 minutes (the tornado would have been spotted and reported 8 minutes after it grounded) to a positive 12 minutes (the tornado is spotted and reported 12 minutes before it grounds).
- Techniques and technology continue to improve; in the near future possibly doubling the positive detection/warning to 20–25 minutes before the tornado grounds.
- By using computer modeling and analysis of weather satellite data, forecasting has become much more reliable. To provide even more data, new weather satellites are being launched more frequently.
- Using those analyses, and weather pattern configurations, the areas potentially covered in forecasting has diminished from points 50 miles apart to points only 10 miles apart.
- Much of the current meteorological research is centered on hurricane detection, how hurricanes interact with the sea surface, and on gathering data to improve prediction about how hurricanes form, gather power, and change configurations.
- Gathering data on the atmosphere/ocean hurricane link at sea is inherently hazardous for both men and equipment, so new data sensors are being designed for deployment as autonomous packages directly into a hurricane. These packages/buoys will then relay real-time data to meteorologists via satellite, on both short- and long-term time frames.
- These data buoys are most significant in their ability to gather and send data not only about the "in-progress" weather but also on the physics and dynamics of ocean/atmosphere interactions between storms, and how new storms form.

3. Discuss some of the roles of water in atmospheric activity.
   - Air always contains some water (up to four percent by volume), usually in its gaseous form (vapor).
   - Water enters the atmosphere through evaporation from the ground, from plants, and from the ocean, and leaves the atmosphere through condensation into its liquid form (rain, fog, snow).
   - Even though water occupies a relatively small volume in an air mass, its presence has a great effect on air's dynamics. Humid air is less dense that drier air at the same temperature, because molecules of water weigh less than the oxygen and nitrogen molecules they displace.
   - Water has a high latent heat of evaporation. Compared to other substances, water can absorb and retain large amounts of heat energy without a change in temperature.
   - At the equator, air heats and expands as it rises, then begins to cool and lose its water as rain.
   - As air moves down toward the equator it warms again, and regains its water through evaporation from Earth's surface.
   - Due to its high latent heat capacity, and its essentially unlimited area in which to expand, Earth's moving air transfers about twice as much heat energy poleward as do the ocean currents.

Lesson 11: Going with the Flow

1. b pages 234–235
2. b page 239, Video
3. c page 244
4. b page 246, Video
5. d page 247, Video
6. b page 239
7. b pages 235–236
8. c page 235
9. b page 235
10. c Video
11. d Video
12. d page 235

Short Answer

1. Ocean current measurements are extremely important in studying atmosphere/ocean dynamics, but obtaining accurate, long-term data remains a formidable problem. Trace the history of ocean current measurements, from the early seafarers to the present.
   - Early methods for measuring current speed and direction were limited to those near or at the sea surface.
   - Objects called drifters were often set afloat and observed, their progress being measured against time.

- Ship drift over time was also used to measure current speed and direction, so much of our early data were from ship-sailing routes.
- Current meters were invented, measuring the speed of rotation of a propeller-like blade in a current.
- Bottom-moored buoy arrays were used, at first serving to hold a number of devices vertically in a water column to measure currents at different depths.
- Next came an Acoustic Doppler Current Profiler, which uses a single deployed acoustic instrument to measure currents throughout a several-hundred meter water column.
- Satellite altimeters measure the changing heights of the sea surface, giving a good estimate of sea surface circulation.
- The NOAA/Japanese buoy/satellite array consists of 70 moored instruments extending from the western Pacific to the coast of South America. These measure temperature, wind, humidity, and other parameters, sending the information to a satellite that relays it to meteorologists for analysis.
- To obtain even more accuracy, the Jet Propulsion Laboratory uses the combined satellite technology of JASON-I and the older TOPEX-POSEIDON, enabling them to double the spatial resolution of circulation details.

2. Describe/discuss the Ekman transport system, and how it affects the configuration of a gyre in a Northern Hemisphere ocean basin.
   - Wind-driven water in the topmost layer of the ocean flows at about 45° to the right of the wind direction.
   - The next layer of water down senses the behavior of the water immediately above, and moves to the right of that layer.
   - While the number of actual layers of water in a column may vary, this mechanism affects each one in the same way, causing it to move to the right of the one above it, forming a spiral pattern called an Ekman spiral.
   - Due to frictional losses caused by those layers impacting each other, each lower layer moves more slowly than the one above it, until water motion at the bottom is nearly negligible.
   - This mechanism normally works down to a depth of about 100 meters (at mid-latitudes), theoretically enabling the water at depth to flow in a direction opposite to that of the surface current.
   - The net motion of water down to about 100 meters (including all of the layers in the spiral) is called Ekman transport. In theory the direction of Ekman transport is about 90° to the right of the wind direction in the Northern Hemisphere, and to the left in the Southern Hemisphere.
   - In the North Atlantic the sea surface is actually raised through wind motion and Ekman transport into a gradually sloping hill about two meters high, maintained by wind energy.
   - This creates a pressure gradient in that area, due to differences in height, and gravity tries to pull that hill back down to level.
   - The net result of all this is that the currents in the gyre continue to move, around the outside edges of the ocean basin.

3. While the El Niño phenomenon remains mysterious and unpredictable, the sequence of events, once it starts, is fairly well known. However, the accompanying return to normal patterns, known as La Niña, which occurs after an El Niño event, is not as well explained or documented. List, in sequence, the events involved in La Niña, and discuss how this also can affect circulation, temperatures, and coastal populations.
   - After an El Niño, the return to normal can often be somewhat vigorous and upsetting.
   - As the normal surface circulation begins to re-form, the easterlies (trade winds) regain their strength and start to reassemble the gyres in the North Pacific basin, causing the waters on the South American coast (particularly Chile and Peru) to cool down.
   - This results in a mass of cold water that tends to deflect winds around it, affecting both local and global weather systems and patterns.
   - As the gyre re-establishes itself, the change in wind strength and direction, and water temperatures at the eastern boundary of the basin can cause strong currents, severe upwelling, storms, and colder-than-normal water and air temperatures.
   - When the eastern boundary cools off, the ocean to the west (north of Australia) can warm rapidly, and the renewed push of the trade winds can begin to pile this water up, causing dramatic anomalies in the temperature and pressure gradients.

- The balance eventually re-establishes itself, although a particularly vigorous La Niña in the late 1990s persisted for nearly a year.

Lesson 12: Deep Connections

1. d page 250
2. d Video
3. b pages 251–252
4. c page 254
5. a page 250, Video
6. b page 257
7. c page 255
8. b Video
9. d pages 256–257
10. d pages 251–254
11. b page 253
12. a Video

Short Answer

1. The Global Heat Circuit has been analogized to a vast conveyor belt that carries surface water to the depths and back again. Describe/discuss that circuit, and the water masses and basins it encounters and affects, beginning with the formation of the North Atlantic Deep Water.
   - North Atlantic Deep Water forms when the warm, salty North Atlantic Ocean cools, as it encounters the cold Canadian winds in the latitude of Iceland.
   - This water mass loses about 8°C, of its heat, and sinks to become deep, cold water.
   - This deep water mass then flows south through the Atlantic, staying cold and salty, until it reaches the Antarctic Deep Water. However, it is not as cold or salty as the Antarctic Deep Water, so it tends to flow over, and to begin to mix with, the Antarctic Deep Water.
   - This combined mass circumnavigates the Antarctic and then moves northward into the Pacific and Indian Ocean basins.
   - Upwelling returns some of this water to the surface, where it continues to move, gain heat, and mix upward, returning to the North Atlantic as warm, shallow currents.

2. We know that some of the global water masses move very slowly, taking many hundreds of years to complete a circuit. We also know that these water masses tend to retain a memory of their initial formation at the sea surface. Because of this historical aspect to thermohaline circulation, scientists are trying to reconstruct some of the possible differences in circulation, and thus in climate patterns, as continental drift has rearranged the land masses over time and, consequently, the ocean boundaries and basins. From the video presentation, recall some of the thoughts, and present concerns, of those scientists as they consider the consequences of long-term continental drift patterns.
   - Thermohaline circulation patterns are greatly influenced by the distribution and configuration of continents and ocean basins.
   - We know that continental drift is a continuing process, and that the landmasses have undergone major changes over geologic time, thus affecting ocean circulation patterns.
   - There is evidence that, in the distant past, nearly all of the ocean bottom waters were warm and salty, with the "haline" part of thermohaline causing increased density and the waters to sink.
   - Because the thermohaline water masses retain hundreds to thousands of years of memory, there may still be remnants of that memory, gradually making their way surfaceward to possibly change present-day climates.
   - Some scientists believe that this formation history could be carried into, and through, all of the ocean basins and, as surface water, cause a rapid, global climate change. This is being tied into concerns about global warming.
   - They propose that if it warms sufficiently at high latitudes to melt ice, leading also to more precipitation at those high latitudes, there could be a layer of fresh, buoyant water which could effectively insulate the sea surface from the atmosphere and basically shut down the Global Heat Circuit, causing drastic and, probably, totally unpredictable changes in the global climate and weather patterns.

3. Discuss the use of chemical tracers (specifically, CFCs) to study ocean circulation.
   - Chlorofluorocarbons (CFCs) were originally manufactured beginning in the 1930s for commercial use as refrigerants, propellants in aerosol cans, and in the making of packing and insulation. There are a number of types of CFCs, but all are man-made—they do not occur naturally.

- CFCs move easily from the atmosphere into the seawater, where they dissolve and spread into the ocean, like a dye.
- During the initial 40 years of their industrial use (1930s to 1970s) their atmospheric concentration increased rapidly, leading to many of them being banned because of their potential to damage Earth's ozone layer.
- The usefulness of these chemicals as tracers, to study ocean circulation, is based on the fact that the history of their production, use, and eventual contact with the ocean through the atmosphere, can be documented.
- CFC-11 and CFC-12 were two of the first to be manufactured and used. By measuring the concentration of these two CFCs in seawater, and comparing their ratios with the ratios of their known changes in the atmosphere, one can estimate the last time that water mass was in contact with the atmosphere.
- The equipment and instrumentation for doing this is quite complex, and is presently being built, tested, and modified, with a lot of sea trials. One of the main requirements is that the sampling and analytic apparatus must be absolutely airtight, to prevent any ambient-air contamination.
- So far, scientists have been able to date the formation of some water masses as far back as the 1920s, and up to the present.

Lesson 13: Surf's Up

1. b page 267, Video
2. b page 266
3. a Video
4. c page 268
5. d page 272, Video
6. c page 272
7. b page 272
8. c pages 276–278, Video
9. d page 269
10. c page 282, Video
11. c page 272
12. b page 281

Short Answer

1. Discuss the so-called rogue wave—what is it, how is it formed, how is it different from other wave forms and patterns, what are some famous rogue wave encounters.
   - Independent wave trains exist simultaneously in the ocean most of the time.
   - Wind waves from different storm systems will have different characteristics, with longer waves outrunning (and interfering with) shorter ones.
   - Destructive interference can have a cancellation effect on the participants, whereas constructive interference can result in waves much larger than the individual participants.
   - Waves of many heights and lengths can converge on an area from different directions, meeting to form a rare, huge wave called a rogue wave, often much larger than is theoretically possible. Such a wave could be 3–4 times the theoretical average.
   - In 1942 the *Queen Mary* was struck broadside by a rogue wave off Scotland, causing it to tilt more than 45 degrees.
   - In 1912 a smaller Cunard liner, the *Carmania*, loaded with tourists, was struck by a rogue wave, sending it into a series of rolls over 50 degrees and causing much damage. A similar wave hit the *Queen Elizabeth II*, in 1996, frightening passengers and crew but causing no injuries or damage.

2. Imagine a large progressive wave train approaching a wide harbor mouth straight-on, entering the harbor and moving through it, assuming there are no major obstructions, to the rear of the harbor and impacting the seawall at the back. If conditions were right, the progressive wave train could be re-formed into a standing wave. Describe the sequence of events.
   - The progressive wave train encounters the harbor mouth straight on, and most, or all of its energy passes into the basin of the harbor.
   - Assuming no major obstructions, the lead wave will move through the harbor, followed by the successive waves in the train, still acting as a progressive wave.
   - When the lead wave hits the seawall at the rear of the harbor, it will be reflected back and move away from the seawall in the direction in which it entered.
   - Meanwhile, the other waves in the train are approaching, and the lead wave, being the first to be reflected back, will impact the following waves and cause constructive interference in the form of vertical oscillations called a standing wave.
   - These waves do not progress but appear as alternating crests and troughs, and because their interference is constructive they can

actually become larger than the original wave.
- This energy will eventually dissipate, because in reality there are no basins or harbors with geometrically square sides, so the waves would assume angles of reflection off the basin walls and the standing wave pattern would disappear.

3. Describe and document the sequence of events occurring as a deep-water wave approaches a shoreline; from the first encounter with the bottom until it breaks as surf, and how the contours and characteristics of the nearshore bottom determine the nature of the breaking wave.
   - A deep-water wind wave moves through water deeper than half its wavelength.
   - As the wave approaches a shoreline it encounters water shallower than half its wavelength but still deeper than 1/20 of its original length. This is a transitional wave, which then continues to move shoreward, into water less than 1/20 of its original wavelength, to become a true shallow-water wave.
   - Here the wave contacts, or "feels" the bottom, and its circular orbit pattern is interrupted, with the part near the bottom flattening into an ellipse and moving more slowly due to friction with the bottom.
   - But the wave still retains most of its forward-moving energy, concentrated in the upper part, causing the formerly rounded wave crests to peak.
   - The following waves, not having felt the bottom yet, continue toward shore at their original speed.
   - As it peaks, the ratio of 1:7 height to length is approached.
   - As the wave form is modified, the water is actually moving faster than the wave form and as the crest moves ahead of its slower-moving base it breaks, when at the ratio of about 3:4 height to length.
   - The break sends this turbulent mass of water onto the beach as surf, and its wind-generated energy is released as random movement, heat, and sound.
   - If the shoreline slopes upward gradually, waves lose their strength because their lower orbit is in contact with the bottom longer. This loss of energy can be further enhanced if the bottom is loose gravel or coral, or if there is floating seaweed or ice in the wave's path.
   - If the bottom is deep until it meets a vertical shoreline or cliff face, the wave may not go through this series of steps leading to the classic break, but will lose its energy as it crashed into the cliff, sending water both downward and skyward.

Lesson 14: Look Out Below

1. d page 284
2. a page 285
3. b page 285
4. c page 286
5. d Video
6. c page 286, Video
7. b pages 287–289
8. d Video
9. a page 286
10. b page 287
11. c Video
12. b Video

Short Answer

1. Describe, in some detail, NOAA's National Tsunami Hazard Mitigation Program.
   - The overall purpose of the Program is to reduce hazards due to tsunami impacts, based on a three-part approach.
   - One element of the Program deals with hazard assessment—on a local/community basis, the recognition and analysis of both the geographical and population aspects of at-risk areas. This can include consideration of past tsunami threats or events, and their impacts on that community, as well as attempting to model/predict reactions to future events.
   - Another component of the program considers hazard mitigation; basically a preparation/evacuation plan. Both state and federal agencies, including FEMA, are involved. Emergency facilities, shelters, food stocks, etc., are documented, as are areas at especially high risk of inundation.
   - In areas at very high risk like Hawaii, the mitigation part of the program also gets involved with building codes and the actual redesigning of shoreside structures to minimize damage in the event of a tsunami.
   - The third element of NOAA's Program is a detection/prediction/warning system, designed to lessen property damage and loss

of life by providing alerts, allowing sufficient preparation and reaction or evacuation time.
- The Warning System receives some of its information from arrays of seismometers, which measure actual ground motion, and a network of tide gauges and deep-water pressure sensors to detect changes in sea level that could result from undersea seismic activity.
- The NOAA system is presently in place but has yet to be tested by a real event.

2. Describe the sequence of events that led to the famous tsunami disaster at Hilo Bay, Hawaii, in April, 1948.
   - At a subduction zone along the Aleutian Trench, the sea floor ruptured.
   - That seafloor displacement was translated to the sea surface above it, disrupting the smooth surface. Gravity immediately tried to restore that smoothness, but the momentum of that pull overcompensated and created a crest/trough oscillation.
   - That oscillation built into a series of progressive waves that radiated out into the ocean in all directions, and one of the first of those waves quickly destroyed a nearby lighthouse in the Aleutian Islands.
   - Five hours later the energy from those waves reached the Hawaiian Islands. While they had lost some of their energy on the trip across the Pacific, they still inundated the Hawaii beaches for more than two hours.
   - Especially hard-hit was Hilo Bay, on the island of Hawaii, because although the seismic sea wave had lost some of its original energy, what remained was concentrated by the funnel-shaped shoreline of the bay.
   - There was much property damage and the loss of at least 150 lives.

3. Describe and discuss the attraction of people to areas of seismic sea wave activity, and the characteristics of those waves that can, in turn, injure or kill those onlookers.
   - Even though a single seismic sea wave is generated at an undersea epicenter, it can quickly be distributed into a series of waves following one another at regular intervals.
   - If the original energy source is far away from the eventual shoreside rendezvous, sea level at that shore will rise and fall as those waves arrive, usually with a period of 15 minutes.
   - Half of the time, a tsunami trough will arrive on shore first, causing sea level to drop dramatically for about 15 minutes.
   - As the water recedes, curiosity-seekers may move out onto the exposed beach to look at fish and other uncovered marine life, or to inspect grounded vessels.
   - Then the first large crest arrives, drowning or injuring the onlookers. When that recedes, would-be rescuers move out onto the beach, only to be hit by later crests.
   - Apparently this is still a common human behavior, so many of the high-risk areas have included information about this phenomenon as a part of their tsunami education and warning programs.

Lesson 15: Ebb and Flow

1. d page 298
2. c pages 304–305, Video
3. d pages 310–311, Video
4. b page 306, Video
5. a page 298
6. b page 299
7. c page 308, Video
8. b page 303
9. d Video
10. d page 307
11. b pages 305–306
12. d page 307

Short Answer

1. Every continental coastline, whether a sandy beach, rocky shore, or a steep cliff, has an area that is covered by water at high tide and exposed to the air at low tide. This is called the intertidal zone, and is best known and studied for the marine organisms that live there. Describe basic physical aspects of the intertidal zone, and how the organisms that live there cope with the changing conditions.
   - The intertidal zone is defined as the area that lies between the high and low tide marks, on any coastline. The surfaces that the organisms live on, in, and under vary with the particular area, and can range from fine sand to rocky cliffs.
   - As water levels rise and fall with the changing tidal cycles, parts of the zone will be covered and uncovered for various lengths of time.
   - Organisms living within those boundaries are also submerged or exposed for various

lengths of time, based on the daily tidal cycles.
- When the tide level falls below their particular habitat or surface of attachment, many of these organisms will be exposed to the air and the chance of drying out, until the water level rises again to cover them. This can also expose them to sudden changes in temperature, between the water and the air.
- Most marine organisms get their food and oxygen from the seawater; during low tide exposure they must cope with the lack of access to these resources until the water returns.
- The organisms that live, and thrive, in the intertidal zone have adapted, over evolutionary time, to survive these periodic extremes in their environment.
- Some species, such as the small grunion fish, actually include the tidal cycles as part of their reproductive strategy. They invade sandy beaches at high tide, and lay their eggs in the sand. The eggs are incubated in the high beach sand, to hatch out two weeks later and be swept out to sea when the high tides return.

2. Tides are defined in several ways—one describes tides as periodic, short-term changes in sea level; another says that tides are a form of wave. Expand upon this second description, listing and discussing the various descriptive components that justify their being called waves.
   - As a wave, tides are the longest of all waves, with a length equal to half of Earth's circumference, or about 20,000 km (12,500 miles).
   - They are also a form of progressive wave, because they have crests, in the form of the tidal bulges. These are separated by half of Earth's circumference.
   - The crests (bulges) move as forced waves, as opposed to free waves (wind waves and tsunami) because they are never free of the forces that cause them.
   - Tides are shallow-water waves, because they will always be in water less than 1/20 of their length (theoretical wavelength is 20,000 km, while the average ocean depth is about 3.8 km).
   - Theoretically, a tidal wave could move across the generalized water-covered Earth surface at about 1600 kilometers per hour. But for the crest to move that fast, the theoretical ocean would have to be 22 kilometers deep, which it is not. So, tidal wave velocity is controlled by the actual depth of the water in which they occur.

3. One of the concepts Newton did not consider in his equilibrium theory of the tides, but which was later used in Laplace's dynamic theory is the rhythmic oscillation of water in an ocean basin. We encountered this standing wave phenomenon earlier, in our lesson on wind waves, and again in our discussion of seiches, but we must now consider it in a larger venue as it relates to ocean basins. Explain the rhythmic oscillation phenomenon, how it relates especially to water movement in the world's ocean basins, and how it affects the tides.
   - When the oscillation begins, there forms a central point of no vertical movement, called a node, around which the wave oscillates. The rhythmic back-and-forth rocking movement in the basin generates alternating crests and troughs, at various distances from the node, as the water encounters the basin boundaries.
   - Even though there are crests and troughs, there is no net movement of water, so this is no longer a progressive wave.
   - The seiche is a slightly different concept, best thought of as occurring in larger, but still enclosed, bodies of water. It was first studied in Switzerland's Lake Geneva, based on wind waves pushing water to one end of the lake, which then recovers to oscillate back to the opposite end. It is a type of standing wave, with the water oscillating about a central, no-motion node.
   - The same basic water dynamics can affect ocean basins, complicated by interference with the landmasses, or continents.
   - Tide wave crests, moving as progressive waves, would encounter continents and, depending on the configuration of the coastline, and the geography of the ocean basin in particular area, could reflect back and begin a resonant oscillation.
   - The shape of the basin has a strong influence on the tidal patterns and tides on the borders of that basin, tending to modify the ideal tidal cycles
   - In a large ocean basin the node, or point of no tide, is called an amphidromic point, around which the tidal crest moves in one tidal cycle. The modifying effect of irregular basin contours can cause the tidal crests and

troughs to cancel each other at those points. There are about a dozen major amphidromic points in the world's ocean basins, with five of those in the Pacific Ocean, accounting for the complex tidal patterns on the Pacific coasts.

Lesson 16: On the Coast

1. c pages 328–329
2. d Video
3. d Video
4. b page 316
5. b page 336
6. a page 323
7. c page 330
8. b pages 330–331
9. a page 319
10. d page 325
11. b page 324
12. b Video

Short Answer

1. While oceanographers and other scientists are busy classifying and describing the world's coasts and their associated features (estuaries, coral reefs, beaches, etc.) sociologists and coastal ecologists are concerned that about most of the U.S. population now lives in what are considered coastal counties, building towns and cities, tapping into coastal fresh water supplies, and impacting other coastal resources. List those concerns, and some of the proposed approaches to solving or addressing them.
   - Since the mid-1950s, there has been a national migration in the United States, away from the middle of the country and toward the coasts (Pacific, Atlantic, Gulf). This has put a large and sudden demand on coastal resources.
   - Much of this human impact has been negative, including the building of homes on unstable coastal bluffs, demands on water supplies and fisheries resources, and recreational impacts on what is considered to be a fairly sensitive marine environment.
   - As sea level continues to rise, many of the homes and other structures that have been on the coast for long periods of time are going to be even more directly impacted by wave activity and continued erosion.
   - Other concerns, on both a national and global scale, include such aspects of a coastal infrastructure as sewage disposal systems, and large coastal highways to serve this growing population.
   - As the coastal population, and its accompanying impacts increased, around the mid-to-late 1960s, there was a growing interest in protecting and conserving the coastal environments and in finding a balance between the natural processes and human activity.
   - To address these issues, coastal zone managers, scientists, and engineers began to develop a master management plan for many areas, but as yet none are fully implemented. Input from scientists is being sought, particularly that which projects natural phenomena such as global warming and sea level fluctuations. State, federal, and local governments and advisory boards are becoming involved in formulating a plan that will consider individual and property rights, as well as the needs of the coastal resources and ecosystems.
2. Coral reefs are a major feature in many tropical oceans, and a significant attraction for both scientific and recreational scuba divers and snorkelers. During his historic voyage on the *Beagle*, Charles Darwin observed and described reefs at many of the areas he visited (from the surface!), and devised a 3-part classification system that we still used today. Compare and contrast the formation and appearance of those three basic types of coral reefs.
   - All coral reefs are made by coral polyps, which are living animals related to sea anemones. Each polyp secretes a cup-shaped skeleton of calcite, forming an aggregation, or colony of living animals. When the polyps die, the calcite skeleton remains, and the next generation of polyps/skeletons grows on top of them, eventually building into large structures or reefs over geologic time. These structures assume different configurations, depending on their location relative to a shoreline.
   - Fringing reefs are connected to a shoreline and extend seaward from it, normally at the downwind edge of tropical islands, in areas of low rainfall.
   - Most new tropical islands will begin to form fringing reefs, which may or may not progress into other configurations.
   - Barrier reefs are separated from the adjacent land by a lagoon. They are normally found in more tropical (lower latitude) locations, and their seaward edges tend to be higher than

those fronting the lagoon, due to better access to normal seawater salinities and food resources.

- Coral growth at the lagoon side of the structure tends to be slow, due to less access to normal seawater and food, and more exposure to land-based sediments and freshwater runoff.
- Australia's Great Barrier Reef is the largest structure of this type on Earth, about 2,500 kilometers (1,560 miles) in length. It is composed of thousands of linked reef segments, covering about 350,000 square kilometers (135,000 square miles).
- Atolls are the third type of reef in Darwin's classification, typified as a ring-shaped island of coral reefs and coral debris enclosing a shallow lagoon, but with no obvious connection to land.
- As the reef matures, coral debris may pile up as sand, and form a protruding landmass, which may then re-form into a type of soil and support palm trees and other land plants. If this structure continues to evolve as an island, birds and other animals from the nearby land may eventually colonize it.
- Atolls can remain as isolated structures or form into loose groups or chains. The Marshall Islands are one of the better known of these formations.
- Atoll formation has been a topic of much discussion. Darwin correctly imagined that they could form around a volcanic island emerging from the sea, gradually increasing in size and depth as the volcanic structure subsides.
- Darwin did not know about plate tectonics and the natural subsidence of volcanic islands, but had correctly observed that the islands tended to sink at a rate comparable to the growth of the coral. So, even though the peak of an island disappeared, to create a true atoll, the ringing coral maintained a growth rate that would keep the top of it at the sea surface.
- Comparing Darwin's three coral reef types, they all form from the successive growth of coral colonies; contrasting them leads to recognition of their different levels of association with the nearby land mass.

3. When one first entertains the idea of playing at the beach one of the main things we are cautioned to watch out for is rip currents. Parents, lifeguards, fellow swimmers tend to be concerned about this narrow, seaward-returning mass of swift-flowing water, and the basic escape routes from it. Characterize and discuss the rip current, as an oceanographic phenomenon and as a concern to swimmers.
   - The correct oceanographic term for this phenomenon is rip current, although rip tide and undertow are more commonly used (descriptive but not accurate).
   - After an incoming wave breaks, some of the remaining energy can push the water further up the beach.
   - When the wave's energy is totally dissipated, which depends upon the composition and steepness of the beach), gravity will begin to pull the water back seaward.
   - If the beach is rather steep, and sandy, the water may actually collect into narrow, high-velocity (up to three meters per second!) masses of seaward-returning current—a rip current.
   - These currents will move back down the beach and, depending on their speed, continue out through the breaker zone, to eventually dissipate. However, successive incoming waves will continue to form these returning currents, until surf conditions change.
   - Often, rip currents carry beach sand back down with them, allowing them to be noticed from the beach, and avoided. They can also form, and flow, nearly undetected, until the swimmer or wader feels a powerful seaward pull.
   - Basic safety considerations are to check with beach guards for current locations of rip currents, and if caught in one, allow it to carry you out with it, and when it loses its energy one can move out of it to either side and back to shore.
   - Small children and weak swimmers are particularly vulnerable to these surprise encounters, and with them the best defense is to avoid contact.

Lesson 17: Due West

1. b Video
2. a Video
3. c Video
4. d Video
5. c Video
6. d Video
7. d Video

8. b Video
9. b Video
10. d Video

Short Answer

1. The instability of the Malibu coastline and the resulting landslide threat has been attributed to a variety of factors. List those factors, and briefly define and discuss them.
   - Major changes in sea level in the recent geologic past, and the resulting coastal erosion at different sites.
   - During periods of heavy rain, especially during El Niño events, water percolates through the ground into the deeper, less geologically stable, strata. This causes the soils to lose strength and move downslope due to gravity.
   - There is still active mountain formation in the area, adding to the instability by faulting.
   - As the population continues to increase, more structures are being built on the prime "view sites," resulting in more roads being cut and the disruption of once-stable hillsides.
   - Some of the road cuts have removed material from the bottom of the slopes, causing the slopes to move downward due to gravity, in an attempt to restabilize.

2. Describe and discuss, in some detail, the Point Mugu wetlands ecosystems, and attempts to restore them.
   - The Point Mugu wetlands area (technically a salt marsh) is one of the most extensive in southern California.
   - The watershed feeding into the wetlands covers about 320 square miles of surrounding urbanized and agricultural lands.
   - Compared to other similar areas, these wetlands are reasonably well protected by federal and state laws, and by the U.S. Navy.
   - The Mugu wetlands are an important area for both resident and migrating waterfowl, including three endangered and one threatened species. The light-footed clapper rail is highest on the endangered list.
   - Restoration already in progress includes recontouring the land to form bird-nesting islands, and creating tidal creeks to bring back fish and invertebrates.
   - Former sewage treatment ponds are being decontaminated or reconstructed to prevent further sludge effluent damage.
   - Special focus is on removal of invasive plants and replanting of more suitable species.
   - With the birds, the primary focus has been on the endangered clapper rail, monitoring their return and attempting to support repopulation through a captive breeding program.
   - Most of this planning, restoration, and monitoring is being directed by a Restoration Advisory Board representing regulatory agencies, the Navy, the public, and cleanup specialists.

3. List, in approximate chronological order, the steps involved in the planning, construction, and maintenance of the Ventura Marina.
   - Started in the 1950s as a private venture, to construct a harbor dedicated to recreational boating.
   - Up to that time there was little precedent or protocol for such a coastal construction, and little if any pre-design was done.
   - As construction progressed, it became evident that the marina jetties would interrupt the natural longshore sand transport system, but there were as yet no precedents for bypassing the sand onto the downcoast beaches. The sand was trapped at the marina entrance, causing significant shoaling and decreasing the depth of the navigation channel.
   - Again, working without the experience of precedent, the constructors built breakwaters and jetties in odd configurations, in an attempt to solve immediate shoaling problems.
   - As the natural longshore drift patterns were interrupted, the beaches downcoast from the jetties were sand-deprived and began to erode.
   - In the 1960s the City of Ventura began a maintenance-dredging program to keep the harbor entrance clear, but did not have the funds to continue it, so in the late 1960s the U.S. Army Corps of Engineers assumed that responsibility. This included building a sand trap upcoast from the jetties to prevent sand from filling the harbor entrance.
   - Presently, the Corps of Engineers and the Ventura Port District are considering a sand-bypass system of pumps and inductors, to continuously move sand from the upcoast trap and bypass it past the harbor and onto the sand-deprived downcoast beaches.

Lesson 18: Building Blocks

1. b page 346
2. c Video
3. d page 366
4. a pages 365–366, Video
5. b page 357
6. c Video
7. a page 353
8. c pages 365–366
9. d Video
10. c page 367, Video
11. b page 369
12. a Video

Short Answer

1. Outline the mechanism of evolution by natural selection, or Darwinian evolution.
   - Darwin observed that organisms reproduce excessively, producing more offspring than the environment can support. He also noticed that natural populations stay relatively constant in number.
   - He concluded, therefore, that the individuals of populations must compete for the limited resources, mates, and refuges.
   - Furthermore, if they are competing and have variations in their traits, individuals better adapted would, on average, survive and reproduce more successfully (natural selection).
   - As natural selection results in particular genetic traits getting passed on and others not, the gene pool changes and evolution occurs. (Darwin did not know of genes but was aware that certain traits are heritable.)
2. Explain the role of biogeochemical cycles in biology and describe the carbon cycle in the ocean.
   - Living things are dynamic entities and continuously cycle matter in and out of their bodies as they live and later decompose. The pathways that matter follows between organisms and their environment are called biogeochemical cycles.
   - One might start the cycle with atmospheric CO_2, which dissolves in seawater by diffusion and physical mixing. It is also released by respiration. It will exist in solution as CO_2, bicarbonate, and carbonate. Photosynthetic organisms convert the CO_2 to organic molecules such as carbohydrates, proteins, and lipids. Via respiration, they release the CO_2 back into the water where it can be reused by another autotroph or diffuse back into the atmosphere.
   - Via consumption or death and decomposition the organic molecules move up food chains being metabolized with further release of CO_2. Death and secretion also contribute particulate organic carbon (detritus) and dissolved organic carbon to the oceanic pool.
   - Carbonate and bicarbonate can also be incorporated into shells and skeletons or precipitate onto sediment. The skeletal and rock carbonate can move in a slow loop on the tectonic conveyor belt to subduct, melt, and later return via outgassing to the atmosphere or ocean.
3. The video states that there are three or four steps necessary for biogenesis. List and describe them.
   - The four steps are 1) produce the small biomolecules such as sugars and amino acids; 2) polymerize smaller molecules to make more complex molecules such as polypeptides and polysaccharides; 3) store genetic information in a polymer such as DNA or RNA; 4) encapsulate it into a cell.
   - Simple biomolecules such as amino acids and monosaccharides can be made *in vitro* from molecules like methane, ammonia, hydrogen, and cyanide under conditions simulating the pre-biotic Earth. This is the Urey-Miller model. Energy sources such as electric discharge, heat, and ultraviolet can stimulate these syntheses.
   - Under a variety of conditions and using a variety of energy sources the simpler molecules polymerize.
   - Among the polymers that self-assemble is RNA, a molecule that can replicate and mutate. The hypothesis that RNA preceded biosynthesis is currently popular. RNA is capable of catalysis and could give rise to both DNA and protein.
   - Although the video does not mention lipid biogenesis, it stresses the importance of compartmentalization and encapsulation of this chemical system as a precursor of the first cell.

Lesson 19: Water World

1. a page 369
2. a page 372, Fig. 13.25

3. d Video
4. c Video
5. d Video
6. c page 359
7. c page 359, Video
8. b page 362, Video
9. c page 371
10. a page 372, Fig. 13.25
11. d Video
12. a page 364

Short Answer

1. Tuna fish, dolphins, and penguins have teardrop-shaped bodies. Discuss this and other adaptations exhibited by marine organisms that swim rapidly, efficiently, and for sustained periods.
   - This general shape is shared by a wide variety of nektonic organisms that require sustained efficient swimming.
   - Its major advantage over other shapes is that it has much less drag because water flows over with less turbulence. The teardrop shape has only about six percent of the drag of a disc of identical cross sectional area.
   - The dolphin is a mammal and the penguin is a bird, so both are homeotherms and can maintain higher metabolic rates because of their higher body temperature. Although most fish are cold-blooded, large tunas like the bluefin can maintain elevated body temperatures because a counter current blood flow through their muscles captures and retains heat. Elevated body temperature contributes to more efficient swimming in all three.
   - Other adaptations discussed in the text or video pertain particularly to the tunas and billfishes and include secretion of mucus or oil that reduces friction and drag, indentation to allow fins to lie flat against the body, reduced scalation and skin that indents to reduce turbulent eddies.

2. List the six kingdoms of life, give examples of each, and briefly compare and contrast them.
   - 1) Archaea: thermophilic, methaogenic, and other extremophile prokaryotes. 2) Bacteria: the "true" bacteria including staph, strep, cyanobacteria, and most of the heterotrophic marine bacteria. 3) Protista: diatoms, dinoflagellates, red green, and brown algaes, foraminifera. 4) Fungi: mold, mildew, yeasts, mushrooms. 5) Plantae: mosses, ferns, vascular plants. 6) Animalia: sponges, jellyfish, worms, mollusks, arthropods, chordates.
   - The Archaea and the Bacteria are the two prokaryotic kingdoms. Until recently they were combined as the Monera, but fundamentally distinctive genes and other features of their biochemistry made it logical to separate them.
   - Among the eukaryotic kingdoms, the Protista are the least natural assemblage. Included in this kingdom are organisms that are among the smallest unicellular eukaryotes and the largest of the multicelled kelps. When they are multicelled, their level of organization barely exceeds the tissue level.
   - The fungi are heterotrophic and may be unicellular or multicellular. They are decomposers or parasites and among the six kingdoms, they are the least well represented in the marine environment.
   - The plants and animals are the most complex eukaryotes. They are always multicellular and organized at the organ and system levels. The plants are attached, immobile, and autotrophic while the animals are sessile or motile and heterotrophic consumers.

3. Explain the meaning of the term "limiting factor" and discuss temperature as an example of a limiting factor.
   - A limiting factor is any factor such as temperature, light, and salinity that limits the survival of organisms.
   - Organisms have a range of tolerance for most limiting factors. That range includes an optimal range in which the organism flourishes, suboptimal high and suboptimal low ranges where they are stressed, and a lethal high value and a lethal low value at which they die.
   - Light, temperature, and salinity are probably the three most important physical factors that limit the abundance and distribution of marine organisms. Ocean temperatures range from approximately −5°C to 40°C. Marine life is found at all these temperatures and certain extreme species are capable of surviving freezing temperatures, while other organisms live in waters exceeding 110°C, such as found in hydrothermal vents.
   - Mammals and birds are true endotherms and maintain constant body temperatures. A few large pelagic fish such as bluefin tuna function at temperatures exceeding ambient temperature by trapping metabolic heat with

a rete-like pattern of blood circulation through their active, warm muscles.

- Endotherms trap heat using blubber, feathers, fur, and circulatory specializations. They benefit by maintaining a constant body temperature and therefore a rather constant and high metabolic rate.
- Ectotherms cannot regulate their body temperatures and therefore have temperatures that reflect the temperature of the environment. Those whose optimal temperatures are higher, such as the tropical species have higher metabolic rates than those whose habitat is colder. Metabolic rates approximately double with every 10°C rise in temperature (within optimal ranges). Because of higher surface area to volume ratios, smaller species have greater difficulty in retaining heat.

Lesson 20: Food for Thought

1. a pages 384–386
2. b pages 388–389, Video
3. b page 392
4. d Video
5. a page 394
6. c page 398
7. a pages 382–383
8. a page 347, Video
9. d page 394
10. c page 389
11. c page 392, Video
12. c page 391

Short Answer

1. A student driving to campus displays a bumper sticker that reads: "Everyone loves dolphins, but who cares about diatoms?" Later, she eats a tuna fish sandwich for lunch. Explain clearly what primary production is and relate it to her bumper sticker and her lunch.
   - Primary production is called primary because it is the way food is manufactured at the base of all food chains. Photosynthesis is the major mechanism of primary production, but some bacteria carry out chemosynthesis.
   - In photosynthesis, light energy is captured by chlorophyll and used to manufacture sugars—primarily glucose—from simple molecules of carbon dioxide and water. Photosynthesis also produces oxygen.
   - The primary producers use most of the sugar they manufacture for their own life processes but the unused portion, typically about 10 percent of the total, is available for the rest of the food web.
   - The layers of a food web can be represented by trophic pyramids. Primary producers form the base. They are consumed by the herbivores or primary consumers. These animals and other consumers are, in turn, fed upon by secondary consumers and higher-level carnivores up to some top carnivore.
   - The diatoms are one of the most abundant and efficient of the ocean's primary producers and are among the organisms at the lowest trophic level in the food web that ultimately leads to the tunas.
   - Tunas are often top carnivores in their ecosystem. Their food web may begin with diatoms and pass through zooplankton, filter feeding fish like anchovies, medium size fish like mackerel and finally to the tuna. If the student eats the tuna sandwich, she becomes the top carnivore.
   - Because only about 10 percent of the energy in a lower trophic level is available to the next higher one it would take about 10,000 kilograms of diatoms to produce one-quarter pound of tuna for the sandwich (10,000 kg of diatoms, 1,000 kg of zooplankton, 100 kg of anchovies, 10 kg of mackerel, 1 kg of tuna and 0.1 kg or about one-quarter pound of tuna for the sandwich.)
2. A significant change in our understanding of marine primary production has emerged with the discovery of the role of picoplankton. Discuss the role of these tiny organisms in ocean ecology.
   - Throughout most of the history of modern oceanography the diatoms and dinoflagellates were considered to be the most important primary producers. Within the past few decades biologists have been uncovering a previously unknown category of primary producers and evaluating its role in total ocean productivity. It seems to be significant.
   - These organisms are the picoplankton. They are prokaryotes and their taxonomy is debated. Many, such as Synechococcus, seem to be cyanobacteria. There may be 100,000 per milliliter and they may account

for up to 70 percent of the primary production in some ocean regions.

- In addition to their tiny size, they may have gone unnoticed because they seem to operate in ecological loops that do not feed significantly to higher levels.
- They are grazed efficiently by microflagellates and the organic material of these tiny autotrophs and the microflagellates as well, are recycled rapidly and efficiently by even smaller heterotrophic bacteria that are involved in decomposition at every trophic level.
- An additional discovery from an even smaller ultraplankton world recognizes the abundance and importance of phycoviruses and bacteriophages that cause picoplankton to burst and release organic molecules into the sea.

3. Describe and explain the general pattern of primary production in temperate oceans.
   - The two factors controlling primary production are nutrients and light. The seasonal changes of temperate oceans produces a characteristic pattern of productivity. This pattern is most pronounced in the temperate North Atlantic.
   - The temperate zone is somewhat like the polar ocean in the winter and somewhat like the tropical ocean in the summer. A seasonal thermocline forms in the spring, strengthens throughout the summer, weakens and finally disappears in the fall.
   - The water is well mixed and nutrients are brought into the photic zone in the fall and mixing is strong throughout the winter. But light is low.
   - As sunlight increases in the spring and couples with the already-present nutrients there is a bloom of phytoplankton. As the water warms and summer approaches, the seasonal thermocline forms and the nutrients tend to get trapped below it.
   - The depletion of nutrients tends to make productivity decline in the summer. In the fall the thermocline breaks down and wind and storms mix the nutrients upward again. As long as this fall upwelling occurs before the sunlight declines too much, there will be a fall bloom as well.
   - The more intense spring bloom is triggered by increasing sunlight and terminated by decreasing nutrients. Conversely, the fall bloom is caused by upwelling nutrients and ends with declining sunlight.

Lesson 21: Survivors

1. b Video
2. a page 404, Video
3. a page 404, Video
4. b pages 405–406
5. d page 406
6. c page 407
7. c page 407
8. b page 417, Video
9. a page 410
10. d page 410
11. c page 412
12. d page 414

Short Answer

1. For purposes of general description, the true animals have been divided into two categories, based on the presence or absence of an internal supporting structure (skeleton). The invertebrates lack such a structure, and comprise roughly 90 percent of all known living and fossil animals, representing 33 phyla. In this lesson we met eight truly invertebrate phyla, as well as the invertebrate members of the phylum Chordata. Those eight invertebrate phyla were chosen to represent the invertebrates, and are presented to us in order of increasing complexity of body structure and function. Beginning with the most primitive list each of those phyla, in ascending order of complexity and, for each, briefly describe its distinguishing features, particularly those that make it primitive or advanced.
   - Porifera (sponges) – simple construction; supporting structure of spicules or fibers; a few types of specialized cells but no tissues, organs or systems; attached to the sea floor; filter feeders.
   - Cnidaria (sea anemones, jellyfish, etc.) – composed of two tissue layers; armed with stinging cells for feeding and defense; primitive nerve network but no brain; radial symmetry; body shaped as polyps (attached) or medusae (free-swimming); feed by trapping and stinging prey.
   - Platyhelminthes (flatworms) – small, wormlike; free-living or parasitic; primitive organs and the beginnings of a central nervous system; composed of three tissue layers; bilat-

eral symmetry; feed as predators, scavengers, or parasites.
- Nematoda (roundworms) – complete digestive system; occupy many habitats; occur in large numbers; free-living or parasitic.
- Annelida (segmented worms) – most advanced of the worm phyla; body composed of similar segments; well-developed systems (nervous, digestive, reproductive, etc.); class Polychaeta the largest in the phylum; move freely over or through sea floor sediments, or live in tubes; feed as predators, scavengers, filter feeders.
- Mollusca (clams, snails, and octopus) – large and diverse phylum; organs and systems well-developed; all have some kind of foot for attaching and moving, and most have a shell; feed as predators, grazers, scavengers; mostly on sea floor, but a few are planktonic.
- Arthropoda (shrimps, crabs, and lobsters) – largest animal phylum in numbers of species and of individuals; well-developed and advanced systems; possess a tough, flexible outer skeleton (exoskeleton), highly-specialized nervous and muscular systems, and jointed appendages; occupy a wide variety of habitats, both on the sea floor and in the midwater; employ a variety of feeding strategies; class Crustacea is the largest marine grouping.
- Echinodermata (sea stars, sea urchins, etc.) – all marine bottom-dwellers; basic body shape is 5-part form with radial symmetry; most systems developed but have a rudimentary nervous system; most move using tube feet powered by a hydraulic water vascular system; feed as grazers, scavengers, predators.

2. The video presentation for this lesson states that "perhaps nowhere on Earth is the ability to survive better demonstrated than in the intertidal zone." List some of the characteristics of the intertidal zone, some of the types of organisms that live there, and the mechanisms and strategies they use to survive.
   - The intertidal zone is that part of the shore alternately exposed and submerged by water.
   - Animals living there are exposed to wave stress, desiccation, exposure to air, rapid changes in temperature and salinity, and ultraviolet radiation from the sun.
   - Rocky shorelines have the most diversity of animal types.
   - Crabs and fishes can move with the changing water levels, staying mostly in the lower intertidal and hiding in crevices and under rocks.
   - Barnacles and mussels are attached to rocks and cannot move, so they close their shells until the water returns.
   - On sandy beaches the small crustaceans and sand crabs bury themselves to avoid exposure.

3. The members of the phylum Chordata are considered the most advanced animals, yet that phylum also includes two interesting types of invertebrate chordates. Briefly describe and discuss those two types of primitive chordates, and the features that place them in the phylum Chordata.
   - The phylum Chordata includes animals which possess at least the following features at some time during their lifecycle: a structural, stiffening notochord, gill slits, and a dorsal nerve cord.
   - Some species lose the notochord during development, and are called invertebrate chordates.
   - The tunicates, so named for their unique body covering (tunic), occupy both bottom and midwater habitats.
   - Tunicates feed by filtering suspended material from the water through their systems.
   - Amphioxus are small, fish-like animals that live in shallow-water sands, and feed by trapping nutrient particles.
   - Amphioxus retain the rudimentary notochord as adults, as well as the dorsal nerve chord, which is thought to be a transitional stage leading to the true vertebrate spinal cord.

Lesson 22: Life Goes On

1. c page 417, Video
2. a page 420, 427–429
3. b Video
4. d page 430, Video
5. c page 424
6. b page 421
7. d page 425, Video
8. c page 420, Video
9. c page 424, Video
10. a Video
11. c page 427, Video
12. d page 422

Short Answer

1. List and discuss four characteristics shared by all marine mammals.
   - Streamlined body shape. Propulsion through water is more efficient if the body is hydrodynamically streamlined. Limbs are reduced or modified into flippers or flukes. The ratio of limb musculature compared to trunk musculature is low. In the whales the tail fluke is horizontal with the greatest mass of muscle along the dorsum for the power stroke. Pinnapeds are fusiform in shape and the seals even lack external ears. Seals' rear flippers are partially fused and work like a tail fluke to provide powerful thrust in water while being virtually useless on land. The front flippers are used for steering. Sea lions do use their front flippers for propulsion. Like cetaceans, sirenians have a single tail fluke.
   - Thermal adaptations. Marine mammals have high metabolic rates and are voracious eaters to fuel this metabolism. Thick layers of blubber and reduced dermal capillaries conserve heat. The sea otter does not have thick blubber but rather has the densest fur of any mammal. It spends considerable effort maintaining its pelt so it traps a lot of air for insulating purposes. Marine mammals are large giving them the heat conservation of a low surface area to volume ratio. Many whales migrate to warmer latitudes in colder seasons.
   - Respiratory modifications. The nostrils are muscular and can be closed when diving. In the Cetacea the nostrils have moved to the top of the head as blowholes and are completely separated from the mouth or pharynx. The muscles are rich in myoglobin and the blood in hemoglobin to collect, transfer, and deliver oxygen efficiently. The percent evacuation of the lungs per breath is high.
   - Osmotic adaptations. Unlike the reptiles and birds that have salt glands the marine mammals rely solely on their kidneys to produce their highly salty and concentrated urine. They swallow very little water, have skin that is not permeable, and rely largely on metabolic water produced from the oxidation of foods.

2. List four problems faced by marine fishes that are different than those faced by land vertebrates. Discuss each problem and its solution.
   - Seawater is more viscous, has a higher density, and less available oxygen than air. Because it is an aqueous saline environment, it presents osmotic problems as well.
   - Viscosity. Swimming through water is fundamentally different from walking on land through air. Fishes have great diversity of body shape, but the streamlined fusiform shape exhibited by many, and is especially well developed in the tunas and billfishes, permits efficient movement through water. Their shape minimizes drag. They reduce friction and eddies by reducing scalation, having grooves into which they can retract fins, having flattened eyes. Their bodies have a high percentage of muscle and most of it is on the trunk and connected through the posterior of the body to a high aspect-ratio tail fin that oscillates rapidly. Some tunas, like the bluefin, and some fast sharks, like the great white and mako, can conserve body heat and function at elevated temperatures. Other shapes vary from the tuna shape, as dictated by particular ways of life. These shapes are less efficient at fast long distance swimming, but can provide different survival advantages. The eel-shape allows movement into cryptic burrows or crevices, the flattened flounder shape is adapted to a relatively motionless existence lying flat on the substrate.
   - Density. Water is denser than air and the ocean is a three dimensional environment often several miles deep. A few fish "walk" on leg-like fins like land vertebrates, but they are rare exceptions. Fish are supported in the dense watery world by the water itself. Most of their tissues, bone muscle, etc. are denser than water. Fat is the exception. Many fish species have a gas filled buoyancy organ called the swim bladder. By regulating the volume of the swim bladder at different depths (pressures), they maintain neutral buoyancy. Most of the musculature on their trunks is used for propulsion, and little is associated with their appendages. Their fins are used primarily for steering. Deep ocean fishes, benthic species and species which travel quickly between depths often lack a swim bladder.
   - Oxygen. Gills are the organs of gas exchange of fishes. They have enormous surface areas, counter-current flow of respiratory water and gill capillary blood, and a one- passage of

water over the gills rather than the tidal flow of air in and out of lungs. These features permit an oxygen extraction ratio of up to 85 percent as compared to the approximately 25 percent efficiency of air-breathing vertebrates.

- Osmoregulation. Marine Osteichthyes are osmotic-regulators. Their blood has a solute concentration roughly equivalent to 8–14‰ while seawater is 35‰. They are hypotonic to their environment and continuously lose water by osmosis from their gills. They compensate by drinking seawater that, of course, creates the problem of excess salt. They actively excrete salt with special gill cells and with their small volumes of salty urine. Marine Chondrichthyes like sharks have a different solution. They maintain elevated levels of urea in their blood, and so are isotonic to the sea.

3. Compare and contrast the tubenoses and the penguins.
   - The tubenoses, order Procellariiformes, and the penguins, order Sphenisciformes, are the two orders of birds most highly adapted to the ocean world. The specializations of the tubenoses are associated with life in the air over the sea while the penguins are specialized for life under the water.
   - The tubenoses include the albatrosses, petrels, and shearwaters. Their name refers to the elongated nostril. This serves as the exit for salty brine from the salt gland, but it also is highly specialized for sensing air speed and is part of these birds' highly developed sense of smell.
   - The speed sensing capabilities of the tubenoses is one of several adaptations to their remarkable powers of flight. The wandering albatross can have a wingspan of 12 feet while weighing a mere 22 pounds. The bones are very light; the total weight of the feathers exceeds the weight of the skeleton. The feet are placed rearward which also improves aerodynamics. The enormous wings have a very high aspect ratio; they are long, thin, and pointed. The albatrosses soar over the ocean for weeks at a time feeding on fish and squids detected by smell and captured by dipping. They rarely land on the water, and come ashore on remote islands only to breed.
   - By contrast the penguins do not fly in the air at all. They "fly" through the water. Penguins are totally flightless. Their wings are short and sickle-shaped. Light weight is less important than insulation as these cold water, southern hemisphere, birds hunt below the surface. They have short, greasy, peg-like feathers, and a thick layer of blubber. Their large size is also a heat-conserving feature.
   - The emperor penguin is the largest species. It is an Antarctic species and one of the few species that can survive the harsh conditions of the Antarctic actually breeding and incubating their eggs in the winter. They mass in tight crowds exhibiting huddle behavior to conserve heat.

Lesson 23: Living Together

1. c page 439
2. b page 441, Video
3. a page 438
4. d page 437
5. b page 454, Video
6. a page 446
7. b page 456, Video
8. c page 450–452
9. a page 449
10. b pages 447–449
11. d page 447, Video
12. a Video

Short Answer

1. Compare and contrast the hydrothermal vent community and the abyssal plain community and explain how the whale fall community might be related to both.
   - Both of these are deep ocean benthic communities and both are rich in life. The hydrothermal vent community is associated with heated plumes of mineral-rich water located near spreading centers. The abyssal plains are regions of constancy. They are perpetually cold, dark, and hyperbaric.
   - Chemosynthetic bacteria are the primary producers of the vent community and many animals including *Riftia* house mutualistic species of these autotrophs. Much of the food of the abyssal plains falls from the photic zone above although chemosynthetic bacteria have been found as deep as 500 meters within abyssal sediments.
   - The animals of the abyssal plains are slow moving, have low metabolic rates, often exhibit gigantism, and often are long-lived.

Standing crops may be high but productivity is low. Brittle stars, sea cucumbers, and other deposit feeders are common.
- *Riftia*, and large bivalves with symbiotic bacteria as well as blind, heat sensitive shrimps are among the animals of the vent community.
- The abyssal plain community seems to be characterized by a slowness and constancy but evidence accumulating as vents get revisited suggests that these are ephemeral communities lasting perhaps a decade or less.
- The whale fall community is the community of organisms colonizing the carcasses of dead cetaceans. The carcasses including their oily vertebrae are sulfur-rich, can support chemosynthetic bacteria, and may serve as stepping-stones for the distribution of larvae of vent organisms across the abyssal plains. Whaling has dramatically reduced whale populations in the past few centuries and probably reduced the availability of these putative stepping-stones. Whales have existed for only the past 50 million years so other stepping-stones would have been necessary before that.

2. Compare the coral reef community with the estuarine community.
   - These two communities are highly productive, coastal, shallow water communities. Coral reefs are limited to clear, tropical, high-salinity, nutrient-poor seas while estuaries are turbid, not limited by latitude, have high nutrient levels and are, by definition, places where seawater is diluted by fresh water from land runoff.
   - Coral reefs have high biodiversity and primary production by sparse phytoplankton in these nutrient-poor seas is enhanced by endosymbiotic zooxanthellae in the coral polyps. Production from adjacent mangroves, sea grasses, and calcareous and coralline algae also contribute.
   - Estuaries have low biodiversity and the food webs are phytoplankton and detritus-based. Salt marshes, mangroves, and sea grasses often supply a rich input of detritus to estuaries. Many species are euryhaline and many species brood their young, lay attached eggs, or have other strategies to minimize larval mortality due to river outflow.
   - Both these coastal communities attract human populations so are heavily impacted by eutrophication, sedimentation, pollution, overfishing, and other anthropogenic assaults.

3. A useful approximation of the pattern of growth of populations is the S-shaped growth curve. Make a drawing illustrating the features of an S-shaped growth curve and explain it.
   - Student illustrates curve including correctly labeled axes, illustration of environmental resistance slowing the growth rate, and leveling of growth at the carrying capacity.
   - The relationship between births and deaths (and immigration and emigration) controls population growth. Populations grow if births exceed deaths and shrink if deaths outrun births.
   - If a new species is introduced into a habitat where it has infinite resources and can flourish it will exhibit J-shaped population growth. The small population will take advantage of the environment and grow exponentially. Growth rate is slow at first because there are few individuals reproducing.
   - As the population grows, the number of individuals producing offspring grows and the population numbers skyrocket.
   - There is no such thing as infinite resources so the population will encounter environmental resistance. Competition for food, mates, and space increase as do disease, parasites, and predation. These factors slow the rate of growth of the population by increasing the death rate.
   - The population levels off at some maximum size that the environment can sustain. At this level, called the carrying capacity, births and deaths are approximately in balance.

Lesson 24: Treasure Trove

1. a page 478, Video
2. d page 466
3. c page 484
4. b synthesis
5. b page 478
6. c pages 473–476
7. d page 463
8. d page 463, Video
9. c page 477
10. b pages 472–473
11. a page 480, Video
12. a Video

Short Answer

1. Outline and discuss several aspects of the fisheries management problems.
   - Most world fisheries resources are being heavily exploited or over exploited. The human population is dramatically increasing; the demand on fisheries far exceeds the supply. It is likely this will worsen.
   - Fisheries are generally a common property resource so without rational management and effective enforcement, overfishing can only be prevented by the self-restraint and altruism of the fishers. These are not universal characteristics of resource exploiters.
   - Management requires sound biological data based on sampling and modeling of the fisheries stocks, and the rate at which fishers exploit stocks can often exceed scientists' capacity to evaluate them.
   - The oceans might seem to have a limitless supply of fish, but this idea is patently false. The difficulty in researching the parameters of fish stocks coupled with growing demand and highly sophisticated and efficient fishing techniques, boats, and strategies make overexploitation common.
   - Exploitation of one species often leads to bykill of others. Gear such as plastic drift nets often are lost at sea. These may be 50 miles long and float in the sea as death traps for decades.
   - The intersection of politics and profit have led to instances in which price supports, fuel tax exemptions, and other government policies have resulted in the cost of fishing far exceeding the profit. In 1995 the commercial fishing industry spent $124 million to catch fish that returned $70 million of profit. This situation has been dubbed "madhouse economics."
   - Marine sanctuaries offer some hope in controlling overfishing and establishing refuge for exploited stocks and bykill species as well.

2. Explain how the ocean's thermal gradient can be used to generate electricity.
   - Most electricity is generated by evaporating a liquid and using the expanding vapor to drive electricity-generating turbines. Most power plants use the heat from burning fossil fuels or from nuclear reactions to convert water to steam to drive this process.
   - Ocean thermal gradient electricity generation utilizes the difference in temperature between colder deep water and warmer surface water to evaporate ammonia gas to drive turbines. The concept was proposed by d'Arsonval in the 1880's and takes advantage of ammonia's low boiling point.
   - Ammonia will boil at the temperature of surface water and can be condensed back to liquid state with cold water pumped from the depths.
   - Tropical regions with narrow continental shelves are generally best suited for thermal gradient exploitation. The tropics have the largest temperature gradient between surface and deep water and a narrow shelf permits easier access to deep water. A small plant operated successfully in Cuba before it was destroyed by a hurricane in the 1930s. The OTEC Project in Hawaii also demonstrated the feasibility of this process but was disassembled in 1997 because it was not competitive with traditional construction.
   - Thermal gradient electricity generation lacks the air pollution of fossil fuel plants and the radioactive wastes of nuclear plants. It does, though, have its own problems including importing deep nutrient rich water into the tropical photic zone. OTEC plants create an artificial upwelling, which can alter local ecosystems. An OTEC plant would have to pump deep water at volumes five times the flow of the Mississippi River to generate as many kilowatts as a large nuclear facility.
   - The efficiency of thermally powered electricity generation depends on the difference between the hot side of the process and the cold side. The difference between the heat generated by fossil fuels or nuclear fusion that evaporates the water in the system and the coolant water or air that condenses the steam back to water is far greater than an ocean thermal gradient. This makes OTEC systems inherently less efficient than traditional ones.

3. Fishing is aquatic hunting while aquaculture is aquatic farming. Discuss ocean aquaculture.
   - Freshwater aquaculture has been practiced for thousands of years. In 1998 worldwide aquaculture produced 28 million metric tons of fish and the industry is growing yearly. The yield of fish from aquaculture may

exceed the yield of meat from cattle ranching by 2010.

- Ocean aquaculture is called mariculture and is also a rapidly growing industry. Asia is the most populous region of the world and also leads the world in aquaculture. China leads in freshwater aquaculture and Japan is mariculture leader.
- Mariculture is conducted near shore with protected bays and estuaries being more suitable than high energy exposed waters. The safety of enclosed bays also reduces the flushing time of the water so the eutrophication associated with intensive culture is exacerbated.
- The utilization of coastal regions for mariculture excludes their use for commercial or sport fishing, recreation, commerce, and other competing uses of the sparse coastal zone. Sometimes mariculture coexists in a symbiotic relationship with power plants where the waste hot water from the plants' cooling systems accelerate the growth of cultured species.
- Shrimp is the fastest growing segment of the mariculture industry. In the U.S.A., oysters and salmon are the leaders. Shrimps, mussels, seaweed, striped bass, abalone, and many other species are being cultured as well.
- Some open water cage-mariculture is being explored. The Japanese are also the leaders in this. Yellowtail and other oceanic species are being raised in suspended cages with some success.
- A variation of mariculture is ocean ranching. Any species is most vulnerable to predation during the earliest stages of life. In ranching, the organisms are spawned and raised through their earliest stages and then released to feed and grow in the wild as late-stage juveniles and adults. Salmon are suited for ranching because their homing instinct returns them from open-ocean feeding grounds to their natal stream at spawning time. They are easily harvested as they crowd the streams on their spawning runs.
- The harvest of algae such as kelp is probably a hybrid between ranching and culture. In southern California and elsewhere, natural kelp beds are harvested by large cutter vessels that shear off the top meter of kelpage. The algae grow back without assistance or maintenance and can be reharvested periodically.
- In addition to food, harvested mariculture species yield other products including cultured pearls, algin, and other phytochemicals useful in making industrial products, food additives, and pharmaceuticals.

Lesson 25: Dirty Water

1. b page 502
2. c page 496, Video
3. c Video
4. d page 498
5. a Video
6. a pages 500–501
7. a page 493
8. b page 499
9. c page 492
10. b Video
11. c Video
12. d Video

Short Answer

1. Using the information from the text chapter, compare the toxicities and potential for environmental damage of crude oil, refined oil, and used refined (motor) oil.
   - Oil is a natural part of the marine environment; large quantities have been seeping into the ocean from natural subsea deposits for millions of years. This natural oil is called crude oil, and is what is drilled for and recovered for eventual use by humans.
   - In 1998 about six million metric tons of oil entered the world oceans, but natural seeps accounted for only about 10 percent of that.
   - Refined oil is crude oil that has been extensively treated and processed to remove and break up the heavier components and to concentrate the lighter (and more biologically active) components. Chemicals are also added during the refining process, making it even more deadly to natural systems.
   - Spills of crude oil are generally larger in volume and more frequent than spills of refined oils. Most components of crude oil do not dissolve easily in seawater, but those that do can damage sea life.
   - Residual, undissolved crude oil on the sea surface can prevent gas exchange at the air-sea interface, clog the respiratory systems of marine organisms, decrease photosynthesis

and, if it comes ashore, cover and smother intertidal populations.
- Crude oil is not highly toxic and will eventually break down (biodegrade) and cause comparatively little biological damage.
- Spills of refined oil can be more destructive for longer periods of time. The refining process removes the more degradable components and concentrates the more harmful ones.
- If crude oil is left undisturbed, natural bacterial activity will eventually consume it. For refined oil, the natural cleanup and dissipation process can take much longer.
- Used motor oil has not only been refined, but through exposure to the heat and pressure of the engines, has added carcinogens and metallic compounds making it even more dangerous to marine organisms.
- Used oil reaches the ocean from street runoff, illegal disposal down city drains, and hidden in trash destined for the dumps of coastal cities. Every year over 240 million gallons of used oil enters the sea.
- The development of recreational harbors and marinas also adds to both the unused (but refined) and the used (leakage from boat motors) oil volumes discharged. In addition, these areas are often somewhat enclosed, concentrating the effects of the oils and further affecting the resident biota.

2. Discuss the process of eutrophication: its definition, potential causes, and types of organisms and habitats affected.
   - Eutrophication has been defined as a set of physical, chemical, and biological changes brought about when excessive nutrients are released into water. In short, the water is too fertile.
   - While the nutrients may not be toxic, they can stimulate the growth of some species to the detriment of others, causing an imbalance in the local ecosystems.
   - Eutrophication is most often connected to the sudden growth of algaes and phytoplankton, seen as red tides, yellow foams, and green slimes.
   - These plants use up disproportionate amounts of oxygen (during the night), depleting the oxygen supply for other organisms. When the excess plant material (and the oxygen-depleted organisms) die, they sink to the bottom and decompose, causing more oxygen to be used up by the decomposing bacteria.
   - The abundant algal growths can release natural toxic substances, which also sicken and kill fish and other species.
   - The "extra" nutrients come from wastewater treatment plants, factory effluents, soil erosion, and land-based lawn and garden fertilizers.
   - Eutrophication is most pronounced at areas around river mouths, and in estuaries and other enclosed bodies of water where there is little natural water circulation.

3. At the beginning of the text chapter for this episode, the tragic history of Easter Island sets the stage for our discussion of human-related environmental disasters. List, describe, and discuss briefly the sequence of events involved in that history.
   - Easter Island (also called Rapa Nui) lies in the mid-Pacific ocean off the coast of Chile. It was first visited by Europeans in 1722, and was later visited and documented by Captain James Cook in 1774.
   - Archaeologists start the history of Easter Island in about 350 A.D. when it was encountered and eventually settled by voyagers from the Marquesas. Their accounts detail fertile volcanic soils, dense forests, and abundant grasslands.
   - These voyagers settled there and began cutting down the palm trees to make canoes, clearing forests to plant crops, and catching the nearshore fishes and dolphins for food.
   - In the year 1400 the population had grown from the few early voyagers to between 10,000 and 15, 000, with the accompanying support activities (agriculture, fishing, lumbering) already stressing the island's resources.
   - Overfishing and harvesting of nearshore intertidal and benthic marine organisms had depleted these food supplies, so fishermen had to build larger canoes so they could go further offshore, using more of the large palm trees.
   - As the natural resources became more inadequate, the people appealed to their gods by carving worshipful images and stone statues. This entailed even more need for large trees and ropes, to roll and maneuver the massive stones into place.
   - The resident and migrant seabird populations had already been consumed for food, as

had the offspring of the rats that had originally hitchhiked to the island on the first canoes.
- Once-abundant plants and seeds were now rare and considered delicacies, and all usable land was under cultivation. Wars broke out over the dwindling food and resources. The nearshore marine organisms had been used up for food, but fishermen could not go further offshore because the large palm trees once used for canoes were gone.
- Many in the starving population resorted to cannibalism just to stay alive. Tribes of the remaining population warred against each other, further destroying the grasslands and forests, and tipping and burying the large stone statues and images.
- When Captain Cook arrived in the late 1700s, the population was about one-tenth of its original peak, and no stone statues stood upright.
- Any survivors of this tragedy could not have left the island, because there were no large trees from which to build ocean-going canoes.
- The successors of the early voyagers eventually died out, leaving the world wondering, for a long period of time, what the stone statues and images were there.

Lesson 26: Hands On

1. a page 56
2. d page 56
3. a Video
4. b Video
5. c Video
6. b page 55
7. a page 56
8. a pages 55–56
9. c Video
10. d Video

Short Answer

1. Remote sensing and sophisticated research submarines have become an integral part of modern ocean science. Review the course and find one example of the use of a manned or unmanned submersible (ROV), a satellite and ocean buoys to study an oceanographic problem. Describe each.
   - *Alvin* is the most famous of the few manned research submersibles presently operating. It has been used extensively to study deep ocean bathymetry, hydrothermal vents, etc. It made its 3000th dive in 1995, carries three people, and can dive to 4,000 meters.
   - TOPEX/Poseidon (1992) and its replacement Jason-1 (2001) measure sea surface topography. They precisely measure ocean surface topography and study global ocean currents and sea surface winds and improve understanding of ocean circulation and its effect on global climate. They have been used successfully to predict and monitor El Niño events.
   - The TAO TRITON array of ocean sensing buoys extends across the equatorial Pacific Ocean and is used primarily to forecast El Niño and La Niña climate events. TAO is The Tropical Atmosphere Ocean array. The anchored buoys monitor atmospheric and ocean temperatures at the surface and at various depths and upload their data to the Argos satellite system that relays it to land bases where it is analyzed. The array is a cooperative venture sponsored by the United States, Japan, and France. The TRITON buoys, manufactured in Japan, have replaced some of the original TAO buoys in the array.

2. Oceanography is a science and at its core it is driven by the innate curiosity of the human species. But several other motives also drive oceanographers in their work. Discuss several of these and give examples.
   - The history of oceanography is closely related to the history of exploration and is driven by innate curiosity of scientists, but the exploitation of profitable resources, the era of colonial expansion, the industrial revolution, military objectives and the space program, to name a few, have all played significant roles in the advancement of ocean science.
   - Some motivations discussed by the oceanographers might be categorized as inspiration by mentors, altruism, spiritualism, romance, adventure, and pragmatism.
   - Several of the scientists discuss experiences, including some from their childhood that sparked their curiosity about the oceans. An encounter with a live octopus and another with bolts of lightning were mentioned.
   - Some were motivated by inspirational mentors and parents. One discusses her high school chemistry teacher and another, her fishing trips with her father.

- The sense of adventure seems to contribute to the drive of some oceanographers. The adventure of diving in the lightless abyss aboard R/V *Alvin*, polar exploration, and the childhood desire to be an astronaut are discussed.
- Practical and pragmatic motivations were seen throughout the course. Improving navigation for navy and merchants also improved it for oceanographers. Improving weather forecasting, warning of impending tsunamis, improving submarine warfare, studying global climate, engineering ship hydrodynamics, preventing beach erosion, improving fisheries management, and an endless list of pragmatic and practical problems have led to a vast array of oceanographic discoveries including very significant ones.
- Pragmatic and altruistic motives often overlap. Improving fisheries while protecting and restoring environmental quality are examples.
- A spiritual or mystical dimension to working in the sea is probably important to all oceanographers to a greater or lesser extent.
- Although it is not a motive, the role of serendipity is worth mentioning. The use of cargo lost at sea to study current patterns and the discovery of the background microwaves that lend support to the big bang theory by scientists at Bell Labs studying telecommunications, are examples.

3. The practicing oceanographers in the video paint a romantic, adventurous, and generally inviting picture of their vocation. The road to a career in oceanography is actually a long one requiring hard work and dedication. Discuss the topic of working in marine science including training required, job opportunities, and factors that affect the employability of prospective oceanographers.
   - Oceanography is a science and it is multidisciplinary. A rigorous and dedicated approach to academics is required.
   - All sciences require math and advanced math. Science students must develop proficiency in the world of mathematics if they are to communicate with their peers and succeed.
   - A career in research requires skills. Students can enter rewarding oceanographic careers with a bachelor's degree if they wish to be technicians and staff at research institutions. People who enjoy hands-on work in the field and laboratory, including the computer lab and those able to operate and maintain equipment and instrumentation are suited for technical positions.
   - Research leaders require graduate degrees, generally the Ph.D. These individuals direct research and may work in the field or in the lab. They must also be skilled at framing questions and convincing funding agencies to support their research. They often work for universities, independent laboratories, government agencies, or private industry.
   - Marine biology is the branch of oceanography that seems most attractive to students, but it is the one with the fewest number of non-academic jobs. Many marine biologists work at colleges, but there are also positions with federal agencies such as the National Marine Fisheries Service and NOAA, for public utilities in environmental monitoring and at aquariums and museums, in aquaculture and in state departments of natural resources.
   - Physical oceanography including marine geology, ocean engineering, and marine chemistry and physics offers more employment opportunities. The oil and mineral industry employs thousands of physical oceanographers, and extracting minerals from the continental shelf is an expanding industry.
   - Other professionals such as computer engineers, lawyers, mathematicians, marine veterinarians, economists work in oceanography and there is always a demand for gifted science teachers at all levels.
   - Experience, good grades, willingness to relocate geographically, and diversification of training and skills are helpful in any field.